AF685371

Novel Adsorbents and Adsorption Methods for Pollutants Removal

Guest Editors

Yuezhou Wei
Yan Wu
Xinpeng Wang

MDPI

Basel • Beijing • Wuhan • Barcelona • Belgrade • Novi Sad • Cluj • Manchester

Guest Editors

Yuezhou Wei
School of Nuclear Science and Technology
University of South China
Hengyang
China

Yan Wu
School of Nuclear Science and Engineering
Shanghai Jiao Tong University
Shanghai
China

Xinpeng Wang
Guangxi Key Laboratory of Processing for Non-Ferrous Metals and Featured Materials
School of Resources, Environment and Materials
Guangxi University
Nanning
China

Editorial Office
MDPI AG
Grosspeteranlage 5
4052 Basel, Switzerland

This is a reprint of the Special Issue, published open access by the journal *Toxics* (ISSN 2305-6304), freely accessible at: https://www.mdpi.com/journal/toxics/special_issues/Adsorbents_Adsorption.

For citation purposes, cite each article independently as indicated on the article page online and as indicated below:

Lastname, A.A.; Lastname, B.B. Article Title. *Journal Name* **Year**, *Volume Number*, Page Range.

ISBN 978-3-7258-4109-7 (Hbk)
ISBN 978-3-7258-4110-3 (PDF)
https://doi.org/10.3390/books978-3-7258-4110-3

© 2025 by the authors. Articles in this book are Open Access and distributed under the Creative Commons Attribution (CC BY) license. The book as a whole is distributed by MDPI under the terms and conditions of the Creative Commons Attribution-NonCommercial-NoDerivs (CC BY-NC-ND) license (https://creativecommons.org/licenses/by-nc-nd/4.0/).

Contents

About the Editors

Yuezhou Wei

Distinguished Professor and Doctoral Supervisor at the University of South China's School of Nuclear Science and Technology, Yuezhou Wei is a recipient of China's National High-Level Talent Recruitment Program. Formerly a professor at Tohoku University (Japan), Shanghai Jiao Tong University, and Guangxi University, Wei has published over 420 SCI papers in journals including the Chemical Engineering Journal and the Journal of Hazardous Materials, authored one monograph, co-authored four books, and holds 40 patents. Recognized with the JNST Most Cited Paper Award, he has chaired six international conferences and serves on editorial boards for multiple SCI journals. His research has been funded by Japan's Major Scientific Research Program, China's NSFC Key Program, and international collaborations. His expertise spans multifunctional composite adsorbents for radionuclide/rare metal/heavy metal separation and nuclear fuel reprocessing and radioactive waste management (low/intermediate/high-level).

Yan Wu

Associate Professor and Doctoral Supervisor at Shanghai Jiao Tong University's School of Nuclear Science and Engineering, Yan Wu earned her Ph.D. in Quantum Science and Energy Engineering from Tohoku University, Japan. She previously served as a Researcher at Mitsubishi Materials Corporation's Central Research Institute and Assistant Professor at Tohoku University's Cyclotron and Radioisotope Center. With over 50 SCI-indexed publications in journals including *Chemical Engineering Journal, Journal of Hazardous Materials*, and *Separation and Purification Technology*, she co-authored one academic book, holds three invention patents, and has one highly cited paper in the *Journal of Nuclear Science and Technology (JNST)*. She has chaired sessions at international conferences and delivered invited lectures. Leading over 10 provincial/ministerial projects funded by China's National Natural Science Foundation (NSFC) and Ministry of Education, her research focuses on the synthesis of inorganic composite functional adsorbents, heavy metal/radioactive pollution control technologies, and resource recovery from solid waste.

Xinpeng Wang

Associate Professor at Guangxi University's School of Resources, Environment and Materials, Xinpeng Wang holds a Ph.D. in Nuclear Science and Engineering from Shanghai Jiao Tong University. Leading two National Natural Science Foundation of China (NSFC) projects and over 10 provincial/ministerial-level and industry-sponsored initiatives, his research focuses on adsorbent material development, radionuclide separation, spent fuel reprocessing, and lead-bismuth fast reactor chemical processes. He has published 40+ SCI-indexed papers in journals including *Chemical Engineering Journal, Journal of Colloid and Interface Science*, and *Journal of Cleaner Production*, with over 1400 total citations. An active contributor to the academic community, he has delivered invited lectures and chaired conference sessions at international forums. His expertise spans adsorbent material synthesis; rare metal separation; environmental pollution remediation; nuclear fuel cycle; and lead-bismuth fast reactor chemical technologies.

toxics MDPI

Editorial

Novel Adsorbents and Adsorption Methods for Pollutant Removal

Yuezhou Wei [1,2,*], Yan Wu [2,*] and Xinpeng Wang [3]

1 School of Nuclear Science and Technology, University of South China, Hengyang 421009, China
2 School of Nuclear Science and Engineering, Shanghai Jiao Tong University, Shanghai 200240, China
3 Guangxi Key Laboratory of Processing for Non-Ferrous Metals and Featured Materials, School of Resources, Environment and Materials, Guangxi University, Nanning 530004, China; wangxinpeng@gxu.edu.cn
* Correspondence: yzwei@usc.edu.cn (Y.W.); wu_yan@sjtu.edu.cn (Y.W.)

Over the past few decades, with the rapid growth of the global population and economy, the increasing levels of various pollutants such as heavy metals, radionuclides, and organic/inorganic/biological toxins from various industries and human activities, which diffuse into aspects of the environment such as the atmosphere, soil, and natural water, have posed a serious threat to human health and the environment. Therefore, these pollutants must urgently be removed in a highly efficient and concise manner. Adsorption, as the main technology for removing pollutants from wastewater, contaminated liquids, and soil pollutants, has attracted extensive attention from researchers. In recent years, many new adsorbent materials, such as highly porous adsorbents, multifunctional hybrid materials, biomass adsorbents, nanostructured materials, and COF/MOF materials, have been developed and applied to the elimination of toxic pollutants.

This Special Issue highlights the recent progress with regard to research on novel adsorbents and adsorption methods for pollutant removal. The authors investigated the synthesis and characterization of novel adsorbent materials targeting specific pollutants in terms of their physical state, content, radiative load, and biogenicity. The evaluation of novel adsorbents was achieved through experiments on the mechanism of adsorption and thermodynamic and kinetic methods, as well as on the separation, concentration, and removal of pollutants.

A study on the separation of actinides from HLLW was investigated by Tianjiao Jiang et al. [1] This paper used an advanced process to successfully separate Am(III) from Ln(III) by setting up two separation columns using CMPO and R-BTP as the core extractants. Mohammed F. Hamza et al. [2] studied novel adsorbents of biopolymers for the mitigation of environmental pollution via heavy metals and radionuclide ions. A biopolymer was prepared as a renewable energy source via the functionalization of chitosan with sulfonic groups. A series of experiments on elemental U were carried out to study its adsorption and desorption properties. Min Huang et al. [3] reported the effect of uranium–thorium leaching from a refractory tantalum–niobium slag via micro-mineralogical analysis and roasting experiments. It is of great practical significance to reduce the radioactive hazard of waste tantalum–niobium slag and to strengthen the sustainable utilization of resources via suitable process improvement techniques.

Two studies [4,5] synthesized an anion-exchange resin and demonstrated the possibility of applying the resin to the removal of ^{99}Tc elements from real radioactive waste liquids. Porous silica gel is a matrix used to support the adsorbent morphology and reduce the loss of high-value adsorbents. Yueying Wen et al. [6] prepared a $KAlFe(CN)_6/SiO_2$ adsorbent for the enrichment of Pd from HLLW. The adsorbent maintained the structural integrity below 100 kGy, and the K_d value of Pd was larger than 1625 cm^3 g^{-1} even in 7 M HNO_3. Xujie Chen et al. [7] presented a SiO_2 nanofiber membrane by using the electrostatic spinning and calcination process for the removal of vaporized ^{210}Po.

Citation: Wei, Y.; Wu, Y.; Wang, X. Novel Adsorbents and Adsorption Methods for Pollutant Removal. *Toxics* **2023**, *11*, 954. https://doi.org/10.3390/toxics11120954

Received: 13 November 2023
Accepted: 16 November 2023
Published: 23 November 2023

Copyright: © 2023 by the authors. Licensee MDPI, Basel, Switzerland. This article is an open access article distributed under the terms and conditions of the Creative Commons Attribution (CC BY) license (https://creativecommons.org/licenses/by/4.0/).

From the perspective of biopurification, three articles [8–10] discussed the purification of dioxins in soil via bioremediation technology, the purification of metal chromium in wastewater via biopolymers, and the degradation of biological pollutants via photocatalytic ternary nanocomposites. The advection and dispersion of tritiated water and pollutants such as iodine species (iodide: I^- and iodate: IO_3^-) and plutonium in deep geological repositories were studied in two respective articles [11,12]. In order to study the purification and detection technology of heavy metals in water, Dengpeng Wang, Chunlin He, and Yuling Zhao et al. [13–15] discussed CdSe/CdS core–shell quantum dot tracing Cd^{2+} in aqueous solutions, carbon adsorbent microwaves with waste amine–oxime removing Pb(II), and advanced Fe_3O_4@SFBC adsorbent removing As(III/V), respectively. Based on waste ore, two articles [16] studied the rational recovery process of radionuclides in tantalum–niobium slag and the adsorption effect of discarded coal gangue on the pollutant tetracycline hydrochloride, which reflected the important idea of turning waste into treasure.

The articles in this Special Issue demonstrate the removal of pollutants via novel adsorbents under severe conditions of different types, states, concentrations, temperatures, irradiation doses, etc., from a variety of application scenarios, such as HLLW, industrial wastewater, and the diffusion of environmental pollutants. Meanwhile, the synthesis and characterization of a variety of novel adsorbents as well as the computational methods of adsorption thermodynamics, kinetics, and product prediction based on the existing equations are also introduced.

The research on the removal methods of pollutants using new adsorbents focuses on selecting suitable methods for the removal of pollutants in complex environments, designing the structures of new adsorbents from their functions, and formulating adsorption programs that are practical and reasonable. These new adsorbents and adsorption methods will play an important role in pollutant removal and environmental remediation.

Conflicts of Interest: The authors declare no conflict of interest.

References

1. Jiang, T.; Ning, S.; Yu, T.; Wang, J.; Wei, Y.; Wu, Y.; He, H.; Chen, F.; Wang, Q. Separation of Minor Actinides from High-Level Liquid Waste Using Novel Silica-Based Butyl-BTP Adsorbents. *Toxics* **2022**, *10*, 741. [CrossRef] [PubMed]
2. Hamza, M.F.; Abu Khoziem, H.A.; Khalafalla, M.S.; Abdellah, W.M.; Zaki, D.I.; Althumayri, K.; Wei, Y. Ecofriendly Composite as a Promising Material for Highly-Performance Uranium Recovery from Different Solutions. *Toxics* **2022**, *10*, 490. [CrossRef] [PubMed]
3. Huang, M.; Hu, K.; Li, X.; Wang, Y.; Ouyang, J.; Zhou, L.; Liu, Z. Mineralogical Properties of a Refractory Tantalum-Niobium Slag and the Effect of Roasting on the Leaching of Uranium-Thorium. *Toxics* **2022**, *10*, 469. [CrossRef] [PubMed]
4. Sang, H.; Mao, C.; Wu, Y.; Wei, Y. Study on the Effect of Gamma-Ray Irradiation on the Adsorption of 99Tc and Re by a Silica-Based Pyridine Resin. *Toxics* **2022**, *10*, 638. [CrossRef] [PubMed]
5. Qin, B.; Hu, Y.; Xie, M.; Xue, L.; Liao, C.; Yang, F. Highly Selective Adsorption of $^{99}TcO_4^-/ReO_4^-$ by a Novel Polyamide-Functionalized Polyacrylamide Polymer Material. *Toxics* **2022**, *10*, 630. [CrossRef] [PubMed]
6. Wen, Y.; Wu, Y.; Xu, L. Radiation Resistance and Adsorption Behavior of Aluminum Hexacyanoferrate for Pd. *Toxics* **2023**, *11*, 321. [CrossRef] [PubMed]
7. Chen, X.; Chen, X.; Zeng, X.; Zhao, Y.; Li, X.; Huang, X.; Fujita, T.; Wang, X. Removal of the Homolog Tellurium of Polonium by SiO_2 Nanofiber Filter for Lead Alloy-Cooled Reactors. *Toxics* **2022**, *10*, 275. [CrossRef] [PubMed]
8. Nhung, N.T.H.; Nguyen, X.-T.T.; Long, V.D.; Wei, Y.; Fujita, T. A Review of Soil Contaminated with Dioxins and Biodegradation Technologies: Current Status and Future Prospects. *Toxics* **2022**, *10*, 278. [CrossRef] [PubMed]
9. Samrot, A.V.; Bavanilatha, M.; Krithika Shree, S.; Sathiyasree, M.; Vanjinathan, J.; Shobana, N.; Thirugnanasambandam, R.; Kumar, C.; Wilson, S.; Rajalakshmi, D.; et al. Evaluation of Heavy Metal Removal of Nanoparticles Based Adsorbent Using Danio rerio as Model. *Toxics* **2022**, *10*, 742. [CrossRef] [PubMed]
10. Afzal, M.I.; Shahid, S.; Mansoor, S.; Javed, M.; Iqbal, S.; Hakami, O.; Yousef, E.S.; Al-Fawzan, F.F.; Elkaeed, E.B.; Pashameah, R.A.; et al. Fabrication of a Ternary Nanocomposite g-C_3N_4/Cu@CdS with Superior Charge Separation for Removal of Organic Pollutants and Bacterial Disinfection from Wastewater under Sunlight Illumination. *Toxics* **2022**, *10*, 657. [CrossRef] [PubMed]
11. Shi, Y.; Yang, S.; Chen, W.; Xiong, W.; Zhang, A.; Yu, Z.; Lian, B.; Lee, C.-P. Study on Tritium and Iodine Species Transport through Porous Granite: A Non-Sorption Effect by Anion Exclusion. *Toxics* **2022**, *10*, 540. [CrossRef] [PubMed]

12. Shi, Y.; Yang, S.; Wu, E.; Wang, L.; Chen, W.; Xiong, W.; Zhang, Y.; Zhang, A.; Lian, B. Advection–Dispersion Behavior for Simulation of H-3 and Pu-238 Transport in Undisturbed Argillaceous Shale of a Near-Surface Repository. *Toxics* **2023**, *11*, 124. [CrossRef] [PubMed]
13. Wang, D.; Gao, F.; Wang, X.; Ning, X.; Wang, K.; Wang, X.; Wei, Y.; Fujita, T. Detection of Cd^{2+} in Aqueous Solution by the Fluorescent Probe of CdSe/CdS QDs Based on OFF–ON Mode. *Toxics* **2022**, *10*, 367. [CrossRef] [PubMed]
14. He, C.; Liu, Y.; Zheng, C.; Jiang, Y.; Liao, Y.; Huang, J.; Fujita, T.; Wei, Y.; Ma, S. Utilization of Waste Amine-Oxime (WAO) Resin to Generate Carbon by Microwave and Its Removal of Pb(II) in Water. *Toxics* **2022**, *10*, 489. [CrossRef] [PubMed]
15. Zhao, Y.; Shi, H.; Tang, X.; Kuang, D.; Zhou, J.; Yang, F. Performance and Mechanism of As(III/V) Removal from Aqueous Solution by Fe_3O_4-Sunflower Straw Biochar. *Toxics* **2022**, *10*, 534. [CrossRef] [PubMed]
16. Wan, H.; Wang, C.; Gong, L.; Zhu, X.; Yan, J.; Lu, J.; Zhang, W. Potential Application of Discarded Natural Coal Gangue for the Removal of Tetracycline Hydrochloride (TC) from an Aqueous Solution. *Toxics* **2023**, *11*, 20. [CrossRef] [PubMed]

Disclaimer/Publisher's Note: The statements, opinions and data contained in all publications are solely those of the individual author(s) and contributor(s) and not of MDPI and/or the editor(s). MDPI and/or the editor(s) disclaim responsibility for any injury to people or property resulting from any ideas, methods, instructions or products referred to in the content.

toxics

Article

Removal of the Homolog Tellurium of Polonium by SiO_2 Nanofiber Filter for Lead Alloy-Cooled Reactors

Xujie Chen [1], Xiyong Chen [1], Xian Zeng [2], Yuan Zhao [2], Xiaoping Li [1], Xi Huang [3,*], Toyohisa Fujita [1] and Xinpeng Wang [1,*]

1 School of Resources, Environment and Materials, Guangxi University, Nanning 530004, China; chanxj00@163.com (X.C.); xiyongchen@gxu.edu.cn (X.C.); Pu976811@163.com (X.L.); fujitatoyohisa@gxu.edu.cn (T.F.)
2 China Nuclear Power Technology Research Institute, Shenzhen 518000, China; zengxian@cgnpc.com.cn (X.Z.); zhaoyuan2@cgnpc.com.cn (Y.Z.)
3 Advanced Nuclear Energy Research Group, College of Physics and Optoelectronic Engineering, Shenzhen University, Shenzhen 518060, China
* Correspondence: xi.huang@szu.edu.cn (X.H.); wangxinpeng@gxu.edu.cn (X.W.)

Abstract: The lead–bismuth eutectic (LBE) can be easily activated by neutron radiation to produce the radionuclide ^{210}Po. It is therefore necessary to establish an effective method to remove vaporized polonium in the cover gas to prevent its release into the air in scenarios of reactor maintenance and coolant leakage accidents. This paper presents a SiO_2 nanofiber membrane prepared based on the electrostatic spinning and calcination process. The SiO_2 nanofiber membrane had the advantages of good flexibility, high-temperature resistance, and corrosion resistance. In the trapping experiments, the SiO_2 nanofiber membrane filters showed excellent filtration performance at 300~400 °C, and the filtration efficiencies for Te, Pb, and Bi could reach 99%, 99%, and 98%, respectively. Proper filtration temperature and gas flow rate are important to maintain high filtration efficiency. After five cycles, the SiO_2 nanofiber membrane filter still exhibited excellent cycle-use performance. In the density functional theory (DFT) calculations, PbPo and PbTe had strong interactions with amorphous SiO_2, having adhesion energies of −2.96 to −2.83 eV/molecule.

Keywords: lead–bismuth eutectic; polonium; SiO_2 nanofiber filter; density function theory

Citation: Chen, X.; Chen, X.; Zeng, X.; Zhao, Y.; Li, X.; Huang, X.; Fujita, T.; Wang, X. Removal of the Homolog Tellurium of Polonium by SiO_2 Nanofiber Filter for Lead Alloy-Cooled Reactors. *Toxics* **2022**, *10*, 275. https://doi.org/10.3390/toxics10060275

Academic Editors: Xin-Gui Li and Mei-Rong Huang

Received: 26 April 2022
Accepted: 19 May 2022
Published: 24 May 2022

Publisher's Note: MDPI stays neutral with regard to jurisdictional claims in published maps and institutional affiliations.

Copyright: © 2022 by the authors. Licensee MDPI, Basel, Switzerland. This article is an open access article distributed under the terms and conditions of the Creative Commons Attribution (CC BY) license (https://creativecommons.org/licenses/by/4.0/).

1. Introduction

With the advantages of a low melting point, high boiling point, small neutron cross-section, and high thermal conductivity, lead–bismuth eutectic (LBE) has become a spallation target material and a coolant for accelerator-driven subcritical systems (ADSs) [1–4]. However, LBE can be easily activated by neutron radiation to produce the radionuclide ^{210}Po, which is a highly toxic radionuclide with a physical half-life of 138.4 days. If the cumulative radiation dosage reaches 4 MBq/kg-body-mass in a short time, it will be fatal to the human body. Therefore, to achieve the wide use of LBE in nuclear energy applications, an effective Po pollution prevention system must be established [5,6]. At present, polonium in molten LBE can be effectively removed by lead–polonium distillation, Po-hydride separation, rare-earth filtration, and alkaline extraction [7–9]. However, these methods are not applicable for removing polonium evaporated from liquid LBE. Without any prevention means, polonium on the surface of the liquid LBE and its vapor in the cover gas will diffuse out of the main system and be released into the atmosphere during reactor maintenance and coolant leakage accidents [10–13]. Therefore, removal of polonium from the gas phase needs to be seriously examined.

In an ADS system, the cover gas typically will have the highest concentration of polonium [14,15]. To reduce the concentration of polonium in the cover gas, efforts have been made to carry out the purification of gaseous polonium compounds and aerosols in the lead stack cover gas. Obara et al. used a stainless-steel mesh filter to remove polonium aerosols in the gas phase under vacuum conditions, and the filtration effect was significant [16,17]. However, the filtration performance of the stainless-steel mesh filter causes a severe decrease or even ineffectiveness due to the corrosion behavior of stainless steel in LBE [18]. Based on thermal chromatography experiments, a high adhesion affinity of polonium was achieved using noble metals (Au, Ag, Pt, and Pd) and 316L stainless steel, as reported by Maugeri et al. [19–21]. Meanwhile, the adhesion energies of Po, Po_2, PbPo, PbBi, H_2Po, and PoOH on noble metals were calculated by using the density functional theory (DFT) for ^{210}Po trapping [22–24]. However, the high price of noble metals limits its large-scale application for polonium removal. The development of polonium aerosol filtration materials with high-temperature resistance, corrosion resistance, and high filtration accuracy is still the main focus of research [25].

Nanofibers with small pore size, high throughput, and high filtration accuracy are regarded as candidate filtration materials in the field of nuclear waste treatment [26–28]. According to the literature, the whole process of evaporating and capturing volatile materials involves sub-processes of volatilization, condensation, and solidification [29–31], which are also applicable for the current study. Polonium-containing vapors migrate from liquid lead–bismuth pools to the cover gas at high temperatures. The supersaturated vapors will thereafter undergo two processes to form aerosol particles: spontaneous nucleation (vapor condensation in the gas phase to form new particles) and nonspontaneous nucleation (vapor condensation and deposition directly on impurity particles). As for aerosol particles, the nanofibers capture these particles by inertial collision, interception, Brownian motion, and gravitational sedimentation [32]. Inorganic nanofibers meet the requirements of high-temperature resistance, corrosion resistance, and high filtration accuracy of polonium filters, and they remove gaseous components and aerosols in the gas phase through condensation and physical filtration to achieve the purification of the cover gas in a lead-based reactor [33]. However, most of the inorganic nanofibers have limited applications in gas purification due to their own brittleness.

Electrospun fibers with small fiber diameter, large aspect ratio, and high porosity have outstanding filtration and barrier properties. Moreover, the SiO_2 material itself has the characteristics of high temperature and corrosion resistance. In this study, the SiO_2 inorganic nanofiber membranes prepared based on electrostatic spinning have excellent flexibility and overcome the disadvantage of brittle inorganic nanofibers, which can be used for the removal of polonium from the cover gas in LBE under high-temperature conditions [34,35]. For the sake of safety in the lab experiments, Te was used in this study as a substitute for the Po radionuclide, which has been proven to be feasible in previous published works [36,37]. In particular, this work focused on the filtration performance and mechanism of the inorganic nanofiber membrane for Te aerosol in the gas. The results provide practical insights for removing Po aerosols in the cover gas in a real ADS system. Furthermore, calculations based on DFT were applied to study the adhesion energies of possible adsorbate molecules (Po, Te, Bi, PbPo, and PbTe) on the surface of amorphous SiO_2 to explain the possible interaction mechanisms. The proposed SiO_2 nanofiber filter has outstanding filtration performance for the removal of polonium under high temperature and airflow conditions and cycle-use performance, so it has the potential to be used in the polonium purification systems.

2. Materials and Methods

2.1. Materials

All chemicals used in this study were of analytical grade and were directly used without further treatment. Phosphoric acid (H_3PO_4, 85 wt.%) was purchased from Xilong Chemical Co., Shantou, China and nitric acid (HNO_3), tetraethyl orthosilicate (TEOS), and poly(vinyl alcohol) (PVA1788) were purchased from Sinopharm Chemical Reagent Co., Ltd., Shanghai, China.

2.2. Preparation of SiO_2 Nanofiber Membrane

The SiO_2 inorganic nanofibers used in the filter were prepared by electrostatic spinning and high-temperature calcination processes. Tetraethyl orthosilicate (10.4165 g), deionized water (9 g), and phosphoric acid (0.577 g) were first mixed and stirred at room temperature for 12 h to obtain a silica sol after full hydrolysis. PVA powder was dissolved in deionized water and stirred in a water bath at a temperature of 90 °C for 4 h to prepare a 10 wt.% PVA solution. The prepared silica sol and the PVA solution were later mixed at a mass ratio of 7:3 and stirred for another 12 h to obtain an electrostatically spun precursor solution. The prepared precursor solution was thereafter loaded into a syringe and fixed on an electrostatic spinning device (Electrospinning Equipment, ET-2535DC).

During the electrospinning process, the charged polymer droplets are accelerated at the apex of the Taylor cone at the syringe needle under the action of an electric field force. When the electric field force is large enough, the polymer droplets overcome the surface tension to form a jet stream. The solvent evaporates or solidifies during the trickle jetting process and finally falls on the receiving device to form a non-woven nanofiber mat [26]. In the current experiments, the electrospinning process parameters were set as follows: the temperature was 40–50 °C, the humidity was 30–40%, the receiving distance was 17 cm, the injection speed was 1 mL/h, and the accelerating voltage was 8 kV. The obtained electrospinning precursor nanofiber membrane was dried in an oven at 80 °C for 6 h. The dried nanofiber membrane was later calcined in a muffle furnace at 800 °C for 2 h to remove the organic matter, thereby yielding flexible SiO_2 inorganic nanofibers. The corresponding specific surface area and the average pore size were generally about 5.39 m^2/g and 15.50 nm, respectively, for all the prepared nanofiber membrane samples, which were determined using the method described in the next section.

2.3. Phase and Microstructure Characterization

The crystal structures of the prepared pure SiO_2 nanofibers before and after the Te trapping experiments (described in Section 2.4) were analyzed by X-ray diffraction (XRD, Rigaku D/MAX 2500V). The N_2 adsorption–desorption curves of the SiO_2 nanofibers were measured by a Tristar II 3020 instrument. The specific surface areas were calculated according to the Brunauer–Emmett–Teller (BET) method, and the pore size distribution was derived from the desorption branch of the isotherm with the Barrett–Joyner–Halenda (BJH) model. Prior to the test, the samples were treated under vacuum for 12 h at 200 °C. The surface morphologies and compositions of the intact SiO_2 nanofibers and the LBE-Te@SiO_2 samples obtained after the Te trapping experiments were analyzed by scanning electron microscopy (SEM, Hitachi SU5000, Tokyo, Japan) and energy-dispersive X-ray spectroscopy (EDS).

2.4. Te Trapping Evaluation Setup and Procedure

The Po evaporation in the lead–bismuth fast reactor was simulated by replacing the radioactive Po with the light element Te within the same main group. Figure 1 shows a schematic of the experimental system that was designed to capture Te with an SiO_2 nanofiber filter (SNF) under gas flow. The SNF, which was composed of a flexible SiO_2 inorganic nanofiber membrane and a stainless-steel mesh and fixed in a stainless-steel sealing ring, was seamlessly installed in the stainless-steel tube. Sufficient layers of the stainless-steel mesh were adopted in the current study to allow complete capture of the compounds of interest. A quartz boat lined with asbestos was preloaded with the volatile sample of LBE-Te (2.85 g of LBE with 0.15 g of pure Te) and placed in the volatilization zone of the stainless-steel tube. The air-exposed part of the stainless-steel tube on the filtering side was insulated with asbestos, and a thermocouple was used to measure the filter temperature in real time. The pressures at the gas inlet and outlet were recorded with pressure-sensing gauges. During the experiment, a reducing atmosphere of 5 wt.% H_2 balanced with Ar was applied as the carrier gas. The flow rate of the carrier gas was controlled with a gas flow controller. Before heating up the LBE-Te sample, the test system was pre-fluxed with the carrier gas for at least 15 min. After all the air in the test tube was replaced with the carrier gas, the volatilization zone was rapidly heated to 690 °C at a rate of 15 °C/min. The trapping experiment was sustained for 5 h after the heating temperature reached 690 °C. The filtered waste gas was discharged after passing through the acid–base tail gas removal device. Table 1 lists the experimental conditions used in this study. The set I experiments explored the influence of the filter temperature on the filtration efficiency of the SNFs for Te, Pb, and Bi. The set II experiments examined the effects of the loading level of the SiO_2 filter. The set III and IV experiments were designed to test the flow rate of the carrier gas for the influence of filtration efficiency. The cycle-use performance of the SiO_2 filter was studied by the set V experiment.

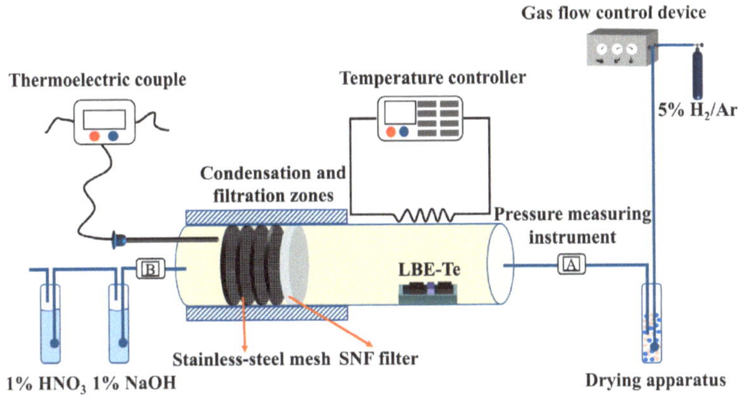

Figure 1. Schematic diagram of aerosol capture system.

Table 1. Experimental conditions for different aerosol capture experiments.

Experiences	Test No.	Mass of Nanofibers (mg)	Trapping Temperature (°C)	Flow Rate (mL/min)
Set I	SNF-1–SNF-5	60	300–500	1000
Set II	SNF-6–SNF-9	20–80	400	1000
Set III	SNF-10–SNF-12	60	400	600–1400
Set IV	SNF-13–SNF-15	60	500	600–1400
Set V	SNF-16	60	400	1000

2.5. Evaluation of Filtration Efficiencies of SiO_2 Nanofiber Membrane Filters

The Te evaporated from the LBE in the current experimental setup could be captured by the SiO_2 nanofiber membrane, collected by the stainless-steel mesh installed after the SiO_2 membrane, and deposited on the stainless-steel tube wall before the filtration membranes. The Te filtration efficiency η by the SNFs can be evaluated with the following equation (Equation (1)):

$$\eta = m_1 \ / \ (m_1 + m_2) \times 100\%, \tag{1}$$

where m_1 and m_2 are the corresponding Te masses captured by the SiO_2 nanofiber membrane and the stainless-steel mesh, respectively. This equation is also applicable for evaluating the filtration efficiency for the evaporated Pb and Bi in the cover gas.

In engineering practice, the filtration factor (QF) has become a meaningful index to identify the filtration performances of materials to balance the filtration efficiency and flow resistance. QF is defined as follows (see Equation (2)):

$$\text{QF} = \frac{-\ln(1-\eta)}{\Delta p}, \tag{2}$$

where Δp is the pressure drop through the membrane, and η is the filtration efficiency.

To obtain the masses of the captured element of interest, the SiO_2 nanofiber membrane and stainless-steel mesh after the experiments were immersed in a 7 mol/L nitric acid solution and ultrasonically oscillated to obtain a solution of Te, Pb, and Bi ions. The ion concentrations in the solution were determined by using an inductively coupled plasma emission spectrometer (ICP) to evaluate the corresponding mass deposition of the volatile elements in both the SiO_2 nanofiber membrane and the stainless-steel mesh.

2.6. Density Functional Theory Calculations

The calculations were performed on the DFT framework of the Gaussian 16 program package. The SiO_2 surfaces were represented using $Si_4O_4(OH)_9$ clusters. The terminal oxygen atoms were saturated with hydrogen atoms [38,39]. First, geometry optimization calculations were performed with the three-parameter hybrid functional (B3LYP) [40]. The geometry optimizations of the Si, O, and H atoms were performed with the standard 6-311G(d) basis set, and the geometry optimizations of the Pb, Po, Te, and Bi atoms were performed with the SDD basis set. Finally, the energy calculations of the optimized structure were based on the def2-TZVP basis set. The adhesion energy was calculated to analyze the direct interactions between molecules and the surface oxygen defects based on the following equation (Equation (3)):

$$E_{ads} = E_{AB} - E_A - E_B, \quad (3)$$

where E_{ads} is the adhesion energy, and E_{AB}, E_A, and E_B are the energies of the complex, monomer A, and monomer B, respectively.

3. Results and Discussion

3.1. Phase and Structural Analysis of LBE-Te

The corresponding phase structures of the unused SiO_2 nanofibers and the LBE-Te@SiO_2 nanofibers were analyzed by XRD, as shown in Figure 2. The SiO_2 nanofibers showed pure amorphous features with only a small envelope of diffraction peaks around 21°. In contrast, the XRD pattern of the obtained LBT-Te@SiO_2 specimen revealed the characteristic reflections of an intermetallic PbTe phase around 23.8°, 27.6°, 39.4°, 46.6°, 48.8°, 57.0°, 64.4°, and 71.5°. The XRD patterns also showed the existence of a pure metallic Bi phase with diffraction angles of 27.2° and 37.9° and overlapping peaks with the PbTe phase at 23.8°, 39.4°, 48.8°, and 64.4°. No elemental lead or Te was detected from the XRD analysis. It is therefore strongly suggested that the Te that vaporized from the LBE was mainly in the form of PbTe, which to some extent confirmed that PbPo would be the main component in the cover gas in a real operating lead-based reactor [10,37,41,42].

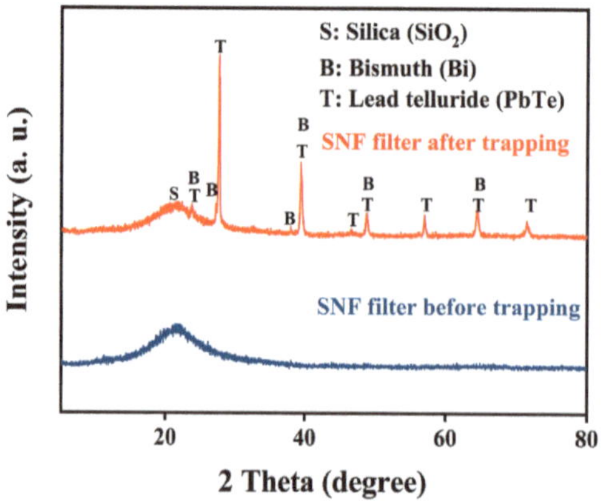

Figure 2. X-ray diffraction (XRD) patterns of the unused SiO_2 nanofibers and the LBE-Te@SiO_2 nanofibers.

By comparing the photographs of the SiO_2 nanofiber membrane samples before and after the Te trapping experiment shown in Figure 3a,d, respectively, the successful capture of either Te, Bi, or Pb was realized. A closer SEM inspection of the samples before and after the experiment (Figure 3b,e, respectively) further proved this conclusion. As shown in Figure 3b, the unused SiO_2 nanofiber membrane possessed well-distributed nanowires with uniform diameters. The nanofiber diameter was controlled to within 500 nm, and the average value was around 322 nm (Figure 3c). However, the sample morphology shown in Figure 3e reveals that the fiber membrane was significantly covered by a large quantity of larger particles, with sizes ranging from 0.5 to 50 μm. The elemental EDS mapping showed the uniformly distributed Te, Pb, Bi, and Si on the sample surface.

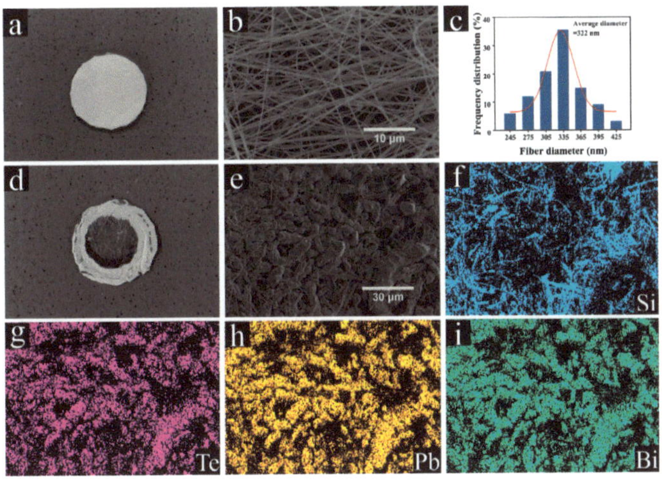

Figure 3. Photographs and scanning electron microscopy (SEM) images of SiO_2 nanofiber membrane (**a**,**b**) before and (**d**,**e**) after the Te capture experiment at the filtration temperature of 400 °C, respectively. (**c**) Fiber diameter distribution of the SiO_2 nanofibers and energy-dispersive X-ray spectroscopy (EDS) elemental mappings of (**f**) Si, (**g**) Te, (**h**) Pb, and (**i**) Bi are also presented.

3.2. Factors Influencing Removal of PbBiTe by SiO_2 Filter

3.2.1. Filter Temperature

ICP was used to measure the concentrations of volatile elements deposited on the filter after the capture experiment and to further calculate the deposition mass of each element on the filters. Figure 4 shows the determined masses of Te, Pb, and Bi deposited on the SiO_2 nanofiber membrane and the stainless-steel mesh at different temperatures. As shown in Figure 4a, the deposited masses of all the elements on the SiO_2 nanofiber membrane tended to decrease as the filtration temperature decreased. More vapor deposited on the tube surface before the vapor arrived at the filter when the filtration temperature was decreased. In addition, Figure 4b shows the captured masses of Te, Pb, and Bi on the stainless-steel mesh as a function of the filtration temperature. The results showed that the amounts of elements permeating through the SiO_2 nanofiber membrane increased with the filtration temperature. In particular, when the filtration temperature changed from 450 °C to 500 °C, the penetration largely increased. Figure 5 shows the respective filtration efficiency of the SNF at different filtration temperatures. When the temperature was lower than 400 °C, the filtration efficiencies for Te, Pb, and Bi could be maintained at 98%, 99%, and 95%, respectively. However, when the temperature was higher than 400 °C, the filtration efficiency for each element began to gradually decrease. The filtration efficiencies for Te, Pb, and Bi dropped down to 85%, 86%, and 47%, respectively, when the temperature reached 500 °C. Among all the elements, the filtration efficiency of the SiO_2 nanofiber membrane for Bi dropped most significantly. In summary, the removal of

Te in the cover gas phase that evaporated from the liquid LBE alloy was affected by the trapping temperature.

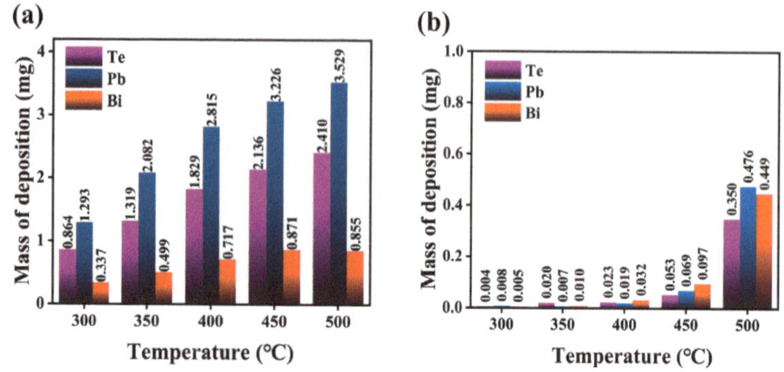

Figure 4. Obtained deposition masses of Te, Pb, and Bi from filters SNF1–SNF5 on (**a**) SiO_2 nanofiber membrane and (**b**) stainless-steel mesh at different temperatures.

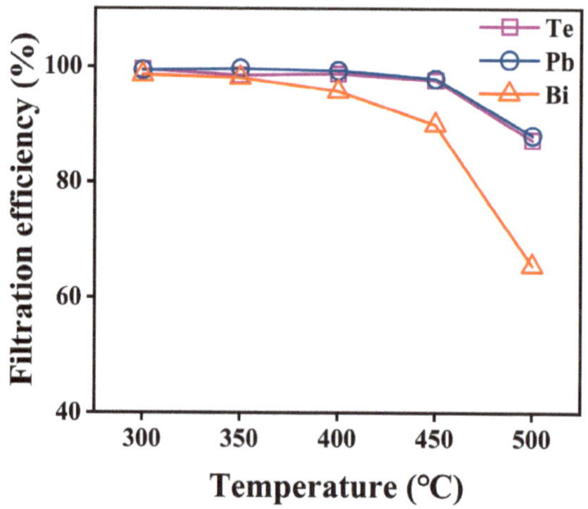

Figure 5. Filtration efficiencies of the SNF for Te, Pb, and Bi at different filtration temperatures.

The reasons for the decrease in the captured mass of the element of interest but the increase in the filtration efficiency of the SiO_2 nanofiber membrane at a decreased filtration temperature can be explained as follows. At a lower filtration temperature, a larger temperature gradient was created between the filter and the vaporization source, which resulted in a large quantity of vapor deposition on the tube wall on the way from the source to the filter, and thus, a lower concentration in the gas phase. It is therefore evident that the captured masses of the elements were lower within a constant capturing timeframe. However, due to the identical and limited capturing capacities of the nanofiber membranes with the same sizes and porosities, a larger fraction of the element vapor could deposit on the membrane, and therefore, a higher filtration efficiency was achieved at a lower filtration temperature. Furthermore, a lower trapping temperature meant that the vapor could nucleate and grow on the nanofiber surface, even with a lower degree of subcooling, which also increased the nucleation rate and accelerated the deposition of vapor on the membrane surface. According to relevant studies, the vapor pressure of PbTe is between those of Pb and Bi, indicating that the volatility of PbTe is higher than that of pure Pb and lower than that of Bi [43–46]. Meanwhile, the melting point of Bi is much lower than that of PbTe; thus, a higher degree of subcooling can be achieved for PbTe. Consequently, capturing Bi is relatively more difficult than capturing PbTe, which results in a lower filtration efficiency for Bi.

3.2.2. Loading Level of SiO_2 Filter

Figure 6a shows the filtration efficiencies of the SNFs with different loading masses on the membrane for the removal of Te, Pb, and Bi at a filtration temperature of 400 °C. The corresponding samples used for this test were labeled as SNF-6, SNF-7, SNF-8, and SNF-9 with 20, 40, 60, and 80 mg loadings of SiO_2 nanofibers, respectively. More generally, an area-specific loading density, defined as the loading mass divided by the cross-sectional area of the filter, can also be adopted as a characteristic structural parameter for the filter design. In this study, the corresponding area-specific loading densities of the filters SNF-6–SNF-9 were calculated to be around 0.76, 1.51, 2.27, and 3.03 mg/cm^2, respectively.

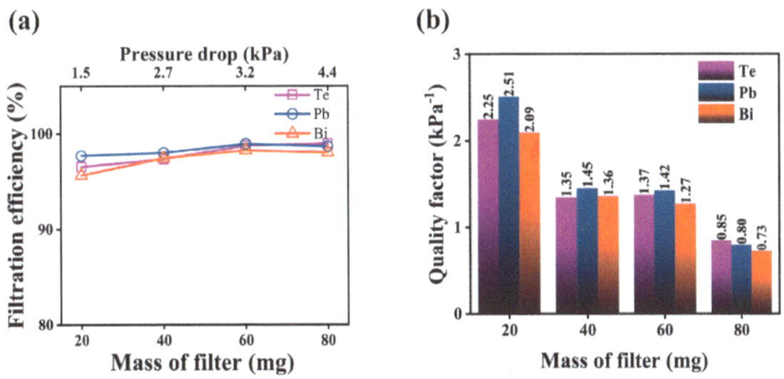

Figure 6. Comparisons of (**a**) filtration efficiency and (**b**) quality factor of the SiO_2 filtration membranes with different SiO_2 nanofiber loading levels for removing Te, Pb, and Bi at a filtration temperature of 400 °C.

The results indicate that the filtration efficiency gradually increased as the loading level or the area-specific loading density of the SiO_2 nanofibers increased. However, higher filtration efficiencies of 96%, 97%, and 95% could still be achieved for Te, Pb, and Bi, respectively, even with the lowest filter loading (20 mg) in the current experiment. In general, the increase in the loading level of the SiO_2 nanofibers would not change the pore size and distribution on the membrane surface. Both the porosity and tortuosity of the obtained membranes remained almost the same for all the SiO_2 nanofiber membranes in the current experiment. However, a higher loading mass of the SiO_2 nanofibers in a membrane would inevitably increase the thickness of the membrane so that the penetration pathway of the gas phase through the membrane would increase. This provided a greater chance for the membrane to capture the elements of interest in the current study.

However, this does not mean that a continuously increased loading level of the SiO_2 nanofibers would always be a better choice. One must consider the pressure drop for the operation. An increase in the membrane thickness would cause a significantly increased flow resistance. More importantly, the experiment revealed that the majority of the vapor deposition occurred at the front surface (feed side) because of the increased surface area due to the accumulated deposits, which decreased the available pore size on the surface for vapor penetration, as evidenced in Figure 3e. Figure 6a shows that the increase in filtration efficiency became insignificant when the loading level of the SiO_2 nanofibers exceeded 60 mg. The calculated quality factor (QF) for the SiO_2 nanofiber membranes is shown in Figure 6b. The quality factor of the membrane decreased with the increase in the area-specific loading density of the SiO_2 nanofibers. Significant changes occurred between the SNF-6 and SNF-7 filters and between the SNF-8 and SNF-9 filters. The filtration efficiencies did not increase significantly from SNF-6 to SNF-9, and thus, the apparently decreased QF indicated a significant increase in the pressure drop when the SiO_2 loading changed from SNF-6 to SNF-7 and from SNF-8 to SNF-9. In particular, the pressure drop increased by 68.8% when the filter loading level changed from SNF-8 to SNF-9. Meanwhile, the QF value did not vary significantly for SNF-7 and SNF-8, with different SiO_2 nanofiber loadings. In summary, an area-specific loading density between 1.51 and 2.27 mg/cm^2 may be the optimal SiO_2 loading level for such nanofiber membranes to achieve the best overall filtration performance to remove Te, Pb, and Bi vapors in the lead-based reactor.

3.2.3. Flow Rate of Carrier Gas

To study the filtration efficiency under different gas flow conditions, flow rates of 600, 1000, and 1400 sccm were selected for testing. The filtration efficiency results of the SNF at different gas flow rates and filtration temperatures of 400 °C and 500 °C are shown in Figure 7. The results revealed that the increase in the gas flow rate generally reduced the efficiency of the filter to remove aerosols in the gas phase. The temperature effect of the reduced filtration efficiency at 500 °C compared to that at 400 °C under the same gas flow rate was consistent with the findings in Figure 5. All these results suggest that a high gas flow rate and filtration temperature are not recommended for the removal of gaseous components and aerosols in the gas phase by the SiO_2 fiber filter. In particular, increasing the gas flow rate would shorten the contact time of the vapor or clusters with the fiber surface and ultimately affect the adhesion and aggregation on the surface, thereby lowering the filtration efficiency of the SiO_2 nanofiber membranes for Te, Pb, and Bi in the gas phase.

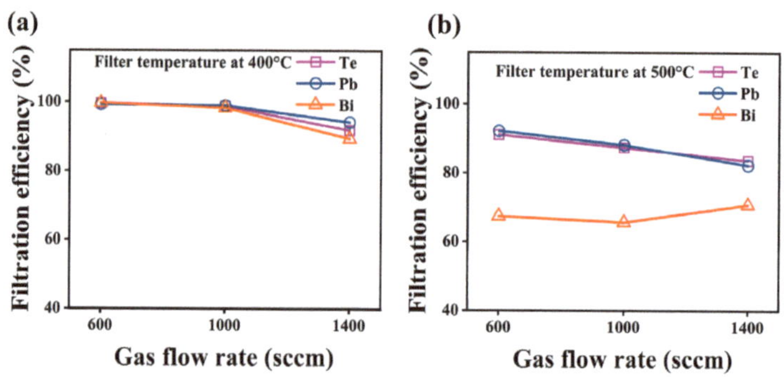

Figure 7. Filtration efficiencies of the SNF for Te, Pb, and Bi at different gas flow rates at filtration temperatures of (**a**) 400 °C and (**b**) 500 °C.

3.2.4. Cycle-Use Performance of SiO_2 Nanofiber Filter

The cycle-use experiments were performed at a filtration temperature of 400 °C and heated for 5 h. The elements of Te, Pb, and Bi on the SiO_2 nanofiber membrane were desorbed by nitric acid solution (7 mol/L). Then, the filter membrane was freeze-dried to obtain a regenerative filter for repeatable experiments. After repeated filtration for five runs, the filtration efficiency of SNF did not decrease significantly (Figure 8). The results showed that the filtration performance of the SiO_2 filter had excellent cycle-use performance, which ensures the long-term stability of purification system at high temperatures.

3.3. DFT Calculation

The adhesion behaviors of the volatile components PbTe and Bi from the LBE-Te sample with amorphous SiO_2 were considered. For comparison, the adhesion properties of Te and Po-containing molecules (Po, PbPo)) were also evaluated. Figure 9 shows the adhesion configuration of monoatomic molecules (Po, Te, Bi) and diatomic molecules (PbTe, PbPo) at the defect sites on the amorphous SiO_2 surface. Table 2 lists the calculated adhesion energies of the different configurations on the amorphous SiO_2 surface defect sites. The results showed that the adhesion energies of all the configurations were negative, indicating that the molecules could be spontaneously adsorbed on the SiO_2 surface. The adhesion energies for monoatomic molecules (Te, Bi, and Po) on the SiO_2 surface were between −3.72 and −4.34 eV, while the adhesion energies for diatomic molecules (PbTe and PbPo) on the SiO_2 surface ranged from −2.0 to −2.96 eV. The adhesion of monoatomic molecules was significantly stronger than that of the diatomic molecules on the SiO_2 surface. However, Te molecules that had strong interactions with SiO_2 were not found in the current Te capture experiment, indicating that the volatile component of Te in the LBE was mainly PoTe. This suggested that the most likely component to contain Po in the cover gas in a real reactor will be PbPo. The calculated adhesion energy results based on different adsorption configurations for diatomic components revealed that configurations e_1 and f_1, with the Pb atom in the PbTe or PbPo molecules binding to the SiO_2 defect sites, had stronger adhesion effects than configurations e_2 and f_2, respectively. This indicated that lead dominated the interactions between PoPb and the filter materials. Based on the experimental and theoretical results, PbTe and Bi were removed from the gas by strong adhesion to the surfaces of the SiO_2 nanofibers. Nonetheless, the filtration efficiency of the SNF for PbTe was significantly better than that for Bi at filtration temperatures above 350 °C, which was inconsistent with the calculation result that the adhesion energy of Bi was greater than that of PbTe. This was because the melting point of Bi was much lower than the filtration temperature used in this study (300–500 °C), which increased the difficulty for capturing Bi.

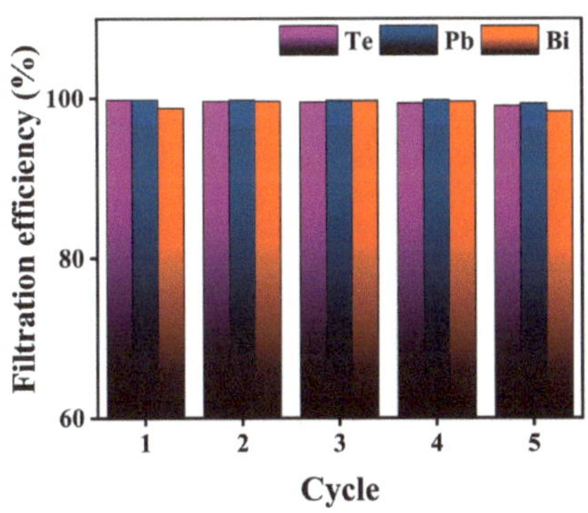

Figure 8. Filtration efficiency in five cycles.

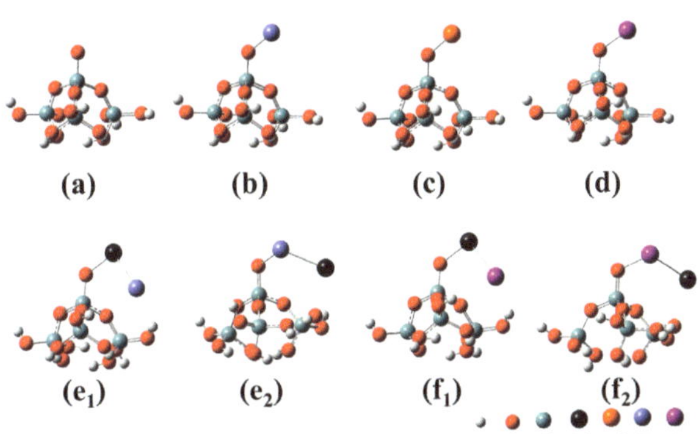

Figure 9. Adsorption configurations of monoatomic molecules (Po, Te, and Bi) and diatomic molecules (PbTe and PbPo) on the ≡Si-O adsorption site used for density functional theory (DFT) calculations: (**a**) SiO_2, (**b**) Te/SiO_2, (**c**) Bi/SiO_2, (**d**) Po/SiO_2, (**e_1,e_2**) PbTe/SiO_2, and (**f_1,f_2**) PbPo/SiO_2.

Table 2. Calculated adhesion energies of monoatomic molecules and diatomic molecules on the ≡Si-O· adsorption sites.

Diatomic Molecules	Configuration	E_{ads} (eV)	Monatomic Molecules	Configuration	E_{ads} (eV)
Te	b	−4.34	PbTe	e1	−2.83
Bi	c	−3.72		e2	−2.00
Po	d	−4.15	PbPo	f1	−2.96
				f2	−2.06

4. Conclusions

In this work, electrostatic spun flexible SiO_2 nanofiber membrane filters exhibited outstanding high temperature and corrosion resistance. This paper investigates the filtration performance of SiO_2 nanofiber membrane filters under different conditions of filtration temperature, loading level of the SiO_2 filter, and flow rate of carrier gas. In general, increasing the filtration temperature and the flow rate of carrier gas has a negative effect on the filtration efficiency of SNF. For example, the filtration temperature and the flow rate of carrier gas in this experiment are not suitable to exceed 400 °C and 1000 sccm. Similarly, the load level of SiO_2 filters should not be too excessive to avoid large pressure drops in the filtration system. Due to its good mechanical properties and acid resistance, the flexible SiO_2 nanofiber membrane could maintain its filtration performance after multiple regenerations. The strong adhesion of amorphous SiO_2 to the volatile components in LBE-Te was demonstrated by DFT calculations. Based on the filtration efficiency of SiO_2 nanofiber membranes for Te aerosols up to 99%, SiO_2 nanofibers are expected to be used as a filtration material to remove Po from lead alloy-cooled reactors.

Author Contributions: Conceptualization, X.W., X.H. and X.C. (Xujie Chen); methodology, X.C. (Xiyong Chen); investigation, X.L.; resources, X.W.; writing—original draft preparation, X.C. (Xujie Chen); writing—review and editing, X.C. (Xujie Chen); supervision, T.F., X.Z. and Y.Z.; funding acquisition, X.W. and X.H. All authors have read and agreed to the published version of the manuscript.

Funding: This research was funded by the China Nuclear Power Technology Research Institute, grant number 008-ZB-B-2018-C30-00223, and the Science and Technology Program of Shenzhen, grant number JCYJ20180305124244969.

Institutional Review Board Statement: Not applicable.

Informed Consent Statement: Not applicable.

Data Availability Statement: Not applicable.

Conflicts of Interest: The authors declare no conflict of interest.

References

1. Gohar, Y.; Cao, Y.; Kraus, A.R. ADS design concept for disposing of the U.S. spent nuclear fuel inventory. *Ann. Nucl. Energy* **2021**, *160*, 108385. [CrossRef]
2. Toshinsky, G.I.; Dedul, A.V.; Komlev, O.G.; Kondaurov, A.V.; Petrochenko, V.V. Lead-Bismuth and Lead as Coolants for Fast Reactors. *World J. Nucl. Sci. Technol.* **2020**, *10*, 65. [CrossRef]
3. Aerts, A.; Gladinez, K.; Prieto, B.G.; Lim, J.; Marino, A.; Rosseel, K. The LBE Coolant Chemistry R&D Programme for the MYRRHA ADS: Chemistry and Control of Oxygen, Corrosion and Spallation Products. In Proceedings of the 14th International Workshop on Spallation Materials Technology, Fukushima, Japan, 11–17 November 2018; Volume 28, p. 071002.
4. Kikuchi, K.; Saito, S.; Kurata, Y.; Futakawa, M.; Sasa, T.; Oigawa, H.; Wakai, E.; Umeno, M.; Mizubayashi, H.; Miura, K. Lead-bismuth eutectic compatibility with materials in the concept of spallation target for ADS. *JSME Int. J. Ser. B* **2004**, *47*, 332–339. [CrossRef]
5. Scott, B.R. Health risk evaluations for ingestion exposure of humans to polonium-210. *Dose-Response* **2007**, *5*, 94–122. [CrossRef]
6. Glasbrenner, H.; Eikenberg, J.; Gröschel, F.; Zanini, L. Polonium formation in Pb–55.5Bi under proton irradiation. *J. Nucl. Mater.* **2004**, *335*, 270–274. [CrossRef]
7. Heinitz, S.; Neuhausen, J.; Schumann, D. Alkaline extraction of polonium from liquid lead bismuth eutectic. *J. Nucl. Mater.* **2011**, *414*, 221–225. [CrossRef]

8. Buongiorno, J.; Loewen, E.P.; Czerwinski, K.; Larson, C. Studies of polonium removal from molten lead-bismuth for lead-alloy-cooled reactor applications. *Nucl. Technol.* **2004**, *147*, 406–417. [CrossRef]
9. Yefimov, E.I.; Pankratov, D.V.; Ignatiev, S.V. Removal and Containment of High-Level Radioactive Polonium from Liquid Lead-Bismuth Coolant. *MRS Online Proc. Libr.* **1997**, *506*, 679–686. [CrossRef]
10. Karlsson, E.; Neuhausen, J.; Aerts, A.; Danilov, I.I.; Eichler, R.; Turler, A.; Vogele, A. Polonium behavior following a vacuum window rupture in a lead-bismuth eutectic based accelerator driven system. *Appl. Radiat. Isot.* **2021**, *168*, 109551. [CrossRef]
11. Prieto, B.G.; Marino, A.; Lim, J.; Rosseel, K.; Martens, J.; Rizzi, M.; Neuhausen, J.; Van den Bosch, J.; Aerts, A. Use of the transpiration method to study polonium evaporation from liquid lead-bismuth eutectic at high temperature. *Radiochim. Acta* **2014**, *102*, 1083–1091. [CrossRef]
12. Mao, L.; Dang, T.; Pan, L.; Zeng, Q.; Bai, Y.; Team, F.D.S. Preliminary analysis of polonium-210 contamination for China LEAd-based Research Reactor. *Prog. Nucl. Energy* **2014**, *70*, 39–42. [CrossRef]
13. Li, N. Lead-alloy coolant technology and materials-technology readiness level evaluation. *Prog. Nucl. Energy* **2008**, *50*, 140–151. [CrossRef]
14. Wang, G.; Niu, S.; Cao, R. Summary of severe accident issues of LBE-cooled reactors. *Ann. Nucl. Energy* **2018**, *121*, 531–539. [CrossRef]
15. Fazio, C.; Sobolev, V.; Aerts, A.; Gavrilov, S.; Lambrinou, K.; Schuurmans, P.; Gessi, A.; Agostini, P.; Ciampichetti, A.; Martinelli, L. *Handbook on Lead-Bismuth Eutectic Alloy and Lead Properties, Materials Compatibility, Thermal-Hydraulics and Technologies*, 2015 ed.; Organisation for Economic Co-Operation and Development: Paris, France, 2015.
16. Obara, T.; Yamazawa, Y.; Sasa, T. Polonium decontamination performance of stainless steel mesh filter for lead alloy-cooled reactors. *Prog. Nucl. Energy* **2011**, *53*, 1056–1060. [CrossRef]
17. Obara, T.; Koga, T.; Miura, T.; Sekimoto, H. Polonium evaporation and adhesion experiments for the development of polonium filter in lead-bismuth cooled reactors. *Prog. Nucl. Energy* **2008**, *50*, 556–559. [CrossRef]
18. Aiello, A.; Azzati, M.; Benamati, G.; Gessi, A.; Long, B.; Scaddozzo, G. Corrosion behaviour of stainless steels in flowing LBE at low and high oxygen concentration. *J. Nucl. Mater.* **2004**, *335*, 169–173. [CrossRef]
19. Maugeri, E.A.; Neuhausen, J.; Prieto, B.G.; Aerts, A.; Mendonca, T.M.; Stora, T.; Eichler, R. Adsorption of volatile polonium species on metals in various gas atmospheres: Part III-Adsorption of volatile polonium on stainless steel 316L. *Radiochim. Acta* **2018**, *106*, 125–134. [CrossRef]
20. Maugeri, E.A.; Neuhausen, J.; Misiak, R.; Eichler, R.; Dressler, R.; Piguet, D.; Vogele, A.; Schumann, D. Adsorption of volatile polonium species on metals in various gas atmospheres: Part II-Adsorption of volatile polonium on platinum, silver and palladium. *Radiochim. Acta* **2016**, *104*, 769–779. [CrossRef]
21. Maugeri, E.A.; Neuhausen, J.; Eichler, R.; Dressler, R.; Rijpstra, K.; Cottenier, S.; Piguet, D.; Vogele, A.; Schumann, D. Adsorption of volatile polonium and bismuth species on metals in various gas atmospheres: Part I-Adsorption of volatile polonium and bismuth on gold. *Radiochim. Acta* **2016**, *104*, 757–767. [CrossRef]
22. Yang, Y.; Cao, Q.; Chen, Y.; Wu, X.; Zuo, W.; Wu, W.; Lu, Y.; Xiong, W.; Liu, Y. Application of density functional theory to the adsorption of Po, Po_2, PbPo, H_2Po, and PoOH on Ag(111) surfaces for ^{210}Po capture in lead–bismuth eutectic coolant environments. *Appl. Surf. Sci.* **2021**, *554*, 149599. [CrossRef]
23. Jiang, M.; Zhao, Z.; Du, H.; Chen, Z.; Ni, M. Role of surface defects in trap of monoatomic polonium on Pd surface. *J. Nucl. Mater.* **2021**, *557*, 153205. [CrossRef]
24. Rijpstra, K.; Van Yperen-De Deyne, A.; Maugeri, E.A.; Neuhausen, J.; Waroquier, M.; Van Speybroeck, V.; Cottenier, S. Ab initio study of the trapping of polonium on noble metals. *J. Nucl. Mater.* **2016**, *472*, 35–42. [CrossRef]
25. Neuhausen, J. Radionuclide Chemistry in Nuclear Facilities Based on Heavy Liquid Metal Coolants: Past, Present and Future. *CHIMIA Int. J. Chem.* **2020**, *74*, 976–983. [CrossRef] [PubMed]
26. Xue, J.J.; Wu, T.; Dai, Y.Q.; Xia, Y.N. Electrospinning and Electrospun Nanofibers: Methods, Materials, and Applications. *Chem. Rev.* **2019**, *119*, 5298–5415. [CrossRef] [PubMed]
27. Yu, M.; Guo, Y.; Wang, X.; Zhu, H.; Li, W.; Zhou, J. Lignin-based electrospinning nanofibers for reversible iodine capture and potential applications. *Int. J. Biol. Macromol.* **2022**, *208*, 782–793. [CrossRef]
28. Liu, H.; Bell, N.S.; Cipiti, B.B.; Lewis, T.G.; Sava, D.F.; Nenoff, T.M. *Functionalized Ultra-Porous Titania Nanofiber Membranes as Nuclear Waste Separation and Sequestration Scaffolds for Nuclear Fuels Recycle*; Sandia National Lab. (SNL-NM): Albuquerque, NM, USA, 2012.
29. Zhang, W.; Tian, Y.; Liu, D.-C.; Wang, F.; Yang, B.; Xu, B.Q. Experimental study on the thermal volatilization and condensation of zinc at 10 Pa and 200 Pa. *J. Mater. Res. Technol.* **2020**, *9*, 3590–3597. [CrossRef]
30. Cai, X.; Huang, Q.X.; Alhadj, M.; Moussa, M.; Chi, Y.; Yan, J.H. Characterization of zinc vapor condensation in fly ash particles using synchrotron X-ray absorption spectroscopy. *J. Zhejiang Univ.-Sci. A* **2015**, *16*, 70–80. [CrossRef]
31. Liu, Z.S. Control of heavy metals during incineration using activated carbon fibers. *J. Hazard. Mater.* **2007**, *142*, 506–511. [CrossRef]
32. Xiao, J.; Liang, J.; Zhang, C.; Tao, Y.; Ling, G.W.; Yang, Q.H. Advanced Materials for Capturing Particulate Matter: Progress and Perspectives. *Small Methods* **2018**, *2*, 1800012. [CrossRef]
33. Wei, Z.; Su, Q.; Wang, X.; Long, S.; Zhang, G.; Lin, Q.; Yang, J. Nanofiber Air Filters with High-Temperature Stability and Superior Chemical Resistance for the High-Efficiency PM2.5 Removal. *Ind. Eng. Chem. Res.* **2021**, *60*, 9971–9982. [CrossRef]

34. Tepekiran, B.N.; Calisir, M.D.; Polat, Y.; Akgul, Y.; Kilic, A. Centrifugally spun silica (SiO_2) nanofibers for high-temperature air filtration. *Aerosol Sci. Technol.* **2019**, *53*, 921–932. [CrossRef]
35. Peng, Y.; Xie, Y.; Wang, L.; Liu, L.; Zhu, S.; Ma, D.; Zhu, L.; Zhang, G.; Wang, X. High-temperature flexible, strength and hydrophobic YSZ/SiO_2 nanofibrous membranes with excellent thermal insulation. *J. Eur. Ceram. Soc.* **2021**, *41*, 1471–1480. [CrossRef]
36. Maugeri, E.A.; Neuhausen, J.; Eichler, R.; Piguet, D.; Schumann, D. Thermochromatography study of volatile tellurium species in various gas atmospheres. *J. Nucl. Mater.* **2014**, *452*, 110–117. [CrossRef]
37. Ohno, S.; Kurata, Y.; Miyahara, S.; Katsura, R.; Yoshida, S. Equilibrium evaporation behavior of polonium and its homologue tellurium in liquid lead-bismuth eutectic. *J. Nucl. Sci. Technol.* **2006**, *43*, 1359–1369. [CrossRef]
38. Neuhausen, J.; Koster, U.; Eichler, B. Investigation of evaporation characteristics of polonium and its lighter homologues selenium and tellurium from liquid Pb-Bi-eutecticum. *Radiochim. Acta* **2004**, *92*, 917–923. [CrossRef]
39. Ferullo, R.M.; Garda, G.R.; Belelli, P.G.; Branda, M.M.; Castellani, N.J. Deposition of small Cu, Ag and Au particles on reduced SiO_2. *J. Mol. Struct. THEOCHEM* **2006**, *769*, 217–223. [CrossRef]
40. Lopez, N.; Illas, F.; Pacchioni, G. Adsorption of Cu, Pd, and Cs atoms on regular and defect sites of the SiO_2 surface. *J. Am. Chem. Soc.* **1999**, *121*, 813–821. [CrossRef]
41. Miehlich, B.; Savin, A.; Stoll, H.; Preuss, H. Results obtained with the correlation energy density functionals of becke and Lee, Yang and Parr. *Chem. Phys. Lett.* **1989**, *157*, 200–206. [CrossRef]
42. Mertens, M.A.J.; Aerts, A.; Infante, I.; Neuhausen, J.; Cottenier, S. Po-Containing Molecules in Fusion and Fission Reactors. *Chem. Phys. Lett.* **2019**, *10*, 2879–2884. [CrossRef]
43. Van Yperen-De Deyne, A.; Rijpstra, K.; Waroquier, M.; Van Speybroeck, V.; Cottenier, S. Binary and ternary Po-containing molecules relevant for LBE cooled reactors at operating temperature. *J. Nucl. Mater.* **2015**, *458*, 288–295. [CrossRef]
44. Sobolev, V. *Database of Thermophysical Properties of Liquid Metal Coolants for GEN-IV*; SCK-CEN: Mol, Belgium, 2011.
45. Huang, Y.; Brebrick, R.F. Partial Pressures and Thermodynamic Properties for Lead Telluride. *J. Electrochem. Soc.* **1988**, *135*, 486–496. [CrossRef]
46. Ubelis, A.P. Temperature dependence of the saturated vapor pressure of tellurium. *J. Eng. Phys.* **1982**, *42*, 309–315. [CrossRef]

toxics

MDPI

Review

A Review of Soil Contaminated with Dioxins and Biodegradation Technologies: Current Status and Future Prospects

Nguyen Thi Hong Nhung [1], Xuan-Tung Tan Nguyen [2], Vo Dinh Long [3], Yuezou Wei [4] and Toyohisa Fujita [1,5,*]

1 School of Resources, Environment and Materials, Guangxi University, Nanning 530004, China; hnhung_2912@yahoo.com
2 Faculty of Environmental and Food Engineering, Nguyen Tat Thanh University, Ho Chi Minh 700000, Vietnam; ntxtung@ntt.edu.vn
3 Institute of Environmental Science, Engineering and Management, Industrial University of Ho Chi Minh City, Ho Chi Minh 700000, Vietnam; vodinhlong@iuh.edu.vn
4 School of Nuclear Science and Technology, University of South China, Hengyang 421001, China; yzwei@usc.edu.cn
5 School of Chemistry and Chemical Engineering, Guangxi University, Nanning 530004, China
* Correspondence: fujitatoyohisa@gxu.edu.cn

Abstract: This article provides a comprehensive assessment of dioxins contaminating the soil and evaluates the bioremediation technology currently being widely used, and also offers recommendations for future prospects. Soil pollution containing dioxins is extremely toxic and hazardous to human health and the environment. Dioxin concentrations in soils around the world are caused by a variety of sources and outcomes, but the main sources are from the consequences of war and human activities. Bioremediation technology (bioaugmentation, biostimulation, and phytoremediation) is considered an optimal and environmentally friendly technology, with the goal of applying native microbial communities and using plant species with a high biomass to treat contaminated dioxins in soil. The powerful bioremediation system is the growth of microorganisms that contribute to the increased mutualistic and competitive relationships between different strains of microorganisms. Although biological treatment technology can thoroughly treat contaminated dioxins in soil with high efficiency, the amount of gas generated and Cl radicals dispersed after the treatment process remains high. Further research on the subject is required to provide stricter control over the outputs noted in this study.

Keywords: dioxins; soil; contamination; bioremediation; toxic

Citation: Nhung, N.T.H.; Nguyen, X.-T.T.; Long, V.D.; Wei, Y.; Fujita, T. A Review of Soil Contaminated with Dioxins and Biodegradation Technologies: Current Status and Future Prospects. *Toxics* **2022**, *10*, 278. https://doi.org/10.3390/toxics10060278

Academic Editor: Łukasz Chrzanowski

Received: 2 May 2022
Accepted: 23 May 2022
Published: 24 May 2022

Publisher's Note: MDPI stays neutral with regard to jurisdictional claims in published maps and institutional affiliations.

Copyright: © 2022 by the authors. Licensee MDPI, Basel, Switzerland. This article is an open access article distributed under the terms and conditions of the Creative Commons Attribution (CC BY) license (https://creativecommons.org/licenses/by/4.0/).

1. Introduction

Dioxins are environmentally stable solid substances with high melting and boiling points and very low vapor pressure [1]. Dioxins are almost insoluble in water, have high thermal stability that can only be completely decomposed at temperatures above 1200 °C, and are resistant to strong acids and alkalis and adhere to the surface of organic resources, especially in soil [2,3]. In addition, dioxins are also substances which are man-made through activities such as the production of 2,4,5-T herbicides, chlorine-containing plant protection agents, combustion processes (the burning municipal waste, medical waste, industrial waste, especially waste containing PVC and metallurgical processes), and pulp bleaching with chlorine substances [4–6]. Dioxins are dangerous threat agents, even at low concentrations (one part per billion), which are enough to wreak havoc on human health and the environment [3]. In humans, dioxin exposure in humans effects the endocrine glands and reproductive functions, causes diseases in the central and peripheral areas of the nervous system, and impedes fetal development, especially in the nervous system and joints [7,8]. Dioxins can be remain in the environment for a long time, seeping deeply into soil and sediments [2]. Therefore, dioxins remaining in the soil and sediments will seep into water sources and ecosystems, including those that produce fish, shrimp, vegetables, and crops, posing a risk of poisoning for future generations [9].

Several studies have been conducted on physical, chemical, and biological dioxin remediation technologies. In physical treatment, physical processes, such as radioactive material degradation using solvent extraction and liquefied petroleum gas extraction methods [10,11], steam distillation [12], thermal desorption [13], and in situ vitrification [14] are used to degrade persistent organic pollutants,. Chemical reactions, such as basic catalytic decomposition, above and below extreme water treatment [15], photolysis at the spot level [16,17], electronic solvation technology [18], and K-polyethylene glycol technology [19] are used in chemical treatments to degrade persistent organic pollutants in the soil. Recently, nanotechnology has also been applied, using nanoscale zero-valent iron [20] to decompose persistent organic pollutants. Bioremediation aims to remove persistent organic compounds from contaminated soil using the anaerobic and aerobic decomposition of microorganisms [21,22]. Indigenous microorganisms enriched from dioxin-contaminated sites are believed to be able to remove dioxins [23,24]. The dechlorination of dioxins by microbial metabolism under anaerobic and aerobic conditions are the two main mechanisms of dioxin degradation [25]. Microbial strains can use dioxins as a carbon and energy source [26] to effectively dechlorinate dioxins against highly chlorinated congeners [27]. Phytoremediation is more effective and widespread, but only with the use of plants that have a large biomass, and is capable of dioxin adsorption in many laboratory and field studies [28,29].

There are many studies on airborne and food dioxins as the main sources in different countries, but there are not many general reports on dioxin-contaminated soil, especially for the bioremediation of dioxins in soil. Furthermore, this paper focuses on analyzing, comparing, and discussing a global overview of the situation of dioxin-contaminated soil. This review provides a comprehensive summary and discussion of relevant studies on dioxin-contaminated soil bioremediation. An assessment will contribute to the optimization parameters of bioaugmentation, biostimulation, and phytoremediation in the remediation of dioxin-contaminated soils. Current knowledge, research challenges/gaps, and prospects for future research are presented in this study.

2. Overview of Dioxins

Dioxins are persistent organic pollutants produced by both natural and human activities [30]. Dioxins are also the common name for a group of hundreds of chemical compounds that persist in the environment, as well as in the human body and other organisms [31]. Dioxins are very stable compounds with low polar, lipophilic, and hydrophobic qualities. They are classified into three groups: polychlorinated dibenzo-p-dioxins (PCDDs, referred to as dioxins), polychlorinated dibenzofurans (PCDFs, referred to as furans), and coplanar polychlorinated biphenyls (dioxin-like PCBs, referred to as dl-PCBs). The chemical structures of PCDDs, PCDFs, and PCBs are shown in Figure 1 [1]. Depending on the number of chlorine atoms and the spatial position, dioxins have 75 congeners PCDD (poly-chloro-dibenzo-dioxins) and 135 congeners PCDFs (poly-chloro-dibenzo-furans) with different toxicity. In 210 congeners, 17 congeners are known to be highly toxic because they can have chlorine atoms at positions 2, 3, 7, and 8, (at least) on the benzene ring [3]. Based on the international toxicity equivalence factor (I-TEF), 2,3,7,8-TCDD/TCDFs are known to be the most toxic compounds [32].

2.1. Sources, Fate, and Transportation of Dioxins in Soil

Dioxin sources are typically found in both natural and anthropogenic environments (Figure 2). Dioxins are a typical examples of persistent organic compounds with extremely complex structures and many different toxicities. Dioxins are naturally emitted from volcanic eruptions [33], forest fires [34], and natural combustion [35]. However, previous studies have shown that the emission of dioxins in the environment is mainly caused by humans. The main anthropogenic origins of dioxins are classified into three sources: incineration, combustion, and industrial processes [30,36,37].

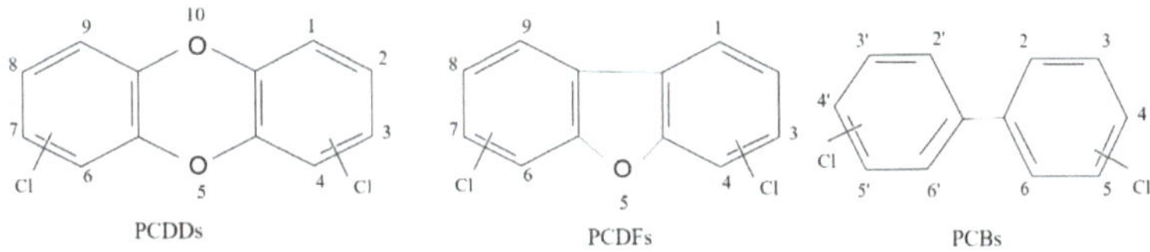

Figure 1. The basic structures of Polychlorinated dibenzo-p-dioxin (PCDD), dibenzofuran (PCDF), and biphenyls (PCBs).

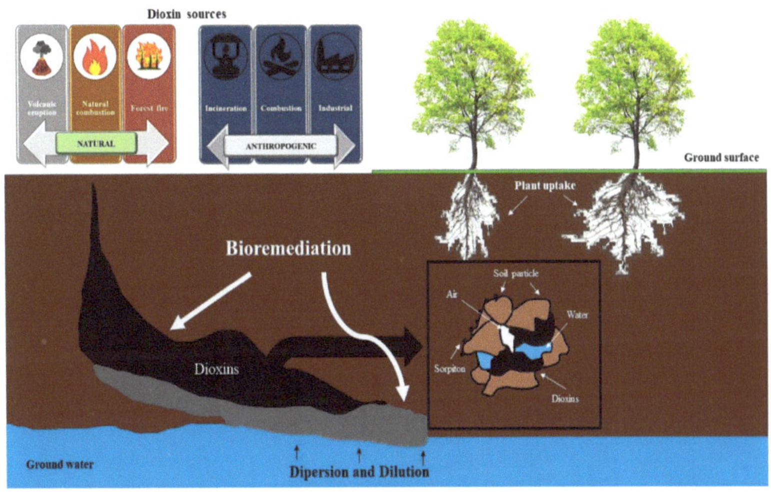

Figure 2. Sources, fate, and transportation of dioxins in soil.

Dioxins in soil are typically solid and cling to soil particles [38]. The fate and transport of dioxins in soil are depicted in Figure 2. Dioxins undergo diffusion and dispersion, as well as biodegradation processes (bioaugmentation, biostimulation, and phytoremediation) [39]. In addition, the fate and transport of dioxins in the soil media is affected by soil characteristics (moisture content, soil texture, pH, and organic matter) and environmental conditions (ground surface, groundwater, plants, weather conditions, and biological activity) [40]. Since soil particles often adsorb dioxins based on their low mobility and biodegradability [38], understanding the fate and transportation of dioxins in contaminated soil is essential in developing bioremediation technologies.

2.2. Toxicity and Health Risk Assessment

The World Health Organization (WHO-2005) has recommended the standard exposure levels of dioxins as 1–4 picograms WHO-TEQ/kg of body weight per one day, or 0.07 ppt in blood; the general environmental limit in most countries is 1200 ppt TEQ in soil and 150 ppt in sediment [3]. The United States Environmental Protection Agency (US-EPA) considers reducing the dioxins limit to 72 ppt TEQ to increase the volume of contaminated soil to be treated [3]. Dioxins, as a dangerous threat agent to the environment and humans, are associated with severe damage to human health, shortening the lives of those exposed and potentially shortening the lives of their children and future generations [3]. When dioxin levels in humans exceed the allowable threshold of 0.0064 picograms/kg of the human body (US-EPA), dark patches of skin appear quickly as a result of cell death, mutated pigment cells, and impaired liver and kidney function [8,41]. If long-term exposure to levels exceeds the threshold, dioxins will affect the immune system, endocrine glands, and reproductive functions, leading to some cancers and diseases of the central and peripheral nervous systems, thyroid disorders, immune system damage, endometriosis, diabetes, etc. [8,42]. As a result, children of exposed parents were born with many tragic deformities, and some children lived in a vegetative state from birth; the association between dioxins exposure and five diseases, such as soft tissue cancer, benign lymphoma, chronic lymphocytic leukemia, hairy leukemia cancer, and chlorosis was also noted [2,43]. Some diseases associated with dioxin exposure, such as acute, chronic, and subacute peripheral neuropathy; chlorosis; type 2 diabetes; liver cancer; lipid metabolism disorders; reproductive abnormalities and birth defects, such as cleft lip and palate; congenital malformations of the legs, hydrocephalus, neural tube defects, adhesions (sticky fingers or toes), muscle malformations, and paralysis [43] have also been observed. The half-life of dioxins in the soil is from 60 to 80 years, and at the same time, it persists for a long time in the environment, seeps into the soil and sediments, and migrates into vegetation and aquatic life, leading to bioaccumulation in the soil and food chain [2,9,44].

3. Situation of Dioxin-Contaminated Soil and Standard Limits

Environmental pollution in general, and dioxins pollution in the soil in particular, are markedly on the rise. The main sources of dioxin emissions are industrial activities (such as combustion) which are an important part of human production. Dioxin emissions from G20 industrial activities account for more than 80% of the estimated annual emissions [41]. According to Miguel Dopico and Alberto Gómez et al., 2015, annual global dioxin emissions were 17,226 kg, equivalent to about 287 kg-TEQ. The main sources of dioxins in the soil environment are fuel combustion, metal production, pesticide production and use, waste incineration, accidental fires, landfill disposal, combustion, and herbicide runoff in agricultural uses [42].

Many countries around the world are currently hotspots for dioxin contamination, including Germany [43], Korea [44], the Mediterranean [45], South Africa [46], Poland [47], China [48], Vietnam [3,49], etc. In China, soil dioxin concentrations are primarily found in soils in the vicinity of production areas, such as urban, agricultural, and mountain soils [48]. Dioxin concentrations in soil in China have varied ranges at the provincial sampling points, listed in the Table 1 below, used to assess the concentration and health risks of people living in the area. According to research by Gene J. Zheng et al. in the mainland of China, Hong Kong, and Taiwan, the main source of PCDD/Fs pollution is from industrial waste activities, with total PCDD/Fs up to 967,500 ppt of dry weight in a soil sample located in the eastern part of Guangdong Province [50]. Soil samples in urban and rural areas in Liaoning province were also compared, and the results revealed that the concentration of dioxin-like PCBs in urban areas is higher than in rural areas [51]. In general, the dioxin concentrations in the east of China are lower than in the rest of China, and the urban and manufacturing areas are higher than the rural areas. While dioxins in the soil are primarily found in China due to industrial activities, high concentrations of dioxins have been found in Vietnam in areas affected by previous wars due to Agent Orange [49]. The majority of research documents on dioxin content in Japanese soil come from agricultural sources and incinerators; there have not been many studies published on this topic in recent years. A study of soil samples taken in China and Korea's coastal areas revealed that the concentrations of PCDD/Fs at the sampling sites in Korea were higher than those in China, but both countries are lower than Japan [52].

Table 1. The average concentration of dioxins (homogeneous unit calculated in ppt TEQ in dry weight) in some different nations in Asia.

Country	Year	Type of Soil	Source Area	Concentration	References
China (Sichuan)	2013	Soil	High mountain area	2.48–4.30 ppt	[53]
China (mainland Hong Kong and Tai wan)	2008	Soil	Schistosomiasis disease area	244.8–33,660 ppt	[50]

Table 1. *Cont.*

Country	Year	Type of Soil	Source Area	Concentration	References
		Soil	E-waste recycling	799,000–967,500 ppt	[50,54]
		Paddy soil	E-waste recycling	2552–2726 ppt	[50]
		Soil	Pentachlorophenol manufacturing factory	606,000 ppt	[50,55,56]
South China	2022	Surface Soil	Municipal solid waste incinerator	114–2440 ppt	[57]
North China	2020	Soil	Urban green space in a metropolis	11.5–91.4 ppt	[58]
Eastern China	2009	Surface Soil	Electronic solid-waste with incinerators	0.017–5.04 ppt	[59]
North China	2011	Topsoil	Coastal areas	6.78–12.3 ppt	[52]
Central Vietnam	2019	Surface soil	The storage of Agent Orange in A-So Airbase during the Vietnam War	2.7 to 746 ppt	[60]
Southern Vietnam	2007	Topsoil	Bien Hoa Airbase was a former storage depot for Agent Orange	4.6–184 ppt	[61]
Japan (Osaka)	2013	Surface soil	Incineration plant	>1000 ppt	[62]
		Paddy field soil	Former herbicide use	38–110 ppt	
Japan (Akita)	2007	Paddy soil	Agricultural area	18,000–540,000 ppt	[63]
		Non-agricultural soil samples	Parks	950–1400 ppt	
South Korea	2021	Soil	Industrial sites	77.73 ppt	[64]
West Korea	2011	Topsoil	Coastal areas	14.2–27 ppt	[52]

Dioxin concentrations have not decreased in Europe in recent years, but instead, have increased significantly due to new dioxins emission sources, such as the illegal disposal of electric transformers in Italy [65] and Sweden [66]. Besides the main sources of pollution from manufacturing industries, motor vehicle emissions are one of the sources of pollution in Russia, because motor fuel combustion is dependent on the dopes used [67]. The results of dioxin concentrations in various regions of European countries are compared in Table 2; it is revealed that the concentration of dioxins in the soil in Spain, Slovakia, and Austria was lower than in other countries. Similarly, another study in Spain on dioxin levels affected by various sources, including municipal solid waste incineration, clinical waste incineration, and industrial areas, found results ranging from 0.45 to 14.41 ppt-TEQ (dry weight), with the effects of uncontrolled combustion processes being the most severe [68]. Research results on dioxin content in the soil in the UK by C.S. Creaser et al. showed that uncontrolled combustion leads to higher results in urban areas than in rural areas [69,70].

Table 2. The average concentration of dioxins (homogeneous unit calculated in ppt TEQ in dry weight) in some different nations in Europe.

Country	Year	Type of Soil	Source Area	Concentration	References
Sweden	2013	Soil	Contaminated sawmill site	0.62–690,000 ppt	[66]
Russia	2011	Soil	Urban site	8.2 ppt	[67]
Poland	2015	Soil	Urban site	475.48–3039.27 ppt	[47]
Germany	2007	Soil	Alluvial flood plain of the river	7680 ppt	[57]
Spain	2006	Topsoil	High industrial activity zones	0.33–9.99 ppt	[71]
Slovakia	2012	Topsoil	Industrial site	0.34 to 18.05 ppt	[72]
Austria	2004	Soil	Agricultural site	0.05–23 ppt	[73]

Similar to the results of searching for research documents on the situation of dioxin pollution in the soils of European countries, there are few data on the situation of soil research in America in recent years. In general, across the United States, dioxin concentrations in urban soil are generally higher than in rural soil, with maximum concentrations reaching 186 ng/kg according to TEQ in urban areas [74]. Another study conducted in Washington state discovered relatively low dioxin concentrations in soils, ranging from

0.14 to 4.1 ng/kg by TEQ, with agricultural land having the lowest dioxin value and urban land having the highest dioxin value, in accordance with other studies [75].

Research data on dioxins in countries in Africa are limited because the cost of dioxin analysis is expensive, and the analytical capacity is limited in this country [76]. According to research by C. Nieuwoudt et al., the dioxin concentrations in the soil of the sampled areas in Africa ranged from 0.34 to 20 ng/kg by TEQ, lower than those in Europe and the US [46,77]. The study also found that dioxin concentrations at industrial sites were higher than in agricultural soils and higher than in non-industrial sites, with combustion sources being the primary polluters [77].

Depending on soil pollution and effective land-use planning, several organizations around the world, such as the US EPA and WHO, have issued regulations on the concentration of pollutants in the soil (e.g., dioxins). However, some countries have developed industries in which the emission of pollutants into the soil is high (or have been affected by war, such as Vietnam), so each country will have its own regulations on PCDD/Fs concentrations in soil. The details are presented in Table 3.

Table 3. Some standards limitations for PCDD/Fs (ppt TEQ) in different nations.

National	Standard Limitation	Comments	Regulation/Guideline Values	References
US EPA Region 5	11 ppt 38.6 ppt	PCDD in soil PCDFs in soil	US EPA Region 5 ecological screening levels	[78]
US EPA Region 9	39 ppt	Residential soil	US EPA Region 9 preliminary remediation goal for 2,3,7,8-TCDD	[79]
China (Taiwan)	1000 ppt	General soil	The standard limit—Taiwan EPA	[80]
Vietnam	100 ppt 300 ppt 1200 ppt	Forest soil Agricultural soil Commercial soil	National technical regulation on the permissible limit of dioxins in soil	[81]
Finland	500 ppt	Agricultural and residential soil	Finland Ministry of the Environment, Department for Environmental Protection	[82]
Sweden	10 ppt 250 ppt	Land with sensitive use, Land with less sensitive use and groundwater extraction	Sweden Generic Guidance Value	[82]
Netherlands	10 ppt 1000 ppt	Dairy farming Agricultural and residential soil	The Netherlands Guidelines	[82]
Germany	5–40 ppt 100 ppt 1000 ppt 10,000 ppt	Agriculture Landscape Residential soil Industrial soil	Germany regulatory limit and recommendation	[82]
New Zealand	100 ppt 1500 ppt 18,000 ppt 90,000 ppt 21,000 ppt	Agricultural soil Residential soil Industrial soil Industrial-paved soil Maintenance	New Zealand Interim Acceptance Criteria	[83]
Canada	4 ppt	Alert soil	Canadian Environmental Quality Guidelines	[84]

4. Biodegradation Technologies of Dioxins

Biological treatment methods to reduce pollutants in different environments are considered effective and environmentally friendly [39]. Bioaugmentation is the addition of a degradation capacity into the soil to increase contaminant degradation, with effective potential for the bioremediation of organic-contaminated soils [85,86]. The addition of nutrients can encourage microbial activity by adjusting soil nutrients, which is known as biostimulation [87]. Composting is used to convert organic waste into simple organic substances. Bio-composting has traditionally been considered an eco-friendly remedy for organic soil contaminants, including petroleum, dioxins, and furans. Composting incubation is divided into mesophilic, thermophilic, cooling, and maturation stages, depending on microbial metabolism and heat production [88]. The mesophilic phase (<45 °C) occurs when the microbial community adapts to the initial conditions, and their numbers increase rapidly due to the readily degradable organic substrates [89]. Phytoremediation is frequently studied for its potential for immediate soil use with persistent organic compounds [28]. The main mechanisms involved in phytoremediation are based on the combined effects of plant uptake and the accumulation of toxic substances [90]. In this paper, bioremediation technology, including bioaugmentation, biostimulation, and phytoremediation, is analyzed and discussed below for the treatment of dioxin contamination in soil.

4.1. Bioaugmentation

Using microorganisms and fungus is currently an active application trend in dioxin-contaminated soil because of its low cost and environmental friendliness. The dechlorination of dioxins by microbial metabolism under anaerobic and aerobic conditions are the two main mechanisms of dioxin degradation in biological treatment [25]. Microbial strains can use dioxins as a carbon and energy source [26] to effectively dechlorinate dioxins from highly chlorinated isomers [27]. Table 4 lists microorganisms strains capable of degrading dioxins. Certain microorganisms such as *Pseudomonas, Mendocino*, and *Dehalococcoides* have been shown to effectively dechlorinate dioxins under anaerobic conditions [25,91,92]. Furthermore, aerobic microorganisms were discovered to degrade dioxins more efficiently and quickly than typical dioxin-contaminated soil anaerobes. Catechol 2,3-dioxygenase (C23O) is an important enzyme that catalyzes the reaction using molecular oxygen to destroy benzene rings [22], and *Bacillus* (*Firmicutes*) is the most dominant strain in aerobic degradation [93]. In addition to the strains of microorganisms that have been found to degrade dioxins with high efficiency, fungi also play a similar role, with their high mass and rapid environmental metabolism. Fungi are a diverse group of organisms, present in most environments and playing an integral role in ecosystems. In addition, fungi can regulate the flow of nutrients and energy through their network [89]. Furthermore, fungi are also unique organisms that can be used in the remediation of persistent organic wastes (POPs) in different environments, such as soil, water, and air [94]. Soils heavily contaminated with dioxins can also use fungi to decompose (with high efficiency) some typical strains such as *Cordyceps sinensis* strain *A* [15], *Phlebia lindtneri* [95,96], *Phanerochaete sordida YK-264* [97], etc. Table 5 presents some fungal strains capable of decomposing dioxins in soil, with high efficiency.

Table 4. Bacterial strains capable of biodegrading dioxins in a soil matrix.

Bacterial Strains	PCDD/Fs Congeners	Concentration	Removal Average (%)	Time	References
Terrabacter sp. strain *DBF63*	2-CDD	10 µg/mL	75	18 h	[98]
	2,3-CDD		80		
	2-CDF		82.5		
	2,8-DCDF		85		
Pseudomonas sp. strain *CA10*	2-CDF		60		
Pseudomonas sp. strain *CA10*	2-CDD	1 µg/mL	97	5 d	[98]
	2,3-CDD		89		
Sphingomonas sp. strain *RW1*	DD	10 ppm	90	24 h	[99]
	2-CDD		90		
Sphingomonas sp. strain *KA1*	2-CDD	1 µg/g	96	7 d	[100,101]
	2,3-DCDD		70		

Table 4. *Cont.*

Bacterial Strains	PCDD/Fs Congeners	Concentration	Removal Average (%)	Time	References
Rhodococcus opacus SAO 101	1-CDD	1 ppm	92	7 d	[102]
	Dioxin (DD)		97		
Pseudomonas aeruginosa	3,6-DCDF	10 mg/L	60	5 d	[103]
	1,2,3,4-TCDD		84		
	DBF		90		
Pseudomonas veronii PH-03	1-MCDD	1 µM	88.3	60 h	[104]
	2-MCDD		78.6		
	DD		90.7		
	DF		79.7		
Sphingomonas sp. wittichi RW1	DD	1 mM	81	72 h	[105]
	PCDD	29 ppt	75.5	15 d	
Pseudomonas resinovorans strain *CA10*	2,3-DCDD	1 µg/kg	90.95	7–14 d	[106]
Pseudomonas resinovorans strain *CA10*	2,3-DCDD	1000 µg/L	100	14 d	[106]
Pseudomonas sp. CA10	2-CDD	10,000 µg/L	98.5	7 d	[99]
Pseudallescheria boydii	2,3,7,8-TCDD	125 ng/g	92	15 d	[107]
Stropharia rugosoannulata	1,2,3,4,6,7,8-HpCDF	200 µg/L	64	3 m	[91]
Bacillus-Firmicutes	2,3,7,8-TCDD	136.33 ng/g	75	42 d	[108]
Bosea BHBi7	2,3,7,8-TCDD	170 ng/g	59.1	21 d	[109]
Hydrocarboniphaga BHBi4					
Pseudomonas mendocina NSYSU	OCDD	20.1 mg/kg	74	60 d	[92]

Table 5. Degradation of dioxins by fungi strains in soil matrix.

Fungi sp. Name	Pollutants Compounds	Nutrients/Conditions	Removal (%)	Time	References
Cordyceps sinensis strain *A*	DD	Glucose or 1,4-dioxane	50	4 d	[15]
	2,3,7-CDD		50		
	octaCDD		50		
Phanerochaete sordida YK-624	2,3,7,8-TetraCDD	Glucose	70	7 d	[97]
	1,2,3,7,7-PentaCDD		70		
	1,2,3,4,7,8-HexaCDD		75		
	1,2,3,4,6,7,8-HeptaCDD		70		
	1,2,3,4,6,7,8,9-OctaCDD		70		
	2,3,7,8-TetraCDF		45		
	1,2,3,7,8-PentaCDF		45		
	1,2,3,4,7,8-HexaCDF		75		
	1,2,3,4,6,7,8-HeptaCDF		70		
	1,2,3,4,6,7,8,9-OctaCDF		70		
Acremonium sp. strain *622*	T4CDD	Activated sludge and effluent	73	24 h	[110]
	P5CDD		85		
	H6CDD		79		
	H7CDD		76		
	O8CDD		88		
	T4CDF		81		
	P5CDF		88		
	H6CDF		84		
	H7CDF		84		
	O8CDF		71		

Table 5. *Cont.*

Fungi sp. Name	Pollutants Compounds	Nutrients/Conditions	Removal (%)	Time	References
Phanerochaete chrysosporium strain *PcCYP11a3*	1-MCDD	Glucose	100	2 h	[111]
	2-MCDD		38.2		
	2,3-DCDD		6.1		
Pleurotus pulmonarius strain *BCRC36906*	HexaCDD/Fs	Solid state fermentation (SSF)	80	72 d	[112]
	HeptaCDD/Fs		97		
	OctaCDD/Fs		90		
Phlebia radiata strain *267*	1,2,3,4,7,8-H6CDD	Laccase, Tween-80 50 mL	28	30 d	[113,114]
	1,2,3,7,8-P5CDF		29		
	2,3,7,8-T4CDF		60		
Phlebia radiata strain *PL1*	1,2,3,7,8-P5CDD	Laccase, 50 mL Tween-80	76.3	30 d	[113,114]
	1,2,3,4,7,8-H6CDD		75.6		
	1,2,3,6,7,8-H6CDD		79.4		
	1,2,3,7,8,9-H6CDD		79.3		
	1,2,3,4,6,7,8-H7CDD		79		
	octaCDD		80		
	1,2,3,4,7,8-H6CDF		100		
	1,2,3,6,7,8-H6CDF		100		
	2,3,4,6,7,8-H6CDF		82.3		
	1,2,3,4,6,7,8-H7CDF		70.2		
	1,2,3,4,7,8,9-H7CDF		100		
	octaCDF		67.4		
P. brevispora strain *BMC3014*	2,7-DiCDD	Glucose and ammonium tartrate	33.8	14 d	[95,114]
	2,3,7-TriCDD		20		
	1,2,8,9-TetraCDD		15		
	1,2,6,7-TetraCDD		18		
P. brevispora strain *BMC9152*	2,7-DiCDD		54		
	2,3,7-TriCDD		30		
	1,2,8,9-TetraCDD		16.5		
	1,2,6,7-TetraCDD		26		
P. brevispora strain *BMC9160*	2,7-DiCDD		40		
	2,3,7-TriCDD		27		
	1,2,8,9-TetraCDD		23		
	1,2,6,7-TetraCDD		16.5		

4.2. Biostimulation

Microbial adsorption highly contributes to the mineralization and co-transformation of organic contaminants during composting [115]. Therefore, microbial activity is the most important factor for the biodegradation of organic contaminants. In addition, the effects of operational parameters, such as humidity, growing conditions, and C/N ratio on the bio-incubation process are very important. Humidity is known as the main factor affecting the biodegradation of organic contaminants because it affects the microbial activity and the physicochemical properties of the contaminants [22]. Oxygen molecules participate in the catabolism and mineralization of hydrocarbon compounds by microbial and fungal activities. In addition, the C/N ratio plays an important role in the biodegradation of contaminants because it controls the composition of the microbial community. Previous studies have shown that the optimal C/N ratio for the biodegradation of organic contaminants by bio-incubation is between 10 and 40 [116]. Bio-composting has been successfully used in the biodegradation of dioxin-contaminated soils on a laboratory scale [25]. Chen et al., 2016, reported that the biodegradation efficiency of PCDD/PCDFs was approximately 95.8–99.7%, from an initial toxic concentration of 1580–3660 μg I-TEQ/kg dw after 42 days of incubation. Table 6 summarizes the organic compositions used to degrade dioxins in the soil by biological composting.

Table 6. Summary of biodegradation statistics for dioxins in contaminated soil.

Initial Concentration	Mechanical Components	Materials	Removal (%)	Time (days)	Conditions	References
16,004 ng-TEQ/kg	Sandy loam	Food waste, sawdust, and compost	75	42	Aerobic	[108]
840–5300 ng-TEQ/kg	Sandy	Wood chips and compost	85	360	Semi-aerobic	[25]
30,000–60,000 ng-TEQ/kg	Sandy loam	Lime granules, Nutrients, and bark	21	175	Anaerobic	[117]
88.8–912.7 μmol/kg	Sandy loam	Sewage sludge	61.2	42	Aerobic	[32]
		Leaves	36.8			
		Animal manure	32.5			
		Sewage sludge and compost	53	280		
		Sewage sludge and animal manure	79			
6048 ng-TEQ/kg	Sandy loam	Food waste, sawdust, and compost	70	49	Aerobic	[118]
300–660 ngTEQ/kg	Sandy loam	Straw manure, bark chips, and wood chips	75	175	Semi-aerobic	[91]

4.3. Phytoremediation

The application of phytoremediation is widely used by plants that grow in the wild, with features such as round and fat stems, many gaseous roots, or large roots, which can crawl on the ground or crawl on the trunk of another tree. These trees can grow, completely covering the ground, and remain green all year, are less deciduous, and yield a large biomass, which can withstand harsh environmental conditions, making it an ideal habitat for microorganisms and fungi in the rhizosphere, etc. They can form a favorable combination that optimizes the absorption and decomposition of toxic chemicals in the soil [119]. In addition, biological products, such as DECOM1 (a mixture of nutrient salts and organic humus), can be applied to the soil to increase the decay time of toxic substances to reduce the concentration of difficult pollutants, degrading, or in other words, helping plants absorb organic toxins more quickly in the soil [120,121]. Especially for the phytoremediation of persistent organic compounds in the soil, it takes a long time considering practical experimental conditions, such as weather, climate, other anthropogenic factors, etc. [122]. Table 7 below reviews some plants with the highest removal performance used for the treatment of dioxins in the soil.

Table 7. Degradation of dioxins by phytoremediation.

Names	Pollutant Compounds	Concentration	Removal (%)	Time	References
Arabidopsis thaliana	TCDD	10 ppt	72	30 d	[123]
		50 ppt	58		
		100 ppt	55		
Black Beauty	Total PCDDs	43 ppt-TEQ	46	32 d	[124]
	Total PCDFs		50		
Gold Rush	Total PCDDs	45 ppt-TEQ	60	32 d	
	Total PCDFs		62		

Table 7. *Cont.*

Names	Pollutant Compounds	Concentration	Removal (%)	Time	References
Spinach	Total PCDDs	3.42 ppt	48.6	ND	[125]
	Total PCDFs	0.519 ppt	37.9		
Garland Chrysanthemum	Total PCDDs	0.543 ppt	36.1		
	Total PCDFs	0.622 ppt	48.8		
Mitsuba	Total PCDDs	0.765 ppt	38		
	Total PCDFs	0.161 ppt	43.8		
Chingentsuai	Total PCDDs	0.268 ppt	39.2		
	Total PCDFs	0.166 ppt	41.6		
Rice leaf and stem	Total dioxins	317 ppt	90	5 m	[126]
Rice paddy chaff	Total dioxins	44 ppt	98		
Atena Polka	PCDD/Fs	7 ppt-TEQ dw	66	5 w	[127]
Zucchini	PCDD/Fs	155 ppt-TEQ dw	37	5 w	[127,128]
Cucumber	PCDD/Fs	122 ppt-TEQ dw	24	5 w	
Zucchini	2,4,8-TrCDF	0.0089 TSCF	64	4 d	[129,130]
	2,3,7,8-TeCDD		70		
Pumpkin	2,4,8-TrCDF	0.0064 TSCF	77		
	2,3,7,8-TeCDD		79		

5. Prospects for Future Research

Bio-composting can degrade dioxins in contaminated soil. However, to achieve high efficiency of the decomposition process, many different strains of microorganisms are required. Therefore, the biodegradation mechanisms of microbial strains are still poorly understood, and further research efforts are needed. To better understand the microbial diversity and structural changes associated with bio-composting, next-generation sequencing is proposed to identify the respective microbial strains. Lastly, the biodegradation process is designed to enhance the biodegradation process and shorten the processing time.

The growth and activity of microorganisms during the composting process determine the efficiency of the biodegradation of dioxins, which can be optimized through operational parameters, such as aeration rate, humidity, incubation time, pH, and C/N ratio. Optimal values of operational parameters vary with laboratory scale, pilot scale, composting material, and soil properties relative to the properties of dioxins. Currently, the knowledge from the literature is not sufficient to achieve commonly used optimal values. In addition, one of the major challenges of composting is that microbial activity is very time-consuming, which demonstrates why field-scale studies are very rare.

Additional studies are required to accurately and completely evaluate dioxin contamination sites of various origins and locations in order to provide effective treatment options. Furthermore, the bioremediation approaches for various contamination sites should be investigated. Other methods currently remain limited, such as hybrid bioremediation strategies in developing some transgenic plants to express dioxin-degrading enzymes, or nano-phytoremediation by combining the nanoparticles and vegetal species, which should be emphasized to improve the biodegradation efficiency of dioxins. In addition, the combination of chemical and biological measures or the combination of physicochemical and biological technologies should be utilized to improve efficiency in the degradation of pollutants.

Overall, future studies should provide more insight into the microbial relationships in the biodegradation of dioxins, in addition to the biodegradation mechanisms outlined above. The performance parameters also need to be studied more deeply for study scaling purposes.

6. Conclusions

With the current rate of industrial development and urbanization, the land area is shrinking, land quality is deteriorating, and land area per capita is decreasing. Currently, there are many hotspots of dioxin pollution in the soils of some countries around the world;

the main source of dioxin-contaminated soil is industrial production activities, followed by the consequences of war, producing high dioxin concentrations and widespread infection. Dioxin contamination in the soil not only has a negative impact on industrial production, agriculture, and service activities, but it also has an indirect impact on human and animal health through food, vegetables, etc. Composting is full of economic benefits, it can treat dioxins in contaminated soil with high efficiency, and it is environmentally friendly. Parameters such as temperature, humidity, pH, oxygen content, aeration rate, and C/N ratio need to be continuously monitored and controlled during the composting process. The microbial community is primarily responsible for the biodegradation of dioxins in the soil. Some microbials can use dioxins as a source of carbon and energy to break down these compounds. The correlation between the microbial communities and the breakdown of dioxins during the composting process needs to be further studied so that the metabolic and congeners mechanisms can be elucidated. Ultimately, current knowledge is insufficient to achieve an optimal set of values for the treatment of soils contaminated with dioxins on a laboratory scale, pilot scale, and field scale.

Author Contributions: N.T.H.N.: writing, investigation, formal analysis; X.-T.T.N.: writing and investigation; V.D.L.: writing and investigation; Y.W.: investigation, validation, writing—review and editing; T.F.: supervision, writing—review and editing. All authors have read and agreed to the published version of the manuscript.

Funding: This research was funded by the Natural Science Foundation, number NSFC 21976039.

Institutional Review Board Statement: Not applicable.

Informed Consent Statement: Not applicable.

Data Availability Statement: Not applicable.

Conflicts of Interest: The authors declare that the research was conducted without any commercial or financial relationships that could be construed as a potential conflict of interest.

References

1. Kulkarni, P.S.; Du, T. Dioxins. In *Encyclopedia of Environmental Health*; Elsevier: Amsterdam, The Netherlands, 2019; Volume 2, pp. 83–92. [CrossRef]
2. Consonni, D.; Sindaco, R.; Alberto, P. Blood levels of dioxins, furans, dioxin-like PCBs, and TEQs in general populations: A review, 1989–2010. *Environ. Int.* **2012**, *44*, 151–162. [CrossRef] [PubMed]
3. USAID. *Dioxin Remediation Bien Hoa Airbase Area Project*; Final Masterplan—Volume 1; USAID: Washington, DC, USA, 2020; Volume 1.
4. Mckay, G. Dioxin characterisation, formation and minimisation during municipal solid waste (MSW) incineration: Review. *Chem. Eng. J.* **2002**, *86*, 343–368. [CrossRef]
5. Stanmore, B.R. An Empirical Model for the De Novo Formation of PCDD/F in Medical Waste Incinerators. *Environ. Sci. Technol.* **2000**, *34*, 4538–4544. [CrossRef]
6. Anderson, D.R.; Fisher, R. Sources of dioxins in the United Kingdom: The steel industry and other sources. *Chemosphere* **2002**, *46*, 371–381. [CrossRef]
7. Ahn, Y.; Liu, F.; Fennell, D.E. Biostimulation and bioaugmentation to enhance dechlorination of polychlorinated dibenzo-p-dioxins in contaminated sediments. *FEMS Microbiol Ecol.* **2008**, *66*, 271–281. [CrossRef]
8. Thi, T.; Hanh, T.; Anh, L.V.; Bich, N.N.; Hung, N.V. Environmental health risk assessment by exposure to dioxin in food at dioxin hot site in Da Nang. *J. Prev. Med.* **2011**, *4*, 4–6.
9. Manh, H.D.; Kido, T.; Okamoto, R.; XianLiang, S.; Anh, L.T.; Supratman, S.; Maruzeni, S.; Nishijo, M.; Nakagawa, H.; Honma, S.; et al. Serum Dioxin Levels in Vietnamese Men more than 40 Years after Herbicide Spraying. *Am. Chem. Soc.* **2014**, *48*, 3496–3503. [CrossRef]
10. Haglund, P. Methods for Treating Soils Contaminated with Polychlorinated Dibenzo-p-Dioxins, Dibenzofurans, and Other Polychlorinated Aromatic Compounds. *Ambio* **2007**, *36*, 467–474. [CrossRef]
11. Marklund, S.; Andersson, R.; Tysidind, M.; Bjrkman, K.E.E.; Grigoriadis, V. Emissions of PCDDs and PCDFs in gasoline and diesel fueled cars. *Chemosphere* **1990**, *20*, 553–561. [CrossRef]
12. Mino, Y.; Moriyama, Y. Possible Remediation of Dioxin-Polluted Soil by Steam Distillation. *Chem. Pharm. Bull.* **2001**, *49*, 1050–1051. [CrossRef]
13. Harjanto, S.; Kasai, E.; Terui, T.; Nakamura, T. Formation and transport of PCDD/Fs in the packed bed of soil containing organic chloride during a thermal remediation process. *Chemosphere* **2002**, *49*, 217–224. [CrossRef]

14. Gray, K.A.; Hilarides, R.J. Radiolytic treatment of dioxin contaminated soils. *Radiat. Phys. Chem.* **1995**, *46 Pt 2*, 1081–1084. [CrossRef]
15. Nakamiya, K.; Hashimoto, S.; Ito, H.; Edmonds, J.S.; Yasuhara, A.; Morita, M. Degradation of dioxins by cyclic ether degrading fungus, *Cordyceps sinensis*. *Fed. Eur. Microbiol. Soc.* **2005**, *248*, 17–22. [CrossRef]
16. Isosaaril, P.; Tuhkanen, T.; Vartiainen, T. Photodegradation of Polychlorinated Dibenzo-p-dioxins and Dibenzofurans in Soil with Vegetable Oil. *Environ. Sci. Pollut. Res.* **2004**, *11*, 181–185. [CrossRef] [PubMed]
17. Isosaari, P.; Laine, O.; Tuhkanen, T.; Vartiainen, T. Photolysis of polychlorinated dibenzo-p-dioxins and dibenzofurans dissolved in vegetable oils: Influence of oil quality. *Sci. Total Environ.* **2005**, *340*, 1–11. [CrossRef] [PubMed]
18. Pittman, C.U., Jr.; He, J. Dechlorination of PCBs, CAHs, herbicides and pesticides neat and in soils at 25 °C using Na/NH_3. *J. Hazard. Mater.* **2002**, *92*, 51–62. [CrossRef]
19. Alassali, A.; Moon, H.; Picuno, C.; Meyer, R.S.A.; Kuchta, K. Assessment of polyethylene degradation after aging through anaerobic digestion and composting. *Polym. Degrad. Stab.* **2018**, *158*, 14–25. [CrossRef]
20. Nam, I.; Kim, Y.; Murugesan, K.; Jeon, J.; Chang, Y.; Chang, Y. Bioremediation of PCDD/Fs-contaminated municipal solid waste incinerator fly ash by a potent microbial biocatalyst. *J. Hazard. Mater.* **2008**, *157*, 114–121. [CrossRef]
21. Suzuki, S.; Hiraishi, A. *Novosphingobium naphthalenivorans* sp. nov., a naphthalene-degrading bacterium isolated from polychlorinated-dioxin-contaminated. *J. Gen. Appl. Microbiol.* **2007**, *228*, 221–228. [CrossRef]
22. Field, J.A.; Sierra-Alvarez, R. Microbial degradation of chlorinated dioxins. *Chemosphere* **2008**, *71*, 1005–1018. [CrossRef]
23. Bunge, M.; Adrian, L.; Kraus, A.; Opel, M. Reductive dehalogenation of chlorinated dioxins by an anaerobic bacterium. *Nature* **2003**, *421*, 357–360. [CrossRef] [PubMed]
24. Barkovskii, A.L. Microbial Dechlorination of Historically Present and Freshly Spiked Chlorinated Dioxins and Diversity of Dioxin-Dechlorinating Populations. *Appl. Environ. Microbiol.* **1996**, *62*, 4556–4562. [CrossRef] [PubMed]
25. Narihiro, T.; Kaiya, S.; Futamata, H.; Hiraishi, A. Removal of polychlorinated dioxins by semi-aerobic fed-batch composting with biostimulation of "Dehalococcoides". *J. Biosci. Bioeng.* **2010**, *109*, 249–256. [CrossRef] [PubMed]
26. Yoshida, N.; Takahashi, N.; Hiraishi, A. Phylogenetic Characterization of a Polychlorinated-Dioxin-Dechlorinating Microbial Community by Use of Microcosm Studies. *Appl. Environ. Microbiol.* **2005**, *71*, 4325–4334. [CrossRef]
27. Wang, X.; Pan, S.; Zhang, Z.; Lin, X.; Zhang, Y.; Chen, S. Effects of the feeding ratio of food waste on fed-batch aerobic composting and its microbial community. *Bioresour. Technol.* **2016**, *224*, 397–404. [CrossRef] [PubMed]
28. Lopez-Echartea, E.; Macek, T.; Demnerova, K.; Uhlik, O. Bacterial Biotransformation of Pentachlorophenol and Micropollutants Formed during Its Production Process. *Int. J. Environ. Res. Public Health* **2016**, *13*, 1146. [CrossRef]
29. Schnoor, J.L. Phytoremediation of Polychlorinated Biphenyls: New Trends and Promises. *Environ. Sci. Technol.* **2010**, *44*, 2767–2776.
30. Mudhoo, A.; Thayalan, G.; Muthoora, N.J.; Muthoora, M.N.; Oozeer, B.Z. *Dioxins and Furans: Sources, Impacts and Remediation*; Springer International Publishing: Cham, Switzerland, 2013.
31. Kulkarni, P.S.; Crespo, J.G.; Afonso, C.A.M. Dioxins sources and current remediation technologies—A review. *Environ. Int.* **2008**, *34*, 139–153. [CrossRef]
32. Chen, W.; Wu, J.; Lin, S.; Chang, J. Bioremediation of polychlorinated-p-dioxins/dibenzofurans contaminated soil using simulated compost-amended landfill reactors under hypoxic conditions. *J. Hazard. Mater.* **2016**, *312*, 159–168. [CrossRef]
33. Feshin, D.; Poberezhnaya, T.; Shelepchikov, A.; Brodsky, E.; Levin, B. PCDD/Fs in emissions of dirt volcano. *Organohalogen Compd.* **2006**, *68*, 2240–2243.
34. Salamanca, M.; Chandía, C.; Hernández, A. Impact of forest fi res on the concentrations of polychlorinated dibenzo-p-dioxin and dibenzofurans in coastal waters of central Chile. *Sci. Total Environ.* **2016**, *573*, 1397–1405. [CrossRef] [PubMed]
35. Reiner, E.J.; Clement, R.E.; Okey, A.B.; Marvin, C.H. Advances in analytical techniques for polychlorinated dibenzo-p-dioxins, polychlorinated dibenzofurans and dioxin-like PCBs. *Anal. Bioanal. Chem.* **2006**, *386*, 791–806. [CrossRef] [PubMed]
36. Rathna, R.; Varjani, S.; Nakkeeran, E. Recent developments and prospects of dioxins and furans remediation. *J. Environ. Manag.* **2018**, *223*, 797–806. [CrossRef] [PubMed]
37. Holtzer, M.; Dańko, J.; Dańko, R. Possibilities of formation of dioxin and furans in metallurgical processes as well as methods of their reduction. *Metalurgija* **2007**, *46*, 285–290.
38. Committee on the Implications of Dioxin in the Food Supply-National Research Council. *Dioxins and Dioxin-Like Compounds*; The National Academies Press: Washington, DC, USA, 2003.
39. Halden, R.U.; Dwyer, D.F.; Halden, R.U.; Dwyer, D.F. Biodegradation of Dioxin-Related Compounds: A Review Biodegradation of Dioxin-Related Compounds: A Review. *Bioremediat. J.* **2008**, *1*, 11–25. [CrossRef]
40. Baveye, P.C.; Otten, W.; Kravchenko, A.; Balseiro-Romero, M.; Beckers, É.; Chalhoub, M.; Darnault, C.; Eickhorst, T.; Garnier, P.; Hapca, S.; et al. Emergent Properties of Microbial Activity in Heterogeneous Soil Microenvironments: Different Research Approaches Are Slowly Converging, Yet Major Challenges Remain. *Front. Microbiol.* **2018**, *9*, 1929. [CrossRef]
41. Booth, S.; Hui, J.; Alojado, Z.; Lam, V.; Cheung, W.; Zeller, D.; Steyn, D.; Pauly, D. Global deposition of airborne dioxin. *Mar. Pollut. Bull.* **2013**, *75*, 182–186. [CrossRef]
42. Dopico, M.; Gómez, A. Review of the current state and main sources of dioxins around the world. *J. Air Waste Manag. Assoc.* **2015**, *65*, 1033–1049. [CrossRef]

43. Nijenhuis, I.; Stollberg, R.; Lechner, U. Anaerobic microbial dehalogenation and its key players in the contaminated Bitterfeld-Wolfen megasite. *FEMS Microbiol. Ecol.* **2018**, *94*, fiy012. [CrossRef]
44. Moon, H.; Choi, M.; Choi, H.; Kannan, K. Severe pollution of PCDD/Fs and dioxin-like PCBs in sediments from Lake Shihwa, Korea: Tracking the source. *Mar. Pollut. Bull.* **2012**, *64*, 2357–2363. [CrossRef]
45. Capanni, F.; Muñoz-Arnanz, J.; Marsili, L.; Fossi, M.C. Assessment of PCDD/Fs, dioxin-like PCBs and PBDEs in Mediterranean striped dolphins. *Mar. Pollut. Bull.* **2020**, *156*, 111207. [CrossRef] [PubMed]
46. Matovu, H.; Mubiru, E.; Ssebugere, P.; Sillanp, M. Human and environmental exposure to PCDD/Fs and dioxin-like PCBs in Africa: A review. *Chemosphere* **2019**, *223*, 483–493. [CrossRef]
47. Zielinski, M.; Czerska, M.; Urbaniak, M.; Gromadzinska, J. Soil contamination of PCDDs/PCDFs and dl-PCBs in the urban catchment in central Poland. *Toxicol. Lett.* **2015**, *238*, S379–S380. [CrossRef]
48. Liu, G.; Cai, Z.; Wu, Y.; Jiang, G. Dioxin analysis in China. *Trends Anal. Chem.* **2013**, *46*, 178–188. [CrossRef]
49. Stellman, J.M.; Stellman, S.D.; Christian, R.; Weber, T.; Tomasallo, C. The extent and patterns of usage of Agent Orange and other herbicides in Vietnam. *Nature* **2003**, *422*, 681–687. [CrossRef] [PubMed]
50. Zheng, G.J.; Leung, A.O.W.; Jiao, L.P.; Wong, M.H. Polychlorinated dibenzo-p-dioxins and dibenzofurans pollution in China: Sources, environmental levels and potential human health impacts. *Environ. Int.* **2008**, *34*, 1050–1061. [CrossRef] [PubMed]
51. Wang, D.G.; Yang, M.; Jia, H.L.; Zhou, L.; Li, Y.F. Levels, distributions and profiles of polychlorinated biphenyls in surface soils of Dalian, China. *Chemosphere* **2008**, *73*, 38–42. [CrossRef]
52. Naile, J.E.; Khim, J.S.; Wang, T.; Wan, Y.; Luo, W.; Hu, W.; Jiao, W.; Park, J.; Ryu, J.; Hong, S.; et al. Sources and distribution of polychlorinated-dibenzo-p-dioxins and -dibenzofurans in soil and sediment from the Yellow Sea region of China and Korea. *Environ. Pollut.* **2011**, *159*, 907–917. [CrossRef]
53. Pan, J.; Yang, Y.; Zhu, X.; Yeung LW, Y.; Taniyasu, S.; Miyake, Y.; Falandysz, J.; Yamashita, N. Science of the Total Environment Altitudinal distributions of PCDD/Fs, dioxin-like PCBs and PCNs in soil and yak samples from Wolong high mountain area, eastern Tibet-Qinghai Plateau, China. *Sci. Total Environ.* **2013**, *444*, 102–109. [CrossRef]
54. Leung, A.O.W.; Wong, A.S. Spatial Distribution of Polybrominated Diphenyl Ethers and Polychlorinated Dibenzo-p-dioxins and Dibenzofurans in Soil and Combusted Residue at Guiyu, an Electronic Waste Recycling Site in Southeast China. *Environ. Sci. Technol.* **2007**, *41*, 2730–2737. [CrossRef]
55. Lee, C.C.; Lin, W.T.; Liao, P.C.; Su, H.J.; Chen, H.L. High average daily intake of PCDD/Fs and serum levels in residents living near a deserted factory producing pentachlorophenol (PCP) in Taiwan: Influence of contaminated fish consumption. *Environ. Pollut.* **2006**, *141*, 381–386. [CrossRef] [PubMed]
56. Lee, C.C.; Guo, Y.L.; Kuei, C.H.; Chang, H.Y.; Hsu, J.F.; Wang, S.T.; Liao, P.C. Human PCDD/PCDF levels near a pentachlorophenol contamination site in Tainan, Taiwan. *Chemosphere* **2006**, *65*, 436–448. [CrossRef] [PubMed]
57. Song, A.; Li, H.; Liu, M.; Hu, J.; Sheng, G.; Ying, G. Polybrominated dibenzo-p-dioxins/furans (PBDD/Fs) in soil around municipal solid waste incinerator: A comparison with polychlorinated dibenzo-p-dioxins/furans (PCDD/Fs). *Environ. Pollut.* **2022**, *293*, 118563. [CrossRef] [PubMed]
58. Die, Q.; Lu, A.; Li, C.; Li, H.; Kong, H.; Li, B. Occurrence of dioxin-like POPs in soils from urban green space in a metropolis, North China: Implication to human exposure. *Environ. Sci. Pollut. Res.* **2021**, *28*, 5587–5597. [CrossRef] [PubMed]
59. Liu, J.; Liu, W. Distribution of polychlorinated dibenzo-p-dioxins and dibenzofurans (PCDDs/Fs) and dioxin-like polychlorinated biphenyls (dioxin-like PCBs) in the soil in a typical area of eastern China. *J. Hazard. Mater.* **2009**, *163*, 959–966. [CrossRef] [PubMed]
60. Le, L.T.H.; Dat, N.D.; Minh, N.H.; Nguyen, K.A. Characteristics of PCDD/Fs in soil and sediment samples collected from A-So former airbase in Central Vietnam. *Sci. Total Environ.* **2019**, *661*, 27–34. [CrossRef]
61. Anh, T.; Vu, T.; Tarradellas, J. Dioxin contamination in soils of Southern Vietnam. *Chemosphere* **2007**, *67*, 1802–1807. [CrossRef]
62. Takeda, N.; Takaoka, M. An assessment of dioxin contamination from the intermittent operation of a municipal waste incinerator in Japan and associated remediation. *Environ. Sci. Pollut. Res.* **2013**, *20*, 2070–2080. [CrossRef]
63. Kiguchi, O.; Kobayashi, T.; Wada, Y.; Saitoh, K.; Ogawa, N. Polychlorinated dibenzo-p-dioxins and dibenzofurans in paddy soils and river sediments in Akita, Japan. *Chemosphere* **2007**, *67*, 557–573. [CrossRef]
64. Cho, H.-K. Distribution of PCDD/Fs and PCBs in Soils and Pine Needles in Ulsan, South Korea. Master's Thesis, Ulsan National Institute of Science and Technology, Ulsan, Korea, 2021.
65. Liberatori, G.; Cotugno, P.; Sturba, L.; Vannuccini, M.L.; Capasso, G.; Velardo, R.; Besselink, H.; Massari, F.; Tursi, A.; Corbelli, V.; et al. Chemosphere Occurrence and spatial distribution of dioxin and dioxin-like compounds in topsoil of Taranto (Apulia, Italy) by GC-MS analysis and DR-CALUX® bioassay. *Chemosphere* **2021**, *279*, 130576. [CrossRef]
66. Henriksson, S.; Hagberg, J.; Bäckström, M.; Persson, I.; Lindström, G. Assessment of PCDD/Fs levels in soil at a contaminated sawmill site in Sweden—A GIS and PCA approach to interpret the contamination pattern and distribution. *Environ. Pollut.* **2013**, *180*, 19–26. [CrossRef] [PubMed]
67. Shelepchikov, A.A.; Brodskii, E.S.; Feshin, D.B.; Zhil'nikov, V.G.; Mir-Kadyrova, E.Y.; Balashova, S.P. Polychlorinated dibenzo-p-dioxins, dibenzofurans, and biphenyls in soils of Moscow. *Eurasian Soil Sci.* **2011**, *44*, 286–296. [CrossRef]
68. Eljarrat, E.; Caixach, J.; Rivera, J. Levels of polychlorinated dibenzo-p-dioxins and dibenzofurans in soil samples from Spain. *Chemosphere* **2001**, *44*, 1383–1387. [CrossRef]

69. Creaser, C.S.; Fernandes, A.R.; Harrad, S.J.; Cox, E.A. Levels and Sources of PCDDs and PCDFs in urban British soils. *Chemosphere* **1990**, *21*, 931–938. [CrossRef]
70. Creaser, C.S.; Fernandes, A.R.; Al-Haddad, A.; Harrad, S.J.; Homer, R.B.; Skett, P.W.; Cox, E.A. Survey of background levels of PCDDs & PCDFs in UK soils. *Chemosphere* **1989**, *18*, 767–776. [CrossRef]
71. Martínez, K.; Abad, E.; Rivera, J. Surveillance programme on dioxin levels in soils in the Campo de Gibraltar (Southwest Spain). *Chemosphere* **2006**, *65*, 382–389. [CrossRef]
72. Dömötörová, M.; Sejáková, Z.S.; Kočan, A.; Čonka, K.; Chovancová, J.; FabiŠiková, A. PCDDs, PCDFs, dioxin-like PCBs and indicator PCBs in soil from five selected areas in Slovakia. *Chemosphere* **2012**, *89*, 480–485. [CrossRef]
73. Ivory, A.; Mobbs, C. Dioxins levels in Australia: Key findings of studies. *Organohalogen Compd.* **2004**, *66*, 3398–3403.
74. Urban, J.D.; Wikoff, D.S.; Bunch, A.T.G.; Harris, M.A.; Haws, L.C. A review of background dioxin concentrations in urban/suburban and rural soils across the United States: Implications for site assessments and the establishment of soil cleanup levels. *Sci. Total Environ.* **2014**, *466–467*, 586–597. [CrossRef]
75. Rogowski, D.L.; Yake, W. Typical dioxin concentrations in agriculture soils of Washington State and potential sources. *Environ. Sci. Technol.* **2005**, *39*, 5170–5176. [CrossRef]
76. Pius, C.; Sichilongo, K.; Mswela, P.K.; Dikinya, O. Monitoring polychlorinated dibenzo-p-dioxins/dibenzofurans and dioxin-like polychlorinated biphenyls in Africa since the implementation of the Stockholm Convention—An overview. *Environ. Sci. Pollut. Res.* **2019**, *26*, 101–113. [CrossRef] [PubMed]
77. Nieuwoudt, C.; Quinn, L.P.; Pieters, R.; Jordaan, I.; Visser, M.; Kylin, H.; Borge, A.R.; Gieasy, J.P.; Bouwman, H. Dioxin-like chemicals in soil and sediment from residential and industrial areas in central South Africa. *Chemosphere* **2009**, *76*, 774–783. [CrossRef] [PubMed]
78. USEPA. *US EPA Region 5 Soil and Sediment Ecological Screening Levels*; USEPA: San Francisco, CA, USA, 2003.
79. USEPA. *USEPA Region 9 Preliminary Remediation Goals*; US Environmental Protection Agency: San Francisco, CA, USA, 2000.
80. Environmental Protection Agency (EPA) of Taiwan. Control Standards of Soil Pollution. 2011. Available online: http://ivy5.epa.gov.tw/%0Aepalaw/search/LordiDispFull.aspx?ltype\protect$\relax\protect{\begingroup1\endgroup\@@over4}$14&lname\protect$\relax\protect{\begingroup1\endgroup\@@over4}$0120 (accessed on 2 February 2022).
81. Ministry of Natural Resources and Environment. *National Technique Regulation on Allowed Limits of Dioxin in Soils*; Ministry of Natural Resources and Environment: Hanoi, Vietnam, 2012.
82. UK Department of the Environment Transport and the Regions (DETR). Compilation of EU Dioxin Exposure and Health Data. Task 1—Member State Legislation and Programmes. Report Produced for the European Commission DG. 1999. Available online: http://ec.europa.eu/environment/dioxin/pdf/task1.pdf (accessed on 6 February 2022).
83. New Zealand Ministry of Environment. Health and Environmental Guidelines for Selected Timber Treatment Chemicals—Chapter 5. New Zealand. 1997. Available online: http://www.mfe.govt.nz/publications/hazardous/timber-guide-jun97/timber-guide-jun97.pdf (accessed on 23 February 2022).
84. Canadian Council of Ministers of the Environment. *Canadian Soil Quality Guidelines for the Protection of Environmental and Human Health—Polychlorinated Dibenzo-p-Dioxins and Polychlorinated Dibenzofurans (PCDD/Fs)*; Canadian Council of Ministers of the Environment: Winnipeg, MB, Canada, 2002.
85. Garbisu, C.; Garaiyurrebaso, O.; Epelde, L.; Grohmann, E. Plasmid-Mediated Bioaugmentation for the Bioremediation of Contaminated Soils. *Front. Microbiol.* **2017**, *8*, 1966. [CrossRef] [PubMed]
86. Coppotelli, B.M.; Ibarrolaza, A.; del Panno, M.T. Effects of the Inoculant Strain *Sphingomonas paucimobilis* 20006FA on Soil Bacterial Community and Biodegradation in Phenanthrene-contaminated Soil. *Microb. Ecol.* **2008**, *55*, 173–183. [CrossRef] [PubMed]
87. Lyon, D.Y.; Vogel, T.M.; de Lyon, U. *6.08 Bioaugmentation as a Strategy for the Treatment of Persistent Pollutants*, 2nd ed.; Elsevier B.V.: Amsterdam, The Netherlands, 2011; Volume 1.
88. Lin, C.; Kao, C.; Grasso, D.; Sheu, D.; Lin, T. Thermophilic Biodegradation of Diesel Oil in Food Waste Composting Processes Without Bioaugmentation. *Environ. Eng. Sci.* **2012**, *29*, 117–123. [CrossRef]
89. Chang, Y. Recent Developments in Microbial Biotransformation and Biodegradation. *J. Mol. Microbiol. Biotechnol.* **2008**, *15*, 152–171. [CrossRef]
90. Xie, H.; Zhu, L.; Wang, J. Combined treatment of contaminated soil with a bacterial Stenotrophomonas strain DXZ9 and ryegrass (*Lolium perenne*) enhances DDT and DDE remediation. *Environ. Sci. Pollut. Res.* **2018**, *25*, 31895–31905. [CrossRef]
91. Tu, Y.T.; Liu, J.K.; Lin, W.C.; Lin, J.L.; Kao, C.M. Enhanced anaerobic biodegradation of OCDD-contaminated soils by *Pseudomonas mendocina* NSYSU: Microcosm, pilot-scale, and gene studies. *J. Hazard. Mater.* **2014**, *278*, 433–443. [CrossRef]
92. Lin, J.L.; Lin, W.C.; Liu, J.K.; Surampalli, R.Y.; Zhang, T.C.; Kao, C.M. Aerobic Biodegradation of OCDD by P. Mendocina NSYSU: Effectiveness and Gene Inducement Studies. *Water Environ. Res.* **2017**, *89*, 2113–2121. [CrossRef]
93. Liu, N.; Hou, T.; Yin, H.; Han, L.; Huang, G. Effects of amoxicillin on nitrogen transformation and bacterial community succession during aerobic composting. *J. Hazard. Mater.* **2018**, *362*, 258–265. [CrossRef]
94. Deshmukh, R.; Khardenavis, A.A.; Purohit, H.J. Diverse Metabolic Capacities of Fungi for Bioremediation. *Indian J. Microbiol.* **2016**, *56*, 247–264. [CrossRef] [PubMed]
95. Kamei, I.; Suhara, H.; Kondo, R. Phylogenetical approach to isolation of white-rot fungi capable of degrading polychlorinated dibenzo-p-dioxin. *Appl. Microbiol. Biotechnol.* **2005**, *69*, 358–366. [CrossRef] [PubMed]

96. Kamei, I.; Kondo, R. Biotransformation of dichloro-, trichloro-, and tetrachlorodibenzo-p-dioxin by the white-rot fungus *Phlebia lindtneri*. *Appl. Microbiol. Biotechnol.* **2005**, *68*, 560–566. [CrossRef] [PubMed]
97. Takada, S.; Nakamura, M.; Matsueda, T. Degradation of Polychlorinated Dibenzo-p-Dioxins and Polychlorinated Dibenzofurans by the White Rot Fungus *Phanerochaete sordida* YK-624. *Appl. Environ. Microbiol.* **1996**, *62*, 4323–4328. [CrossRef] [PubMed]
98. Habe, H.; Chung, J.S.; Lee, J.H.; Kasuga, K.; Yoshida, T.; Nojiri, H.; Omori, T. Degradation of Chlorinated Dibenzofurans and Dibenzo-p-Dioxins by Two Types of Bacteria Having Angular Dioxygenases with Different Features. *Appl. Environ. Microbiol.* **2001**, *67*, 3610–3617. [CrossRef] [PubMed]
99. Halden, R.U.; Halden, B.G.; Dwyer, D.F. Removal of Dibenzofuran, Dibenzo-p-Dioxin, and 2-Chlorodibenzo-p-Dioxin from Soils Inoculated with *Sphingomonas* sp. Strain RW1. *Appl. Environ. Microbiol.* **1999**, *65*, 2246–2249. [CrossRef]
100. Habe, H.; Ide, K.; Yotsumoto, M.; Tsuji, H.; Yoshida, T.; Nojiri, H.; Omori, T. Degradation characteristics of a dibenzofuran-degrader *Terrabacter* sp. strain DBF63 toward chlorinated dioxins in soil. *Chemosphere* **2002**, *48*, 201–207. [CrossRef]
101. Habe, H.; Ashikawa, Y.; Saiki, Y.; Yoshida, T.; Nojiri, H. *Sphingomonas* sp. strain KA1, carrying a carbazole dioxygenase gene homologue, degrades chlorinated dibenzo-p-dioxins in soil. *FEMS Microbiol. Lett.* **2002**, *211*, 43–49. [CrossRef]
102. Kimura, N.; Kitagawa, W.; Mori, T.; Nakashima, N.; Tamura, T.; Kamagata, Y. Genetic and biochemical characterization of the dioxygenase involved in lateral dioxygenation of dibenzofuran from *Rhodococcus opacus* strain SAO101. *Appl. Microbiol. Biotechnol.* **2006**, *73*, 474–484. [CrossRef]
103. Ishiguro, T.; Ohtake, Y.; Nakayama, S.; Inamori, Y.; Amagai, T.; Soma, M.; Matsusita, H. Biodegradation of Dibenzofuran and Dioxins by *Pseudomonas aeruginosa* and *Xanthomonas maltophilia*. *Environ. Technol.* **2000**, *21*, 1309–1316. [CrossRef]
104. Hong, H.; Nam, I.; Murugesan, K.; Kim, Y. Biodegradation of dibenzo-p-dioxin, dibenzofuran, and chlorodibenzo-p-dioxins by *Pseudomonas veronii* PH-03. *Biodegradation* **2004**, *15*, 303–313. [CrossRef] [PubMed]
105. Nam, I.; Hong, H.; Kim, Y.; Kim, B.; Murugesan, K.; Chang, Y.S. Biological removal of polychlorinated dibenzo-p-dioxins from incinerator fly ash by *Sphingomonas wittichii* RW1. *Water Res.* **2005**, *39*, 4651–4660. [CrossRef] [PubMed]
106. Widada, J.; Omori, T. Enhanced degradation of carbazole and 2, 3-dichlorodibenzo-p-dioxin in soils by *Pseudomonas resinovorans* strain CA10. *Chemosphere* **2002**, *49*, 485–491. [CrossRef]
107. Ishii, K.; Furuichi, T. Development of bioreactor system for treatment of dioxin-contaminated soil using *Pseudallescheria boydii*. *J. Hazard. Mater.* **2007**, *148*, 693–700. [CrossRef]
108. Huang, W.Y.; Ngo, H.H.; Lin, C.; Vu, C.T.; Kaewlaoyoong, A.; Boonsong, T.; Tran, H.-T.; Bui, X.-T.; Vo, T.-D.-H.; Chen, J.-R. Aerobic co-composting degradation of highly PCDD/F-contaminated field soil. A study of bacterial community. *Sci. Total Environ.* **2019**, *660*, 595–602. [CrossRef]
109. Huy, P.Q.; Thoa, N.K.; Thi, D.; Ha, C. Degradation of 2,3,7,8-TCDD by a consortium of bacterial strains isolated from heavil herbicide/Dioxin contaminated soil in Bienhoa airbase. *J. Biotechnol.* **2018**, *16*, 777–784.
110. Nakamiya, K.; Furuichi, T.; Ishii, K. Isolation of a fungus from denitrifying activated sludge that degrades highly chlorinated dioxins. *J. Mater. Cycles Waste Manag.* **2002**, *4*, 127–134.
111. Kasai, N.; Ikushiro, S.; Shinkyo, R.; Yasuda, K.; Hirosue, S. Metabolism of mono- and dichloro-dibenzo-p-dioxins by *Phanerochaete chrysosporium* cytochromes P450. *Appl Microbiol Biotechnol.* **2010**, *86*, 773–780. [CrossRef]
112. Kaewlaoyoong, A.; Cheng, C.; Lin, C.; Chen, J.; Huang, W.; Sriprom, P. Science of the Total Environment White rot fungus *Pleurotus pulmonarius* enhanced bioremediation of highly PCDD/F-contaminated fi eld soil via solid state fermentation. *Sci. Total Environ.* **2020**, *738*, 139670. [CrossRef]
113. Tachibana, M.K.S.; Kiyota, Y. Bioremediation of Dioxin-Contaminated Soil by Fungi Screened from Nature. *Pak. J. Biol. Sci.* **2007**, *10*, 486–491. [CrossRef]
114. Pinedo-Rivilla, C.; Aleu, J.; Collado, I.G. Pollutants Biodegradation by Fungi. *Curr. Org. Chem.* **2009**, *13*, 1194–1214. [CrossRef]
115. Zafar, U.; Nzeram, P.; Langarica-Fuentes, A.; Houlden, A.; Heyworth, A.; Saiani, A.; Robson, G.D. Bioresource Technology Biodegradation of polyester polyurethane during commercial composting and analysis of associated fungal communities. *Bioresour. Technol.* **2014**, *158*, 374–377. [CrossRef]
116. Obuekwe, C.O.; Al-jadi, Z.K.; Al-saleh, E.S. International Biodeterioration & Biodegradation Hydrocarbon degradation in relation to cell-surface hydrophobicity among bacterial hydrocarbon degraders from petroleum-contaminated Kuwait desert environment. *Int. Biodeterior. Biodegrad.* **2009**, *63*, 273–279. [CrossRef]
117. Laine, M.M.; Ahtiainen, J.; Wågman, N.; Öberg, L.G.; Jørgensen, K.S. Fate and toxicity of chlorophenols, polychlorinated dibenzo-p-dioxins, and dibenzofurans during composting of contaminated sawmill soil. *Environ. Sci. Technol.* **1997**, *31*, 3244–3250. [CrossRef]
118. Lin, C.; Kaewlaoyoong, A.; Vu, C.T.; Huang, W.Y. Treatment of dioxin-contaminated soil by organic waste co-composting system. *Springer Proc. Phys.* **2018**, *207*, 619–623. [CrossRef]
119. Alkorta, I.; Garbisu, C. Phytoremediation of organic contaminants in soils. *Bioresour. Technol.* **2001**, *79*, 273–276. [CrossRef]
120. Macek, J.K.T.; Macková, M. Exploitation of plants for the removal of organics in environmental remediation. *Biotechnol. Adv.* **2000**, *18*, 23–34. [CrossRef]
121. Reichenauer, T.G.; Germida, J.J. Phytoremediation of Organic Contaminants in Soil and Groundwater. *ChemSusChem* **2008**, *8*, 708–717. [CrossRef]
122. Kurzawová, M.M.V.; Uhlík, O.; Macek, T. Interactions of microorganisms and plants in soil contaminated with PCBs and their importance in phyto/rhizoremediation. *Remediat. Phytoremediation Technol.* **2010**, *126*, 396–398.

123. Hanano, A.; Almousally, I.; Shaban, M. Chemosphere Phytotoxicity effects and biological responses of *Arabidopsis thaliana* to 2, 3, 7, 8-tetrachlorinated dibenzo-p-dioxin exposure. *Chemosphere* **2014**, *104*, 76–84. [CrossRef]
124. Inui, H.; Wakai, T.; Gion, K.; Kim, Y.; Eun, H. Chemosphere Differential uptake for dioxin-like compounds by zucchini subspecies. *Chemosphere* **2008**, *73*, 1602–1607. [CrossRef] [PubMed]
125. Nakagawa, R.; Hori, T.; Tobiishi, K.; Iida, T.; Tsutsumi, T.; Sasaki, K.; Toyoda, M. Levels and tissue-dependent distribution of dioxin in Japanese domestic leafy vegetables—From the 1999 national investigation. *Chemosphere* **2002**, *48*, 247–256. [CrossRef]
126. Uegaki, R.; Seike, N.; Otani, T. Dioxin-like polychlorinated biphenyls in rice plants: Possible contaminated pathways. *Chemosphere* **2006**, *65*, 1537–1543. [CrossRef] [PubMed]
127. Urbaniak, M.; Zieliński, M.; Wyrwicka, A. The influence of the Cucurbitaceae on mitigating the phytotoxicity and PCDD/PCDF content of soil amended with sewage sludge. *Int. J. Phytoremediation* **2017**, *19*, 207–213. [CrossRef]
128. Campanellaand, B.; Paul, R. Presence, in the Rhizosphere and Leaf Extracts of Zucchini (*Cucurbita pepo* L.) and Melon (*Cucumis melo* L.), of Molecules Capable of Increasing the Apparent Aqueous Solubility of Hydrophobic Pollutants. *Int. J. Phytoremediation* **2006**, *2*, 145–158. [CrossRef]
129. Hulster, A.; Muller, J.F.; Marschner, H. Soil-Plant Transfer of Polychlorinated Dibenzo-p-dioxins and Dibenzofurans to Vegetables of the Cucumber Family (Cucurbitaceae). *Environ. Sci. Technol.* **1994**, *28*, 1110–1115. [CrossRef]
130. Zhang, H.; Chen, J.; Ni, Y.; Zhang, Q.; Zhao, L. Chemosphere Uptake by roots and translocation to shoots of polychlorinated dibenzo-p-dioxins and dibenzofurans in typical crop plants. *Chemosphere* **2009**, *76*, 740–746. [CrossRef]

toxics

MDPI

Article

Detection of Cd^{2+} in Aqueous Solution by the Fluorescent Probe of CdSe/CdS QDs Based on OFF–ON Mode

Dengpeng Wang [1,2], Feng Gao [1,2,*], Xianran Wang [1,2], Xiaomei Ning [1,2], Kaituo Wang [1,2], Xinpeng Wang [1,2], Yuezhou Wei [3] and Toyohisa Fujita [1,2,*]

1 Guangxi Key Laboratory of Processing for Non-Ferrous Metals and Featured Materials, MOE Key Laboratory of New Processing Technology for Non-Ferrous Metals and Materials, Guangxi University, Nanning 530004, China; 1915391031@st.gxu.edu.cn (D.W.); 2015301071@st.gxu.edu.cn (X.W.); 2115391053@st.gxu.edu.cn (X.N.); wangkaituo@gxu.edu.cn (K.W.); wangxinpeng@gxu.edu.cn (X.W.)

2 School of Resources, Environment and Materials, Guangxi University, Nanning 530004, China

3 School of Nuclear Science and Technology, University of South China, Hengyang 421001, China; yzwei@gxu.edu.cn

* Correspondence: gaofeng@gxu.edu.cn (F.G.); fujitatoyohisa@gxu.edu.cn (T.F.); Tel.: +86-151-774-620-06 (F.G.); +0771-323-2200 (T.F.)

Abstract: The detection of heavy metals in aqueous solutions has always attracted much attention from all over the world. A fluorescent probe of CdSe/CdS core-shell quantum dots (QDs) was designed to detect trace Cd^{2+} in aqueous solutions using the OFF–ON mode rapidly and efficiently, likely based on adsorption and desorption reactions between ethylenediaminetetraacetic acid disodium salt (EDTA) and CdSe/CdS QDs. In the OFF mode, the optical shielding function of EDTA results in fluorescence quenching owing to the strong adsorption ability of EDTA with Cd^{2+} on the sites of CdSe/CdS QDs surface. In the ON mode, the introduction of Cd^{2+} promotes the desorption of EDTA from the EDTA-CdSe/CdS QDs and restores the fluorescence intensity. There were two linear response ranges which were 0.1–20 μmol/L and 20–90 μmol/L for the EDTA-CdSe/CdS system to detect Cd^{2+}. The detection limit was 6 nmol/L, and the standard deviation was below 4% for the detection of Cd^{2+} concentration in tap water.

Keywords: QDs; fluorescence quenching; fluorescence restore; Cd^{2+} detection; adsorption; desorption

Citation: Wang, D.; Gao, F.; Wang, X.; Ning, X.; Wang, K.; Wang, X.; Wei, Y.; Fujita, T. Detection of Cd^{2+} in Aqueous Solution by the Fluorescent Probe of CdSe/CdS QDs Based on OFF–ON Mode. *Toxics* **2022**, *10*, 367. https://doi.org/10.3390/toxics10070367

Academic Editor: Roberto Rosal

Received: 13 May 2022
Accepted: 29 June 2022
Published: 3 July 2022

Publisher's Note: MDPI stays neutral with regard to jurisdictional claims in published maps and institutional affiliations.

Copyright: © 2022 by the authors. Licensee MDPI, Basel, Switzerland. This article is an open access article distributed under the terms and conditions of the Creative Commons Attribution (CC BY) license (https://creativecommons.org/licenses/by/4.0/).

1. Introduction

Water pollution includes heavy metal pollution [1–3] and radionuclide pollution [4,5]. Due to the non-biodegradable nature, heavy metal pollution has always been a problem, until now. The influence of Cd^{2+} pollution should not be underestimated [6]. As a heavy metal ion, Cd^{2+} has a long biological half-life period of 20–30 years and is accumulated in the human body via polluted water, air, soil, or other ways, causing many diseases of the kidney, liver, heart, lung, or other organs. Studies have shown that it can cause a series of health problems, including several fatal diseases such as diabetes, cancer, and chondropathy, even if the accumulated Cd^{2+} content in the body is very low [7].

Currently, there are several effective techniques to detect Cd^{2+}: atomic absorption spectrometry (AAS) [8], atomic fluorescence spectrophotometry (AFS) [9], inductively coupled plasma mass spectrometry (ICP-MS) [10], the electrochemical method [11,12], and the fluorescence probe method [13–15]. Compared to the fluorescent probe method, AAS cannot be used for simultaneous analysis of multiple elements, as AFS and ICP-MS require expensive and complex instruments, complex sample preparation, and the electrochemical method, which has the disadvantage of poor selectivity.

The biggest advantage of the fluorescent probe method is its rapid response, visibility, and high sensitivity. In addition, low cost, simple operation, and a wide linear range for detecting heavy metal ions are also obvious advantages. These advantages make for attracting much attention from researchers, and they have been widely used in the biomedical and analytical chemistry fields [16,17].

Fluorescence probes can be divided into organic fluorescence probes and inorganic fluorescence probes [18]. Inorganic quantum dots (QDs) have been widely used in fluorescence detection in recent years on account of the following merits: high fluorescence quantum yield, size-tunable fluorescence emission spectrum, and visibility. There were several reports about fluorescent QDs probes for detecting Cd^{2+}, such as CdX (X = Te, Se, S) QDs, ZnS QDs, C QDs, and Au QDs [19–22]. According to the spectral characteristics, some QDs fluorescent probes are based on the intensity change of a single fluorescence peak, while others are ratiometric fluorescent probes based on the relative intensities of double emission peaks. According to the structure of the QDs, the probes have single crystal, core-shell, or composite crystal [23–25]. Usually, the detection of Cd^{2+} by QDs has two methods, i.e., TURN–OFF and OFF–ON [16,26]. So far, a considerable number of defects were caused by the TURN–OFF mode. In comparison, there are few detections from OFF–ON mode. However, many results show that the OFF–ON mode is more accurate than the ON–OFF mode in detecting Cd^{2+} [23].

In this work, a novel QDs fluorescence probe based on OFF–ON mode was developed. Considering that the single-core QDs have many defects on their surface which can affect the luminescence efficiency, CdSe/CdS QDs were designed and prepared in a core-shell structure. A schematic diagram of Cd^{2+} detection in aqueous solution is shown in Figure 1. Under UV excitation, an obvious emission band of 500–750 nm was observed in the photoluminescence emission spectrum of the CdSe/CdS QDs. EDTA molecules were partially dissociated into anions and cations when they were added in the aqueous solution of CdSe/CdS QDs, and then some of the $EDTA^-$ ions were adsorbed on the surface of QDs through electrostatic force between $EDTA^-$ and Cd^{2+}, shielded the fluorescence excitation and emission energy, thus resulting in the quenching of fluorescence emission. At that moment, the detecting system turned out to be the OFF mode. When Cd^{2+} was added, the $EDTA^-$ ions chelated with the Cd^{2+} and reduced the number of $EDTA^-$ in the solution system. In order to maintain the chemical balance, some $EDTA^-$ ions would desorb from the surface of the CdSe/CdS QDs and were released into the solution again, leading to their fluorescence restoration.

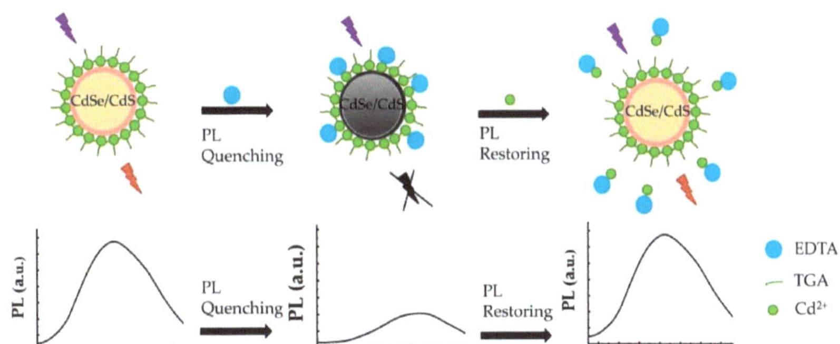

Figure 1. A schematic diagram of the detection of Cd^{2+} in an aqueous solution by CdSe/CdS QDs based on OFF–ON mode through absorption and desorption processes.

2. Materials and Methods

2.1. Apparatus and Reagents

Fluorescence spectra were taken on a FL3C-111 TCSPC luminescence spectrometer equipped with a 20-kW xenon discharge lamp as a light source (Horiba, Kyoto, Japan). X-ray powder diffraction (XRD) spectra were taken on a Rigaku D/MAX2500V X-ray diffractometer (Rigaku, Tokyo, Japan). The microstructural features of the samples were characterized by a F200X (Semerfeld, Seattle, WA, USA) transmission electron microscope (TEM).

Thioglycolic acid (TGA), $NaBH_4$, $CdCl_2{\cdot}2.5H_2O$, $Na_2S{\cdot}9H_2O$, and ethylenediaminetetraacetic acid disodium salt (EDTA) in analytical grade and various standard solution of metal ions (K^+, Na^+, Mg^{2+}, Ba^{2+}, Al^{3+}, Mn^{2+}, Fe^{3+}, Ca^{2+}, Hg^{2+}, Pb^{2+}, Cu^{2+}, Ag^+, Ni^+, Zn^{2+}, Cd^{2+}) were purchased from Sinopharm Chemical Reagent Co., Ltd. Analytically pure tris (hydroxymethyl) methyl aminomethane (Tris) was purchased from Aladdin company. Ultra-pure water was used in the whole experiment except for particular notations in text, and the resistivity of ultra-pure water is above 18 MΩ·cm (25 °C).

2.2. Synthesis of CdSe/CdS Core-Shell QDs

The fabrication method of CdSe/CdS QDs has already been reported elsewhere [27]. In brief, a certain amount of Se, $NaBH_4$, and 10 mL ultrapure water were added into a three-necked flask under the N_2 atmosphere, and then stirred vigorously until the solution became colorless and clarified. The solution at this time was a NaHSe solution, which would be used as a precursor of Se for the next step. A certain amount of $CdCl_2$ was dissolved in 100 mL of ultrapure water, and then a certain volume of TGA solution was dropped into it, and the solution changed from colorless to cloudy rapidly. The molar ratio of Cd:Se:TGA in the reaction system was 1:0.5:2.5. The $CdCl_2$ solution became clear again when the pH value of the solution was adjusted to 11 with 1 mol/L NaOH solution, and before it, N_2 had been introduced for 30 min to exclude oxygen. The prepared NaHSe solution was quickly transferred to the $CdCl_2$ solution, and the mixture was stirred vigorously and heated under 80 °C for 30 min under N_2 atmosphere to get CdSe solution. After cooling to room temperature, a certain amount of $CdCl_2$ solution and Na_2S solution were prepared according to the molar ratio of CdSe:CdS = 1:1. The CdSe solution was added drop by drop under intense stirring, and the reaction system was heated to 80 °C and refluxed for 30 min. The final solution obtained by the above process was an orange-red solution, which was washed with anhydrous ethanol, centrifuged 3 times, and subsequently dispersed in ultrapure water. The solution is an as-prepared TGA-capped CdSe/CdS QDs solution.

2.3. Fluorescence Quenching Method of CdSe/CdS by EDTA

For the study on the fluorescence quenching of CdSe/CdS QDs by EDTA, the following series of solutions were prepared, i.e., 300 µL CdSe/CdS QDs solution, 2.4 mL Tris-HCl buffer (10 mmol/L, pH = 8.0), 300 µL EDTA with various concentrations. All these solutions were added into a colorimetric dish in turn to form a 3 mL solution system. After incubation for 10 min, the photoluminescence (PL) spectra of the solution system were tested. Fluorescence quenching degree is expressed by I/I_0, in which I and I_0 represent the PL intensities of QDs with the various concentrations of EDTA and without EDTA, respectively.

2.4. Fluorescence Restore Method of EDTA-CdSe/CdS by Cd^{2+}

In order to investigate the effect of Cd^{2+} on the fluorescence intensity of the EDTA-CdSe/CdS system, the following series of solutions were prepared: 300 µL CdSe/CdS QDs solution, 2.1 mL Tris-HCl buffer, 300 µL EDTA, and 300 µL Cd^{2+} aqueous solution with various concentrations were added into a colorimetric dish in turn to form a 3 mL solution. After incubation for 10 min, the PL spectra of the solution system were tested. The fluorescence restoring effect is expressed by I/I_0, in which I and I_0 represent the PL intensities of EDTA-QDs with the various concentrations of Cd^{2+} and without Cd^{2+}, respectively.

All PL spectra were measured under the same conditions: the excitation and emission slit were set to 3 nm, and the excitation wavelength was set to 397 nm. The monitoring emission range was 420–780 nm. The fluorescence intensity values were not corrected for inner-filter effects.

3. Results and Discussion

3.1. Characterization of CdSe/CdS QDs

Figure 2a shows the X-ray diffraction (XRD) pattern of the CdSe/CdS QDs. Three wide diffraction peaks centered at $2\theta = 25.8°$, $43.2°$ and $50.5°$ were observed, corresponding to the characteristic peaks of (111), (220) and (311) lattice planes of cubic CdSe or CdS. There were mutual stresses between the core of CdSe and the cell of CdS that caused the lattice parameters to change, and thereby the shifts of the diffraction peaks happened specifically for the two peaks at a higher angle. A similar diffraction pattern was observed in CdSe/CdS nanoparticles in which CdS was epitaxially grown on a CdSe core [28]. These obvious wide peaks reflect the basic characteristics of nanoparticles. The micro-morphology of TGA-capped CdSe/CdS QDs is shown in Figure 2b,c. These QDs, with a nearly spherical shape, display a good dispersion property. The insert graph in Figure 2c shows the HRTEM (high-resolution transmission electron microscopy) image obtained by focus on a nanoparticle within the view field and the spacing of the neighbouring lattice fringe is 0.35 nm, corresponding to the (111) lattice plane of CdSe, which further proves the final prepared products in this work to be CdSe/CdS nanoparticles. By counting the particle size of all the nanoparticles in Figure 2c, the size distribution of the QDs is shown in Figure 2d. The particle size is mainly in the range of 4–20 nm, and the average particle size is about 12 nm.

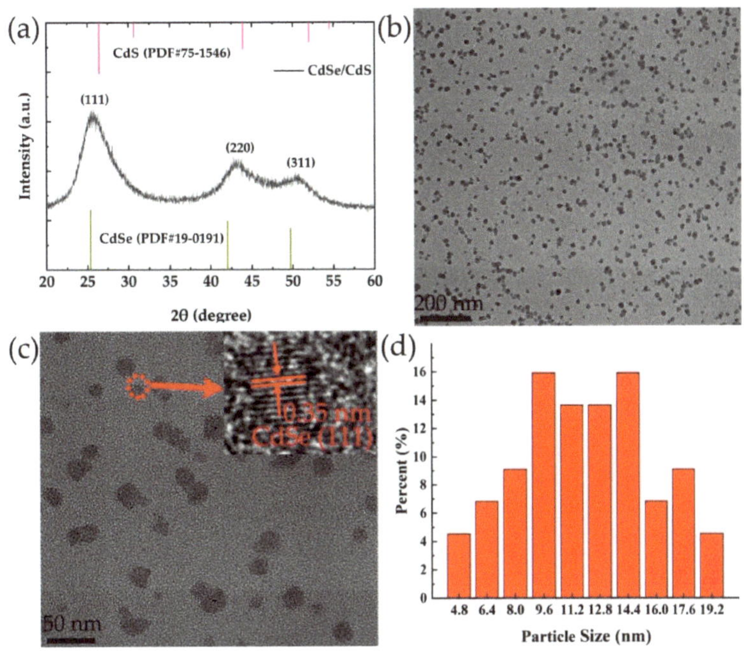

Figure 2. (**a**) XRD of CdSe/CdS QDs, (**b**,**c**) TEM images of CdSe/CdS QDs in different scale, and (**d**) size distribution of CdSe/CdS QDs.

3.2. Fluorescence Quenching Effect of EDTA on CdSe/CdS QDs

The introduction of EDTA can effectively reduce the PL intensity of CdSe/CdS QDs, as shown in Figure 3a. There is a broad emission band centered at about 600 nm for all the CdSe/CdS QDs systems with various EDTA concentrations. As the concentration of EDTA increases, the PL intensity of the EDTA-CdSe/CdS QD system decreases gradually, with a red shift of the emission band in the spectra. This is because EDTA was chemically absorbed on the surface of QDs, which caused QDs to cluster [29]. The change trend of PL intensity can be expressed by the relationship between I/I_0 and the concentration of

EDTA in Figure 3b, in which I and I_0 represent the PL intensities of QDs with the various concentrations of EDTA and without EDTA, respectively. It can be seen that the PL intensity decreases gradually with the increase of EDTA concentration. The fluorescence quenching is fast when the EDTA concentration is less than 35 μmol/L, but becomes slow when the EDTA concentration is more than 35 μmol/L. This is because some Cd^{2+} sites on the surface of CdSe/CdS QD were occupied with the increasing concentration of EDTA, and the more the combination between EDTA and CdSe/CdS QD becomes difficult, the more the fluorescence quenching slows down. The fluorescence intensity response curve of CdSe/CdS QDs can be divided into two stages, as shown in Figure 3c,d, respectively. In the concentration range of 0–35 μmol/L EDTA, the linear relationship between I/I_0 and the EDTA concentration (C_{EDTA}) can be expressed by Equation (1),

$$I/I_0 = 0.98788 + (-0.02182)C_{EDTA} \quad (1)$$

and the correlation coefficient (R^2) of Equation (1) is 0.994. In the range of 35–60 μmol/L EDTA, the correlation equation can be expressed as Equation (2),

$$I/I_0 = 0.39829 + (-0.00582)C_{\mathrm{EDTA}} \quad (2)$$

and the correlation coefficient (R^2) is 0.976.

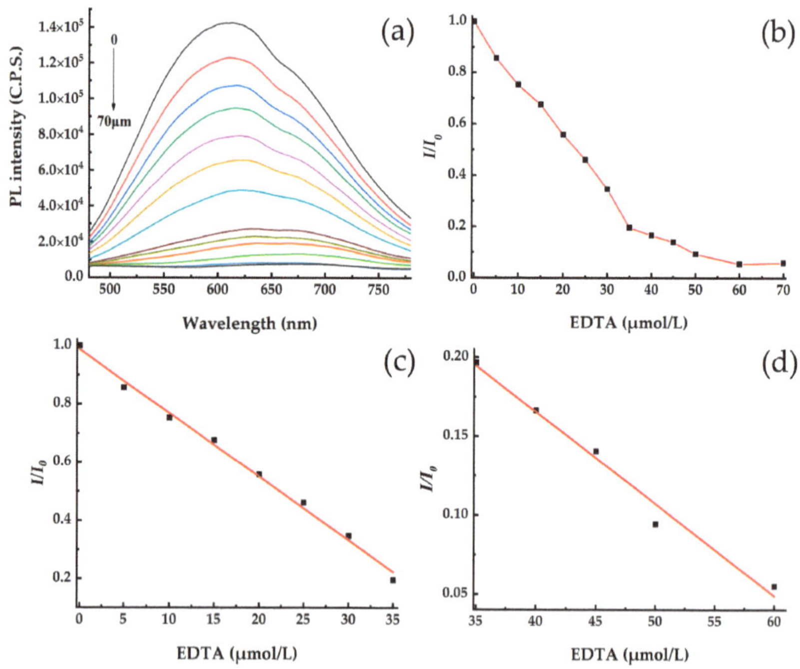

Figure 3. (**a**) Fluorescence spectra of the CdSe/CdS QDs system with various EDTA concentrations (0–60 μmol/L) in 300 μL volume, and (**b**) the relationship between fluorescence intensity ratio (I/I_0) of CdSe/CdS QDs and EDTA concentration, (**c**) 0–35 μmol/L EDTA, and (**d**) 35–60 μmol/L EDTA.

As a common metal chelating agent, EDTA will be chemically adsorbed on the surface of QDs to chelate with Cd^{2+} sites of CdSe/CdS QD when it was added to the CdSe/CdS QD solution, and a large area of optical active sites on the surface of these fluorescent CdSe/CdS QDs were masked, resulting in the fluorescence quenching. It caused a blue shift in the absorption spectra, as shown in Figure 4. After the introduction of Cd^{2+}, the absorption peak was red shifted, which was due to the partial leakage of the photon into the shell matrix [30].

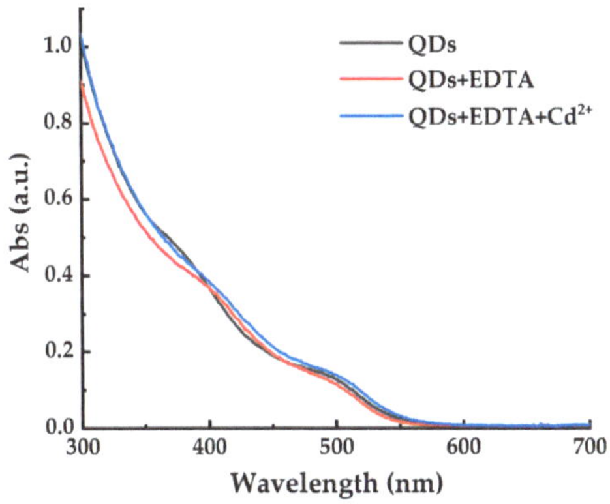

Figure 4. The absorption spectra of the systems of QDs, EDTA-QDs, EDTA-QDs+50 μmol/L Cd^{2+}.

In order to further explore the quenching mechanism of QDs by EDTA, temperature experiments and measurement of the fluorescence lifetime were conducted. The results are shown in Figure 5. The fluorescence quenching data were analyzed by Stern–Volmer Equation (3) [29],

$$I_0/I = 1 + K_{sv}[Q] \tag{3}$$

where K_{sv} is the quenching constant and $[Q]$ is the concentration of quenching agent. As shown in Figure 5a, when the temperature is 298 K and 308 K, the relationship between I_0/I and Q conforms to the Stern–Volmer equation. The value of slope decreases with the increase of temperature, indicating that static quenching occurs between EDTA and QDs. As shown in Figure 5b, the fluorescence lifetimes of QDs, QDs + 10 μmol/L EDTA and QDs + 20 μmol/L EDTA are 27.3 ns, 24.4 ns and 23.7 ns, respectively. The change of the lifetime is not obvious with the increase of Cd^{2+} concentration, consistent with the characteristics of static quenching. When the temperature is 298 K and 308 K, the quenching constants K_{sv} were calculated to be 4.49×10^4 L/mol and 3.82×10^4 L/mol. According to the Formula (4) [31],

$$K_q \tau_0 = K_{sv} \tag{4}$$

where, τ_0 is the lifetime of CdSe/CdS QD, the bimolecular quenching rate constant (K_q) is 1.64×10^{12} L·mol/s and 1.39×10^{12} L·mol/s, respectively, much higher than the maximum dynamic quenching rate 2.0×10^{10} L·mol/s. Therefore, it can be concluded that the quenching mechanism of the QDs by EDTA belongs to a static quenching process.

The concentration of EDTA has great influence on the sensitivity of EDTA-CdSe/CdS QDs to Cd^{2+} detection when Cd^{2+} concentration lies in a proper range. The quenching effect is not obvious when the concentration of EDTA is too low, while the detection of Cd^{2+} was not accurate when the concentration of EDTA is excessive. The fluorescence quenching efficiency $(1 - I/I_0)$ for the CdSe/CdS QDs system reaches 90% when 50 μmol/L EDTA is added. Furthermore, the quenching efficiency of CdSe/CdS QDs increases to 99.5% when the EDTA concentration increases to 60 μmol/L, meaning that the fluorescence is almost completely quenched. For comparison, 50 μmol/L EDTA was selected for subsequent fluorescence restoration experiments to detect Cd^{2+}.

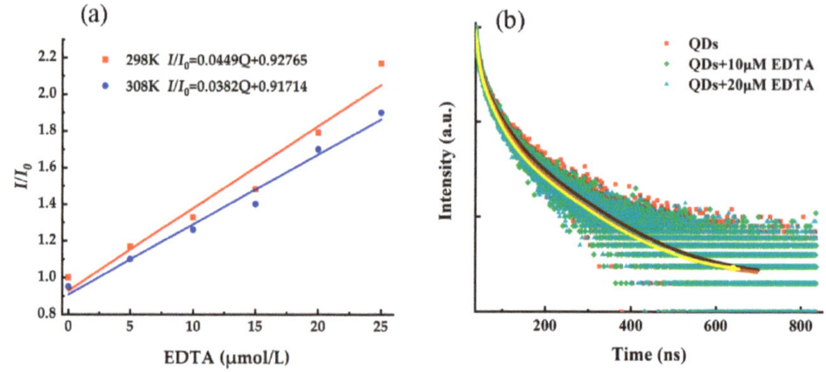

Figure 5. (**a**) Stern–Volmer plot for CdSe/CdS QDs system at 298 K and 308 K quenched by EDTA, and (**b**) fluorescence lifetime of the systems of QDs, QDs + 10 μmol/L EDTA, and QDs + 20 μmol/L EDTA.

3.3. Relationship between Fluorescence Intensity and Incubation Time and pH of Solution

The fluorescence stability of CdSe/CdS QDs, the quenching rate of CdSe/CdS QDs by EDTA, and the fluorescence restoration efficiency for the EDTA-CdSe/CdS system by Cd^{2+} were determined by the changes of PL intensity of these three systems over time. The experimental results are shown in Figure 6. The fluorescence intensity of CdSe/CdS QDs remains stable with the prolongation of incubation time, indicating a good fluorescence stability for the QDs system. After the addition of EDTA, the fluorescence quenching of CdSe/CdS QDs is very significant, and the PL intensity begins to be stable within 5 min, indicating that the reaction between EDTA and CdSe/CdS QDs is rapid, and that the fluorescence quenching is very effective. After Cd^{2+} was introduced into the EDTA-CdSe/CdS QDs system, the fluorescence began to be restored, and the fluorescence was almost completely restored within 3 min, remaining stable after that.

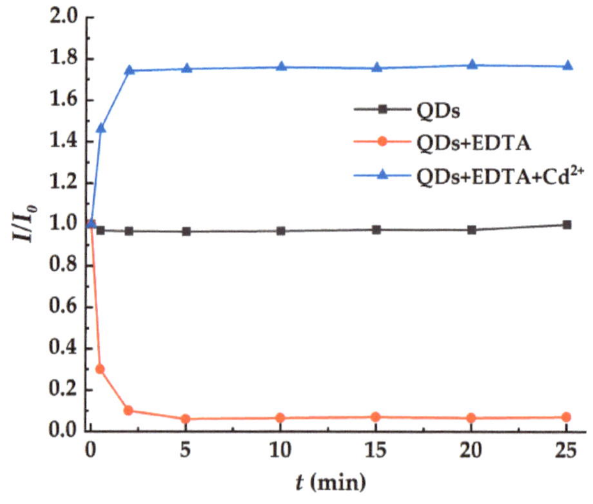

Figure 6. The change trend of I/I_0 for the system of QDs, EDTA-QDs, EDTA-QDs + 20 μmol/L Cd^{2+} with incubation time (t).

The pH in the solution system can affect the fluorescence intensity of QDs, as well as the sensitivity and selectivity of detected substances [32]. Figure 7a shows the influence curve of solution pH on PL intensity of TGA capped CdSe/CdS QDs. When the pH increases from 5.5 to 8.0, the PL intensity of the QDs increases gradually, and after that, the PL intensity tends to be stable with continual increase of pH value. This is because the mercaptan groups of TGA capped QDs are not stable under acidic conditions, which enhances the direct contact frequency between the QD surface and the aqueous solution. At this time, the fluorescence is weaker. Figure 7b shows the effect of solution pH on the fluorescence quenching of TGA capped CdSe/CdS QDs induced by EDTA, and on the fluorescence restoration for the EDTA-CdSe/CdS QDs system by Cd^{2+}. With the increase of the pH value, I/I_0 first decreases and then increases until the solution pH reaches 9.0, and at last decreases again to a higher pH range. The mechanism of the solution pH on the fluorescence intensity is very complicated. As a weak acid, the dissociation equilibrium constant of EDTA becomes smaller with the decrease of the solution pH value from 7.5 to 6, thus reducing the total number of $EDTA^-$ ions in the solution, which causes more ions of $EDTA^-$ to desorb from the surface of the fluorescent QDs and re-enter into the solution. This leads to a weaker quenching effect of EDTA, thus enhancing the fluorescence restoration rate of the QDs system in a lower pH solution by the introduction of Cd^{2+}. This explains why I/I_0 decreases with the increase of pH values in the range of pH 6–7.5. However, just like the discussion regarding the results of Figure 5a, higher pH is benefit for the stability of the mercaptan groups of TGA. The competition between these two factors determines the increasing change trend of fluorescence intensity in pH 7.5–9.0. However, if the solution pH value becomes too high and surpasses 9.0, Cd^{2+} in the solution system tends to react with OH^- to form $Cd(OH)_2$ precipitation [30], and thus I/I_0 begins to once again decrease.

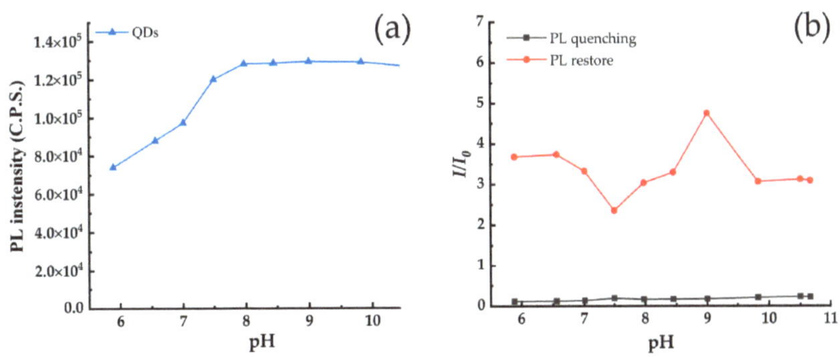

Figure 7. (**a**) The relationship between PL intensity of CdSe/CdS QDs and solution pH; (**b**) the effect of solution pH on the PL quenching and PL restoration.

3.4. Detection of Cd^{2+} in Ultrapure Water Solution

The absorption of EDTA on the surface of CdSe/CdS QDs produced a shielding function for the optical absorption and emission of QDs, which resulted in fluorescence quenching for the QDs system. Then Cd^{2+} was introduced to restore fluorescence of EDTA-CdSe/CdS QDs, and the restoration efficiency depends on the Cd^{2+} concentration in the detected water sample. As shown in Figure 8a, there is a broad emission band centered at about 600 nm for all the EDTA-CdSe/CdS QDs systems added with various Cd^{2+} concentrations. Figure 8b shows the change trend of PL intensity of EDTA-CdSe/CdS system with the increase of Cd^{2+} concentration in the detected solution, and the fluorescence restoration increases continually in the concentration range of 0.1–90 μmol/L Cd^{2+} of ultrapure water.

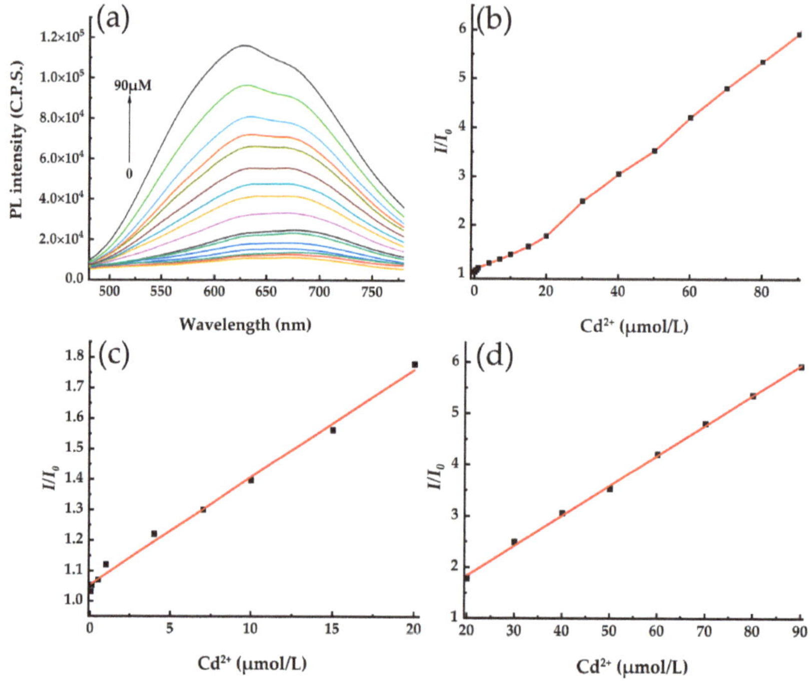

Figure 8. (**a**) The fluorescence spectra of EDTA-CdSe/CdS+Cd^{2+} QDs system with the various Cd^{2+} concentrations, and the relationship between fluorescence intensity ratio (I/I_0) and Cd^{2+} concentration in the detected solution, (**b**) 0–90 μmol/L, (**c**) 0–20 μmol/L, and (**d**) 20–90 μmol/L.

Figure 8c,d show the linear fitting results of the data of I/I_0 and Cd^{2+} concentration in two regions: 0.1–20 μmol/L and 20–90 μmol/L, corresponding to the linear Equations (5) and (6), respectively. The correlation coefficients (R^2) are all above 0.99, suggesting a good linear relationship between I/I_0 and Cd^{2+} concentration. According to the Equation (7) [33], where δ is the standard deviation of blank measurements (n = 11) and S is the slope of calibration graph. The detection limit (LOD) was calculated to be 6 nmol/L.

$$I/I_0 = 1.05536 + 0.03529C_{Cd^{2+}} \ 0.1\text{–}20\ \mu mol/L \quad (5)$$

$$I/I_0 = 0.6664 + 0.05869C_{Cd^{2+}} \ 20\text{–}90\ \mu mol/L \quad (6)$$

$$LOD = 3\delta/S \quad (7)$$

The selectivity of EDTA-CdSe/CdS QDs system to Cd^{2+} in an aqueous solution was evaluated in comparison with 13 other metal ions (K^+, Na^+, Mg^{2+}, Ba^{2+}, Al^{3+}, Mn^{2+}, Fe^{3+}, Ca^{2+}, Hg^{2+}, Pb^{2+}, Cu^{2+}, Ag^+, Zn^{2+}) under the optimal fluorescence restoration conditions. In order to reflect the selectivity of the QD system to various impurities adequately, the concentration of these interfering ions is set to 500 μmol/L except for Zn^{2+}, and both of the Zn^{2+} and Cd^{2+} concentrations are set to be 50 μmol/L. Figure 9 shows the effects of these interfering ions on fluorescence restoration efficiency of the EDTA-CdSe/CdS fluorescence probe. Cu^{2+}, Ag^+, Hg^{2+}, and Pb^{2+} lead to fluorescence quenching of the system completely. They could be adsorbed on QDs surface and quenched PL due to electron transfer from QDs to Ag^+, Cu^{2+}, Hg^{2+}, and Pb^{2+}. In addition, a chemical displacement of surface Cd^{2+} by Hg^{2+}, Cu^{2+}, and Ag^+ occurred due to the extremely low solubility of CuSe, HgSe, and Ag_2Se. Their formation would cause the PL quenching by facilitating non-radiative electron/hole (e^-/h^+) annihilation for the QDs system [34–36]. Ni^+, Fe^{3+}, and Mn^{2+} also have a certain fluorescence quenching effect on the system. K^+, Na^+, Al^{3+}, Ba^{2+}, and

Ca^{2+} have little effect on fluorescence. However, Cd^{2+} shows a significant fluorescence restoration effect on the system, and I/I_0 increases by 3.6 times with the addition of Cd^{2+}. However, Zn^{2+} also shows an obvious fluorescence restoration effect, and I/I_0 increases by 2.0 times with the addition of Zn^{2+}, likely owing to similar chemical properties between Zn^{2+} and Cd^{2+}. Zn-mercaptan forming on the surface of QDs can lead to fluorescence restoration for the system [30]. In conclusion, the CdSe/CdS QDs system has a high selectivity for the detection of Cd^{2+} in an aqueous solution, but it is improper to detect Cd^{2+} in a solution containing Cd^{2+} and Zn^{2+}.

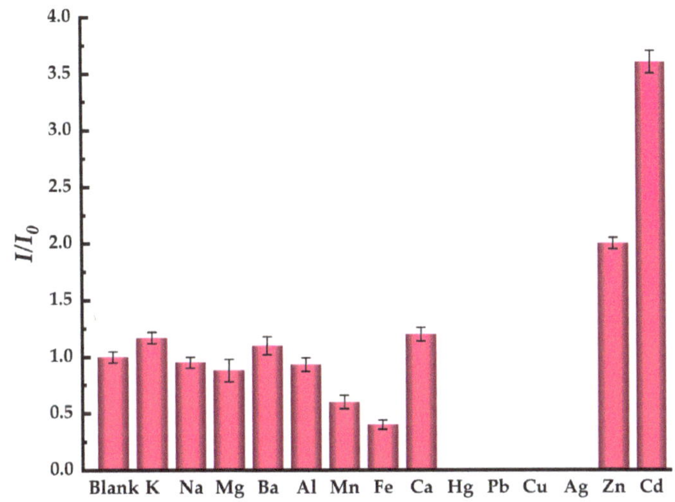

Figure 9. The effect of interfering ions on fluorescent restoration for the system of EDTA-CdSe/CdS QDs.

3.5. Detection of Cd^{2+} in Tap Samples

The detection experiments of Cd^{2+} were conducted in tap water to evaluate the practicability and reliability of the EDTA-CdSe/CdS QDs fluorescent probe. The tap water is a living water from Nanning city of China. The solutions with three different concentrations of Cd^{2+} (10, 20 and 30 μmol/L) were introduced into the tap water, respectively, and the experimental results are shown in Table 1. It can be seen that the measured values are very close to the actual concentration of Cd^{2+} in the aqueous solutions, with the fluorescence restoration efficiency above 96% and the relative standard deviation (RSD) below 4%. It suggests that the inherent ions such as Na^+, Ca^{2+}, Mg^{2+}, and Mn^{2+} in the tap water cannot constitute a barrier in the detection of Cd^{2+} for the EDTA-CdSe/CdS QDs fluorescent probe. Note that the actual Cd^{2+} concentration in the tap water was considered to be equal to the dosage of Cd^{2+} because it is not detectable in pristine tap water.

Table 1. A comparison between actual Cd^{2+} concentration and the measurasdfed values, and the fluorescence restoration efficiency and relative standard deviation for the detection of Cd^{2+} in the tap water.

Sample (No.)	Cd^{2+} Concentration (μmol/L)	ICP-MS Method (μmol/L)	Proposed Method (μmol/L)	Recovery (%)	RSD (%, n = 3)
1	10	9.65	9.68	96.8	2.6
2	20	20.56	20.95	104.8	1.8
3	30	30.08	29.12	97.1	3.8
Mean value	/		/	99.6	2.8

4. Conclusions

A CdSe/CdS QD fluorescent probe with core-shell structure was successfully synthesized via solution reaction method. EDTA was proved to be an efficient fluorescence quenching agent to realize the OFF function in the detection process of Cd^{2+} based on an OFF–ON mode. Under proper conditions, the fluorescence quenching efficiency reached 99.5% within 5 min. The fluorescence of the EDTA-CdSe/CdS system was restored more effectively when Cd^{2+} was immigrated in the solution system, realizing the ON function in the detection process. The relationship between fluorescence restoration efficiency and Cd^{2+} concentration was well expressed by two linear equations, which can be used for the accurate calculation of Cd^{2+} concentration in the 0.1–100 μmol/L range. The QD fluorescent probe shows a good selectivity for Cd^{2+} in the aqueous solution containing twelve kinds of interfering ions such as Na^{+}, Ca^{2+}, Mg^{2+}, Mn^{2+}, and so on. However, the presence of Zn^{2+} would cause serious interference for the detection of Cd^{2+} in an aqueous solution. Finally, the experimental results of Cd^{2+} detection in tap water further proved the practicability and reliability of the EDTA-CdSe/CdS QDs fluorescent probe.

Author Contributions: Performing the experiments, Data collection and analysis, Writing the initial draft, D.W.; Conceptualization, Methodology, Resources, Supervision, Validation, Review and Revised manuscript, Acquisition of the financial support, F.G.; Data collection, X.W. (Xianran Wang) and X.N.; Guiding experiments, K.W., X.W. (Xinpeng Wang) and Y.W.; Part of the financial support, T.F. All authors have read and agreed to the published version of the manuscript.

Funding: This research was funded by the Programs for Natural Science Foundation of Guangxi (2019GXNSFAA185013), Science and Technology Major Project of Guangxi Province (AA18118030), Guangxi Key Laboratory Development Foundation (GXYSSF1802), and National Natural Science Foundation of China (21976039).

Institutional Review Board Statement: Not applicable.

Informed Consent Statement: Not applicable.

Data Availability Statement: Data is contained within the article. The data presented in this study are available in this published article.

Acknowledgments: The authors thanks financial support from the Programs for Natural Science Foundation of Guangxi (2019GXNSFAA185013), Science and Technology Major Project of Guangxi Province (AA18118030), Guangxi Key Laboratory Development Foundation (GXYSSF1802), and National Natural Science Foundation of China (21976039).

Conflicts of Interest: The authors declare no conflict of interest.

References

1. Zhang, X.; Zhang, M.; Liu, H.; Gu, J.; Liu, Y. Environmental sustainability: A pressing challenge to biological sewage treatment processes. *Curr. Opin. Environ. Sci. Health* **2019**, *12*, 1–5. [CrossRef]
2. Yan, F.; Niu, Z.G. Evaluation model of major heavy metals pollution factors in coastal waters and sediments. *Desalination Water Treat.* **2019**, *149*, 335–340. [CrossRef]
3. Huang, X.S.; Zhang, R.J.; Cui, M.J.; Lai, H.J. Experimental Investigation on Bioremediation of Heavy Metal Contaminated Solution by Sporosarcina pasteurii under Some Complex Conditions. *Water* **2022**, *14*, 595. [CrossRef]
4. Zhang, H.; Li, C.; Chen, X.; Fu, H.; Chen, Y.; Ning, S.; Fujita, T.; Wei, Y.; Wang, X. Layered ammonium vanadate nanobelt as efficient adsorbents for removal of Sr^{2+} and Cs^{+} from contaminated water. *J. Colloid Interface Sci.* **2022**, *615*, 110–123. [CrossRef] [PubMed]
5. Luo, X.; Zhang, G.H.; Wang, X.; Gu, P. Research on a pellet co-precipitation micro-filtration process for the treatment of liquid waste containing strontium. *J. Radioanal. Nucl. Chem.* **2013**, *298*, 931–939. [CrossRef]
6. El-Saadani, Z.; Mingqi, W.; He, Z.; Hamukwaya, S.L.; Abdel Wahed, M.S.M.; Abu Khatita, A. Environmental geochemistry and fractionation of cadmium metal in surficial bottom sediments and water of the Nile River, Egypt. *Toxics* **2022**, *10*, 221. [CrossRef]
7. Barone, G.; Storelli, A.; Garofalo, R.; Mallamaci, R.; Storelli, M.M. Residual levels of mercury, cadmium, lead and arsenic in some commercially key species from italian coasts (Adriatic Sea): Focus on human health. *Toxics* **2022**, *10*, 223. [CrossRef]
8. Smichowski, P.; Londonio, A. The role of analytical techniques in the determination of metals and metalloids in dietary supplements: A review. *Microchem. J.* **2018**, *136*, 113–120. [CrossRef]

9. Haribala; Hu, B.T.; Wang, C.G.; Gerilemandahu; Xu, X.; Zhang, S.; Bao, S.H.; Li, Y.H. Assessment of radioactive materials and heavy metals in the surface soil around uranium mining area of Tongliao, China. *Ecotox. Environ. Safe.* **2016**, *130*, 185–192. [CrossRef]
10. He, D.; Zhu, Z.L.; Miao, X.; Zheng, H.T.; Li, X.L.; Belshaw, N.S.; Hu, S.H. Determination of trace cadmium in geological samples by membrane desolvation inductively coupled plasma mass spectrometry. *Microchem. J.* **2019**, *148*, 561–567. [CrossRef]
11. Lee, W.; Kim, H.; Kang, Y.; Lee, Y.; Yoon, Y. A biosensor platform for metal detection based on enhanced green fluorescent protein. *Sensors* **2019**, *19*, 1846. [CrossRef] [PubMed]
12. Yao, Y.; Wu, H.; Ping, J.F. Simultaneous determination of Cd(II) and Pb(II) ions in honey and milk samples using a single-walled carbon nanohorns modified screen-printed electrochemical sensor. *Food. Chem.* **2019**, *274*, 8–15. [CrossRef] [PubMed]
13. Petryayeva, E.; Algar, W.R.; Medintz, I.L. Quantum dots in bioanalysis: A review of applications across various platforms for fluorescence spectroscopy and imaging. *Appl. Spectrosc.* **2013**, *67*, 215–252. [CrossRef]
14. Lin, L.P.; Wang, Y.H.; Xiao, Y.L.; Liu, W. Hydrothermal synthesis of carbon dots codoped withnitrogen and phosphorusas a turn-on fluorescent probe for cadmium(II). *Microchim. Acta* **2019**, *186*, 147. [CrossRef] [PubMed]
15. Pan, Y.C.; Xiao, W.S.; Fan, M.; Chen, W.X.; Xu, S.H. Detection of Cd(II) ions by a self-tuning method using quantum dot fluorescence sensing. *Mater. Res. Express.* **2017**, *4*, 105008. [CrossRef]
16. Lin, W.X.; Xie, X.C.; Wang, Y.J.; Chen, J.J. A new fluorescent probe for selective Cd^{2+} detection and cell imaging. *Z. Anorg. Allg. Chem.* **2019**, *645*, 645–648. [CrossRef]
17. Su, W.; Yuan, S.Z.; Wang, E.J. A rhodamine-based fluorescent chemosensor for the detection of Pb^{2+}, Hg^{2+} and Cd^{2+}. *J. Fluoresc.* **2017**, *27*, 1871–1875. [CrossRef]
18. Pourreza, N.; Mousavi, H.Z. Determination of cadmium by flame atomic absorption spectrometry after preconcentration on naphthalene-methyltrioctylammonium chloride adsorbent as tetraiodocadmate (II) ions. *Anal. Chim. Acta.* **2004**, *503*, 279–282. [CrossRef]
19. Aldewachi, H.; Chalati, T.; Woodroofe, M.N.; Bricklebank, N.; Sharrack, B.; Gardiner, P. Gold nanoparticle-based colorimetric biosensors. *Nanoscale* **2018**, *10*, 18–33. [CrossRef]
20. Liu, J.; Zhao, X.Y.; Xu, H.Y.; Wang, Z.Y.; Dai, Z.H. Amino acid-capped water-soluble near-infrared region $CuInS_2/ZnS$ quantum dots for selective cadmium ion determination and multicolor cell imaging. *Anal. Chem.* **2019**, *91*, 8987–8993. [CrossRef]
21. Wang, S.N.; Zhu, J.; Li, X.; Li, J.J.; Zhao, J.W. Fluorescence turn-on sensing of trace cadmium ions based on EDTA-etched CdTe@CdS quantum dot. *Spectrochim. Acta A* **2018**, *201*, 119–127. [CrossRef] [PubMed]
22. Niu, W.J.; Shan, D.; Zhu, R.H.; Deng, S.Y.; Cosnier, S.; Zhang, X.J. Dumbbell-shaped carbon quantum dots/AuNCs nanohybrid as an efficient ratiometric fluorescent probe for sensing cadmium (II) ions and l-ascorbic acid. *Carbon* **2016**, *96*, 1034–1042. [CrossRef]
23. Jing, W.; Chenxing, J.; Xiaoqi, W.; Ligeng, W.; Aimin, C.; Jun, H.; Zhihui, L. Fabrication of an "ion-imprinting" dual-emission quantum dot nanohybrid for selective fluorescence turn-on and ratiometric detection of cadmium ions. *Analyst* **2016**, *141*, 5886.
24. Wang, Z.; Yang, Y.; Zou, T.; Xing, X.; Zhao, R.; Wang, Y. Novel method for the qualitative identification of chromium ions (III) using l-aspartic acid stabilized CdS quantum dots. *J. Phys. Chem. Solids* **2020**, *136*, 109160. [CrossRef]
25. Liu, F.; Li, S.; Hu, R.; Shao, N. Core-shell structured $CdTe/CdS@SiO_2@CdTe@SiO_2$ composite fluorescent spheres: Synthesis and application for Cd^{2+} detection. *Luminescence* **2017**, *32*, 723–729. [CrossRef]
26. Yin, Y.; Yang, Q.L.; Liu, G. Ammonium pyrrolidine dithiocarbamate-modified CdTe/CdS quantum dots as a turn-on fluorescent sensor for detection of trace cadmium ions. *Sensors* **2020**, *20*, 312. [CrossRef]
27. Wang, J.; Zhou, X.; Ma, H.; Tao, G. Diethyldithiocarbamate functionalized CdSe/CdS quantum dots as a fluorescent probe for copper ion detection. *Spectrochim. Acta A* **2011**, *81*, 178–183. [CrossRef]
28. Talapin, D.V.; Mekis, I.; Götzinger, S.; Kornowski, A.; Benson, O.; Weller, H. CdSe/CdS/ZnS and CdSe/ZnSe/ZnS core−shell−shell nanocrystals. *Spectrochim. Acta A* **2004**, *108*, 18826–18831. [CrossRef]
29. Mohammad-Rezaei, R.; Razmi, H.; Abdolmohammad-Zadeh, H. d-penicillamine capped cadmium telluride quantum dots as a novel fluorometric sensor of copper (II). *Luminescence* **2013**, *28*, 503–509. [CrossRef]
30. Xu, H.; Miao, R.; Fang, Z.; Zhong, X. Quantum dot-based "turn-on" fluorescent probe for detection of zinc and cadmium ions in aqueous media. *Anal. Chim. Acta* **2011**, *687*, 82–88. [CrossRef]
31. Jiang, C.Q.; Gao, M.X.; He, J.X. Study of the interaction between terazosin and serum albumin: Synchronous fluorescence determination of terazosin. *Anal. Chim. Acta* **2002**, *452*, 185–189. [CrossRef]
32. Bian, W.; Wang, F.; Zhang, H.; Zhang, L.; Wang, L.; Shuang, S. Fluorescent probe for detection of Cu^{2+} using core-shell CdTe/ZnS quantum dots. *Luminescence* **2015**, *30*, 1064–1070. [CrossRef] [PubMed]
33. Punithakumari, G.; Wu, S.P.; Velmathi, S. Highly selective detection of Cr^{3+} ion with colorimetric & fluorescent response via chemodosimetric approach in aqueous medium. *J. Fluoresc.* **2018**, *28*, 663–670. [PubMed]
34. Xia, Y.S.; Cao, C.; Zhu, C.Q. Two distinct photoluminescence responses of CdTe quantum dots to Ag (I). *J. Lumin.* **2008**, *128*, 166–172. [CrossRef]
35. Xie, H.Y.; Liang, J.G.; Zhang, Z.L.; Liu, Y.; He, Z.K.; Pang, D.W. Luminescent CdSe-ZnS quantum dots as selective Cu^{2+} probe. *Spectrochim. Acta Part A Mol. Biomol. Spectrosc.* **2004**, *60*, 2527–2530. [CrossRef]
36. Susha, A.S.; Javier, A.M.; Parak, W.J.; Rogach, A.L. Luminescent CdTe nanocrystals as ion probes and pH sensors in aqueous solutions. *Colloids Surf. A Physicochem. Eng. Asp.* **2006**, *281*, 40–43. [CrossRef]

toxics

MDPI

Article

Mineralogical Properties of a Refractory Tantalum-Niobium Slag and the Effect of Roasting on the Leaching of Uranium-Thorium

Min Huang [1,2], Ke Hu [1,3], Xiang Li [1,2], Yun Wang [1,3], Jinbo Ouyang [1,2], Limin Zhou [1,2] and Zhirong Liu [1,2,*]

[1] State Key Laboratory of Nuclear Resources and Environment, East China University of Technology, Nanchang 330013, China
[2] School of Chemistry, Biological and Materials Science, East China University of Technology, Nanchang 330013, China
[3] School of Nuclear Science and Engineering, East China University of Technology, Nanchang 330013, China
* Correspondence: zhrliu@ecit.cn

Abstract: In order to realize sustainable development, it is beneficial to explore an appropriate process to recover the radionuclides contained in tantalum-niobium slag. By micro-mineralogical analysis and roasting experiments, the effect of uranium-thorium leaching from a refractory tantalum-niobium slag is investigated. The uranium and thorium content in the slag is 2.26×10^3 mg/kg and 7.84×10^3 mg/kg, which have large recovery value. As the surface area and pore size of the slag are very small, the leaching agent cannot fully penetrate the particles. Various methods of characterization are used to analyze the mineralogical properties of roasted slag at different temperatures. The leaching ratio of U-Th is 90.84% and 96.62% at the optimum roasting temperature of 500 °C, which are about 39% and 27% higher than original samples. The oxidants Fe^{3+}, O_2 and Mn can also promote the conversion of insoluble U(IV) to soluble U(VI). Roasting reduces the content of organic C and S, thereby preventing reduction of U(VI), and increasing pore size as well as specific surface area also promote radionuclide leaching. Thus, the roasting method at 500 °C can destroy the surface wrapping structure of radionuclides, reduce the internal density of minerals, and improve uranium-thorium leaching ratio significantly. It is of great practical significance to reduce the radioactive hazard of waste tantalum-niobium slag and to strengthen the sustainable utilization of resources by suitable process improvement techniques.

Keywords: tantalum-niobium slag; mineralogical properties; roasting; uranium; thorium; leach

Citation: Huang, M.; Hu, K.; Li, X.; Wang, Y.; Ouyang, J.; Zhou, L.; Liu, Z. Mineralogical Properties of a Refractory Tantalum-Niobium Slag and the Effect of Roasting on the Leaching of Uranium-Thorium. *Toxics* **2022**, *10*, 469. https://doi.org/10.3390/toxics10080469

Academic Editors: Yuezhou Wei, Yan Wu and Xinpeng Wang

Received: 15 July 2022
Accepted: 9 August 2022
Published: 12 August 2022

Publisher's Note: MDPI stays neutral with regard to jurisdictional claims in published maps and institutional affiliations.

Copyright: © 2022 by the authors. Licensee MDPI, Basel, Switzerland. This article is an open access article distributed under the terms and conditions of the Creative Commons Attribution (CC BY) license (https://creativecommons.org/licenses/by/4.0/).

1. Introduction

Both tantalum (Ta) and niobium (Nb) belong to group VB elements, which are transition metals with similar physicochemical properties [1–3]. They often exist as associated minerals in geological environment. Ta and Nb can form a thin oxide layer with high corrosion resistance on surface, so they are considered as excellent biologically inert materials [4]. Since the contraction of lanthanides, tantalum and niobium are related to each other, as well as occur in pairs. Furthermore, they form tantalates and niobates that contain large amounts of uranium (U) and thorium (Th) [5]. Tantalum-niobium minerals associated with uranium-thorium elements exist mainly in the form of multiple oxides. China is rich in reserves of uranium-thorium minerals, tantalum-niobium ores. Many minerals such as tantalum-niobium ore, biotite and monazite often contain certain amounts of radionuclides (especially U and Th) [6,7]. However, these rare heavy metal tailings have complex composition, low content of useful components, and are difficult to extract of heavy metals [8,9]. Additionally, many mines do not handle the waste properly after processing, leaving a great number of abandoned tailings [10]. The harmful substances rich in NH_4^- and F^- in tailings cause secondary pollution [9]. This not only makes the utilization ratio of tailings low, but also endangers the ecological environment [11,12]. Thus, finding appropriate methods to solve the problem of comprehensively utilizing waste slag has become an urgent matter [13].

In recent years, the demand for Ta and Nb products have increased dramatically around the world, and the high-grade tantalum-niobium hard ores could no longer meet the market demand [14,15]. However, the decomposition of tantalum-niobium ore treated with H_2SO_4, HNO_3 or HF is insufficient, resulting in a serious waste of resources. Moreover, the residual HF in the slag is volatile and easy to cause environmental problems [16]. Although tantalum-niobium resources in China are very rich, the ore grade is low, with many associated impurities, which leads to separate difficultly, low recovery ratio and poor occurrence state of useful components [17]. Tantalum-niobium ore undergoes a series of high-temperature as well as strong acid reactions in smelting process, during which the uranium-thorium remains in the slag and at a high content [9]. The risk of heavy metals contamination in mine waste cannot be ignored, especially when the tailings are affected by human activities [18]. After the smelting and processing of tantalum-niobium ore, the uranium-thorium elements in ore are largely retained in the tailings. These toxic elements are spread and remain by natural media, and eventually enter our bodies through the biological chain or biological cycle, posing a threat to human health [19,20]. Due to the radioactive nature of uranium and thorium, the extraction of radionuclides from mine solid waste is critical to environmental sustainability and human health [21,22].

With the development and utilization of nuclear energy, the market demand for nuclear fuel increased. If uranium and thorium in tantalum-niobium tailings can be recovered, the shortage of uranium resources in China can be alleviated. The first step in recovering uranium-thorium from tantalum-niobium slag is to extract the target elements from the hard-to-decompose slag, so that the radionuclides can be initially separated from the slag. The high intensity reaction occurring during smelting makes it difficult to leach uranium-thorium from tantalum-niobium tailings directly with acid. The original mineral structure is broken, resulting in a more complex composition [23]. Therefore, this work selected a refractory tantalum-niobium mine waste residue in south China as the research object, and analyzed the relevant mineralogical properties of tantalum-niobium slag by various mineralogical characterization methods. Additionally, the research elaborated the mineralogical microstructure characteristics of the slag, and explored the composition and distribution characteristics of uranium and thorium in the slag. It is one of the focuses of this work to explain the reason why uranium and thorium are difficult to leach from tantalum-niobium slag by micro-mineralogical perspective. Considering the presence of reductive organic C and S in the slag, and the possibility of some special inclusions to encapsulate radionuclides, so the slag was roasted. The conversion of insoluble U(IV) to U(VI) was promoted by roasting, thus improving the leaching ratio of radionuclides. The ultimate purpose of this research is to improve the recovery of uranium-thorium from tantalum-niobium slag and provide ideas for the treatment of waste slag in industrial applications.

2. Experiments and Methods

2.1. Instrumental Techniques

The characterization methods of tantalum-niobium slag are as follows. The microstructure of tantalum-niobium slag was observed by Scanning Electron Microscope (SEM) using a Nova-Nano SEM450 (FEI, Hillsboro, OR, USA). The types, content and distribution of the main elements in slag were detected by Energy Dispersive Spectroscopy (EDS) using a model x-max 50 spectrometer (Oxford Instruments, Abingdon, UK). Inductively Coupled Plasma-Optical Emission Spectrometer (ICP-OES) technique allows qualitative and quantitative analysis of the elements contained in slag samples. The content of various elements such as Ta, Nb, U, Th and Fe in slag, and the content of U as well as Th in the leaching solution during the detection experiments, were accurately detected by ICP-OES using an Avio200 type spectrometer (PerkinElmer, Waltham, MA, USA). In this work, the material structure of slag was analyzed by X-ray Diffractometer (XRD) of a D8-Advance device

(Bruker, Karlsruhe, Germany). The types and content of elements in slag were detected by X-ray Fluorescence Spectrometry (XRF) using a model Axios-mAX spectrometer (PANalytical B.V., Almelo, The Netherlands). The elemental content of Fe and U in different valence states in slag was measured by X-ray Photoelectron Spectrometer (XPS) using a K-Alpha spectrometer (Thermo Fisher Scientific, Waltham, MA, USA). The microscale specific surface area and pore size of slag were characterized by specific surface area test (BET) using a BSD-PS1 type specific surface pore size analyzer (Beishide, Beijing, China). Also, the content of organic carbon and sulfur elements in slag was detected by organic carbon and sulfur analysis (TOC/TS) of a CS844 type analyzer (Laboratory Equipment Corporation, Livonia, MI, USA).

2.2. Roasting and Leaching Experiments

Considering the presence of reductive organic carbon and sulfur in tantalum-niobium slag, and the possibility of some special inclusions to encapsulate uranium and thorium, so the slag was roasted to improve the leaching ratio of radionuclides [24,25]. The changes of leaching ratio in slag before and after roasting was analyzed by ICP-OES technique to investigate the effect of roasting on leaching efficiency. In addition, the tantalum-niobium slag after roasting at 200–600 °C was analyzed to explore the optimal roasting temperature. In the roasting experiments, the insoluble U(IV) can be converted into soluble U(VI) by roasting, thus increasing the leaching ratio [26,27].

The tantalum-niobium slag should be pretreated before the experiments. The samples were dried in a blast drying oven at 50 °C for 24 h. The coarse samples were ground and sieved with a pulverizer and stored in a sealed bag. Samples with a particle size of −200 mesh (−75 µm) were selected for all experimental procedures. During roasting experiments, the pretreated slag was poured into the crucible, which was placed in muffle furnace at 200–600 °C (interval 100 °C) for 24 h. After cooling, the roasted slag was removed and stored in a sealed bag to keep dry and avoid contact between the sample and water vapor in the air, which may affect the experimental results.

In previous study, the effects of the type of acid used, the concentration of acid, leaching time, liquid-solid ratio and other conditions on the leaching ratio during the leaching process have been investigated [28]. In this work, we discussed the effect of roasting conditions on the leaching ratio of uranium-thorium from tantalum-niobium slag, using H_2SO_4 as leaching agent, and the original pH of the slag was 1.79.

A certain amount of −200 mesh raw slag and the samples roasted at 200–600 °C were respectively put into polytetrafluoroethylene beakers. The H_2SO_4 with the same concentration of 0.75 mol/L was added, the liquid-solid ratio was 10:1, the leaching temperature was 90 °C, and the magnetic stirrer was used for stirring and leaching for 4 h at the speed of 200 r/min. The leachate was obtained by filtration three times and analyzed by ICP-OES for U and Th content in the leachate. The leaching ratio was calculated by the equation (1), in this equation, p_i (%) was the leaching ratio under condition i, C_i (mg/L) was the element concentration in the leachate under condition i, V (L) was the volume of leaching agent, W_i (mg/kg) was the content of uranium or thorium in the slag, M (kg) was the total mass of slag.

$$p_i\ (\%) = \frac{C_i \times V}{W_i \times M} \times 100 \quad (1)$$

3. Results and Discussion

3.1. Characterization Results of Mineralogical Properties of Slag

The tantalum-niobium slag was analyzed by SEM and EDS, and the results were shown in Figures 1 and 2.

Figure 1. SEM image of tantalum-niobium slag.

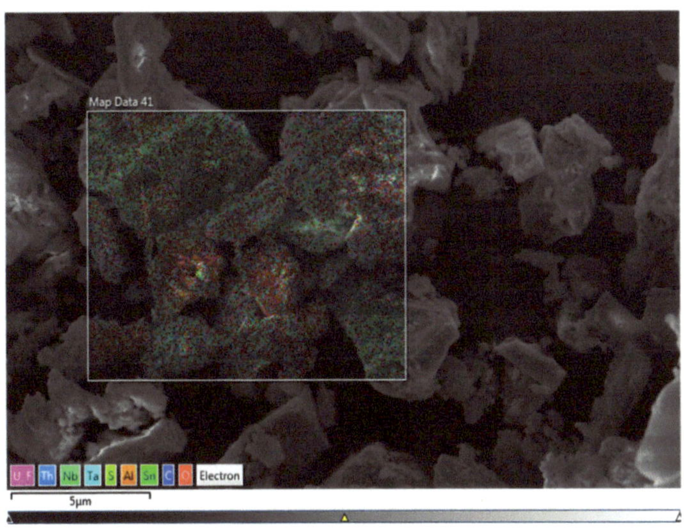

Figure 2. EDS image of tantalum-niobium slag.

Figure 1 was the SEM image of the tantalum-niobium slag at 2000×. It showed that many irregular particles in the slag, the surface between the particles was relatively dense, moreover, there were voids of different sizes between the particles.

The EDS results (Figure 2 and Table 1) of tantalum-niobium slag showed that the slag samples contain Al, F, Mn, Fe, Ta, Nb, Sn, U, Th, S and C. The content of F was 2.55% and S was 0.87%, which was due to the addition of large amounts of H_2SO_4 and HF during the smelting of tantalum-niobium ore, resulting in a small portion of F and S remaining in the slag. Besides uranium-thorium in the slag, there were also reducing substances such as metallic manganese, organic carbon, and sulfur, which could affect the redox atmosphere conditions in the leaching process. Furthermore, a certain amount of divalent iron was also present in the slag with a relatively uniform distribution, which can promote the leaching efficiency of U from slag. The addition of high concentrations of H_2SO_4 and HF destroyed the mineral structure, and made radionuclides enriched in slag.

Table 1. Elemental percentage of tantalum-niobium slag by EDS.

Element	C	O	F	Al	Mn	Fe	Nb	Sn	Ta	Th	U	Total
Wt %	2.18	31.20	1.44	6.44	0.24	1.08	12.88	37.54	5.58	0.04	1.38	100
Atomic %	6.12	65.86	2.55	8.06	0.14	0.66	4.68	10.68	1.04	0.01	0.20	100

3.2. Elemental Content in Tantalum-Niobium Slag

Table 2 showed that the tantalum-niobium slag was rich in Sn, additionally, the content of Ta, Nb and Fe was also relatively high, while the slag contained rare earth elements and radionuclides. These proved that the recovery of radioactive elements in tantalum-niobium slag was of significance.

Table 2. Detection of composition content in tantalum-niobium slag by XRF.

Composition	Sn	Fe	U	Th	O	Al	S	Ta	Nb
Content (wt.%)	8.38	28.5	0.38	1.22	28.5	3.36	16.8	0.08	0.799
Composition	Nd	P	Sc	Si	Yb	Pb	Cr	Ca	Ti
Content (wt.%)	0.55	2.73	0.041	2.58	0.59	0.099	1.27	0.71	0.79

In Table 3, the exact content of Ta, Nb, U, Th and Fe in the slag was obtained by ICP-OES analysis, with 2.26×10^3 mg/kg of U and 7.84×10^3 mg/kg of Th. The tantalum-niobium slag had a high content of U and Th, which was valuable for recycling. Additionally, the Fe element was rich to promote the oxidation of refractory U(IV) to soluble U(VI) and to improve the leaching efficiency. By recovering the uranium-thorium from tantalum-niobium slag, we can not only alleviate the shortage of nuclear fuel resources, but also reduce the pollution of slag on the surrounding environment.

Table 3. ICP–OES element analysis diagram of tantalum niobium slag.

Ta-Nb Slag	U	Th	Ta	Nb	Fe
Content (mg/kg)	2.26×10^3	7.84×10^3	0.88×10^3	2.01×10^3	1.66×10^5

3.3. Elemental Valence Analysis in Tantalum-Niobium Slag

The pretreated slag was taken and analyzed by XPS for the content of two different valence states of U and Fe in the slag, and the results were shown in Table 4.

Table 4. XPS element analysis diagram of tantalum-niobium slag.

Valence States	Fe^{2+} (%)	Fe^{3+} (%)	U(IV) (%)	U(VI) (%)
Original slag	62.61	37.39	63.83	36.17

The content of U and Fe in both valence states was in the form of divalent and trivalent compounds (Table 4). The content of ferrous compounds was 62.61%, and that of ferric compounds was 37.39%, which proved that most of Fe in slag existed in the form of ferric. Uranium in slag mainly existed in the form of tetravalent uranium dioxide, the content of U(IV) was 63.83%, and the content of soluble U(VI) was only 36.17%. In the original sample, the content of Fe^{2+} and U(IV) were higher than Fe^{3+} and U(VI). Although the existence of iron was beneficial to the subsequent leaching, the U(IV) was insoluble. Therefore, it was necessary to promote the conversion of U(IV) into soluble U(VI) by other means.

3.4. Specific Surface Area and Pore Size Analysis of Tantalum-Niobium Slag

The Figure 3 showed that the pore size between the slag particles was small, and mainly distributed between 5 nm, thus the pore size in tantalum-niobium slag is mainly mesoporous (2–50 nm). This result was consistent with the data in the relevant literature.

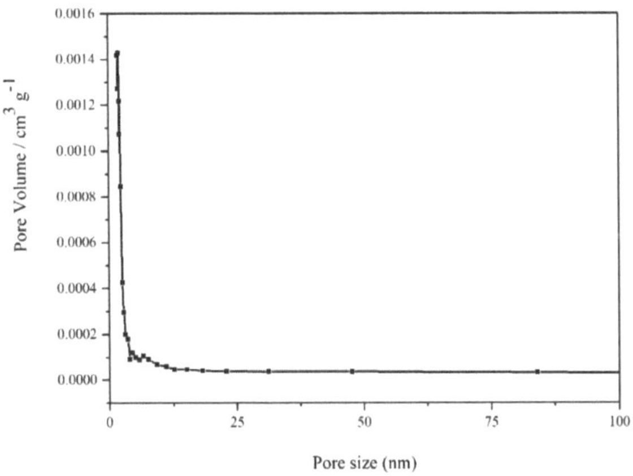

Figure 3. Pore size distribution of tantalum-niobium slag.

There were many narrow pore sizes in the pore channels inside the slag, and the specific surface area, the average pore size as well as pore volume of the slag were relatively small (Table 5). The average pore size of the tantalum-niobium slag was 5.5453 nm, which verified that the pore size was mainly distributed around 5 nm, and the pore volume is 3.12×10^{-3} cm^3/g. This was also consistent with the pore size distribution map of the slag. Nevertheless, the dense void distribution encased the radionuclides inside the slag, thus H_2SO_4 was not able to fully react with the slag particles. The following process was undertaken in order to investigate how roasting affects leaching, and to enhance the recovery effect by roasting experiments.

Table 5. Tantalum-niobium slag performance parameters.

Materials	Pore Volume (cm^3/g)	Average Pore Size (nm)	BET Surface Area (m^2/g)
Ta-Nb slag	3.12×10^{-3}	5.5453	99.9×10^{-2}

The leaching ratio of U and Th of the raw slag was 51.86% and 69.64%. Similarly, the leaching ratio differed depending on the roasting temperatures (Figure 4). The leaching ratio of U and Th at 200 °C were 76.23% and 90.61%. The ratio increased with roasting temperatures from 200 to 500 °C. When the temperature was 500 °C, the leaching ratio reached 90.84% and 96.62%. When the temperature was from 500–600 °C, the ratio has no obvious change trend. Therefore, the roasting experiments could increase the leaching ratio of uranium-thorium. Moreover, 500 °C was chosen as the optimal roasting temperature to leach as much radionuclides as possible from tantalum-niobium samples.

3.5. Mechanism of the Effect of Roasting on the Leaching of Uranium-Thorium

Comparing the leaching results of raw slag and roasted samples, the leaching ratio of U increased about 39% as well as Th increased about 27% after roasting at 500 °C. This showed that roasting method had a facilitating effect on the extraction of radionuclides. Then, the slag roasted at different temperatures was analyzed by SEM, EDS, ICP-OES, TOC/TS, XRD, XPS and BET to explore the reasons for the increased leaching ratio of uranium-thorium from the slag.

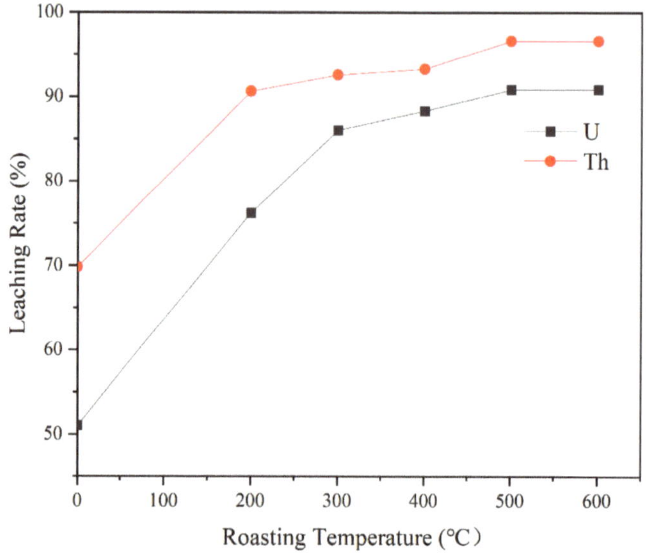

Figure 4. Effects of roasting temperature on uranium and thorium leaching.

The content of uranium-thorium in slag after roasting at different temperatures was measured by ICP-OES (Table 6).

Table 6. The content of U and Th in roasted slag.

Roasting Temperatures (°C)	U Content (mg/kg)	Th Content (mg/kg)
0	2.26×10^3	7.84×10^3
200	2.31×10^3	7.88×10^3
300	2.42×10^3	7.96×10^3
400	2.46×10^3	8.04×10^3
500	2.54×10^3	8.10×10^3
600	2.59×10^3	8.12×10^3

The U-Th content in slag after roasting were increased. With the increase of roasting temperature, the uranium-thorium content in slag were gradually increased. Due to the decomposition of some components and water in slag during roasting, it reduced the overall quality of slag causing the raise of target elements content.

Figure 5 showed the SEM images of the slag after roasting at 200–600 °C. As the roasting temperature increases, the slag presents irregular agglomeration and the voids between slag particles gradually increase, which was beneficial to the leaching of radionuclides. Under the condition of high temperature roasting, the surface impurities were dissolved and removed. The roasting method not only increased the particle void, but also dredged the pore size, thus promoting the elements leaching. When the temperature was 500 °C, the gap between slag particles was large enough and the dispersion was the best.

Scanning results of slag roasted at 200–600 °C were shown in Figure 6.

After roasting at different temperatures, the content of C and S in slag changed, but the EDS technology did not measure the exact content of the elements. Thus, the organic carbon and sulfur analysis was used to obtain the exact content of organic carbon and sulfur elements.

Compared to the original samples, the organic C and S contents decreased at 200 °C. This indicated that the organic C and S within the slag reacted with oxygen in the air during the roasting process. Due to the reducibility of substances, the content of organic

C-S reached the minimum at 500 °C and was conducive to leaching (Table 7). As reducing agents in the reaction, organic C and S restored the soluble U(VI) in the slag back to the insoluble U(IV). By roasting to raise the temperature, the reduction of U valence could be effectively controlled. Therefore, roasting could reduce the presence of organic C and S to improve the leaching ratio.

Table 7. TS/TOC element analysis diagram of tantalum niobium slag.

Elements (%)	Raw Slag	200 °C	300 °C	400 °C	500 °C	600 °C
C	0.1010	0.0924	0.0845	0.0698	0.0654	0.0762
S	0.0354	0.0321	0.0306	0.0241	0.0208	0.0256

The original slag and the 200–600 °C roasted slag were taken and analyzed by XRD for the composition of the physical phases, and the results were shown in Figure 7.

(a)

(b)

Figure 5. *Cont.*

(c)

(d)

(e)

Figure 5. SEM images of tantalum-niobium slag roasted at (**a**) 200 °C; (**b**) 300 °C; (**c**) 400 °C; (**d**) 500 °C; (**e**) 600 °C.

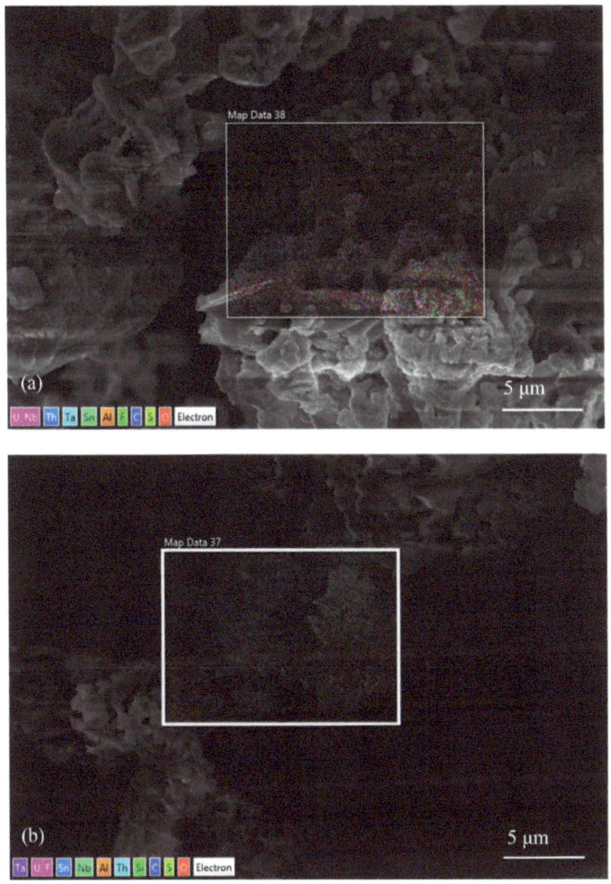

Figure 6. *Cont.*

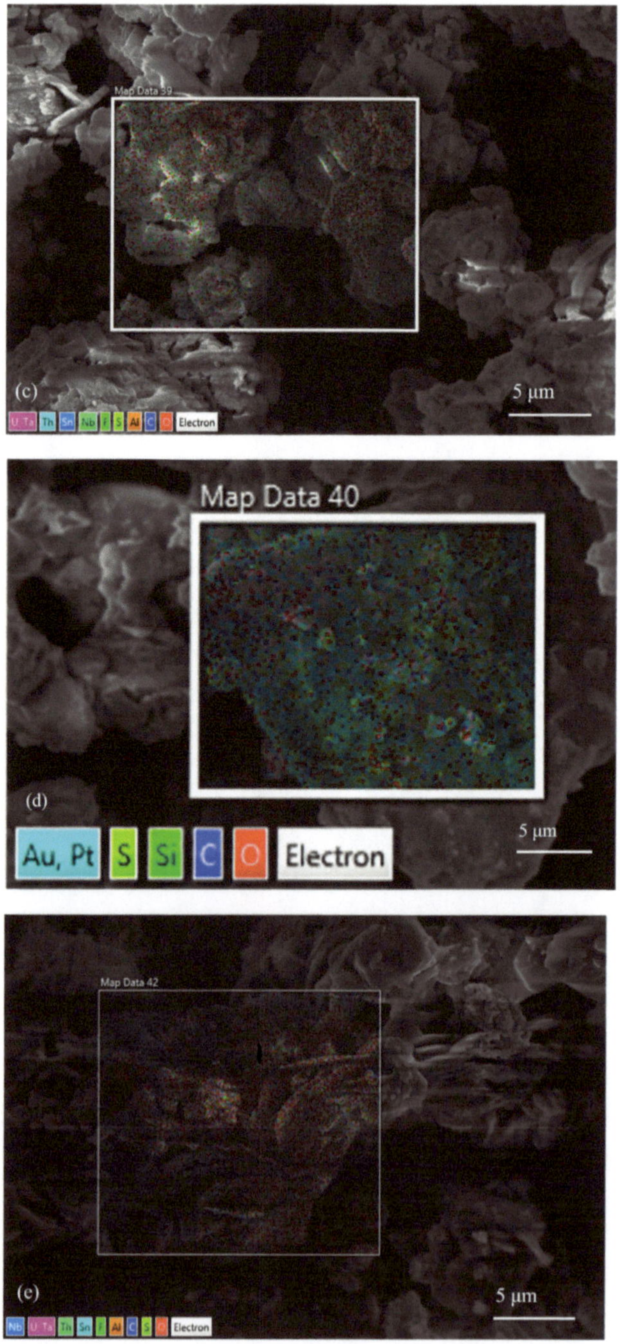

Figure 6. EDS results of tantalum-niobium slag roasted at (**a**) 200 °C; (**b**) 300 °C; (**c**) 400 °C; (**d**) 500 °C; (**e**) 600 °C.

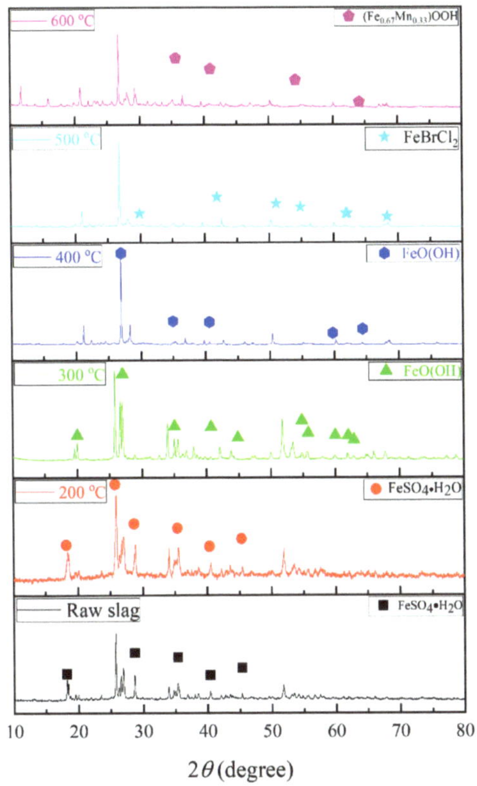

Figure 7. XRD of tantalum niobium slag.

As shown in Figure 7, the XRD spectra of both the original slag and the 200 °C roasted slag matched with the standard PDF card for PDF #45-1364($FeSO_4{\bullet}H_2O$). The XRD spectra of the 300 °C and 400 °C roasted slag with high similarity to the standard PDF card for PDF #89-6096(FeO(OH)). The XRD spectra of the 500 °C and 600 °C roasted slag with high similarity to the standard PDF cards PDF #07-0284($FeBrCl_2$) and PDF #14-0557(($Fe_{0.67}Mn_{0.33}$)OOH). The comparison results confirmed that there were not only tantalum niobium oxide, but also bromine, iron, manganese and other compounds in the slag samples. Iron in the original slag mainly existed in the form of divalent salt compounds. Due to the complex composition of the tantalum-niobium slag and the variety of elements, the phase and the corresponding peaks of uranium-thorium were not detected from Figure 7. With the increase of roasting temperature, the divalent iron in slag gradually changed to higher valence. Therefore, the roasting method had oxidation effect on iron in tantalum-niobium slag.

The Fe^{3+} and U(VI) content in the tantalum-niobium slag reached 52.11% and 37.46% at 200 °C (Table 8). When the temperature was increased from 200 °C to 500 °C, the content of Fe^{2+} and U(IV) in the samples gradually decreased, and the Fe^{3+} and U(VI) reached a maximum of 89.43% and 66.79%. When the roasting temperature was 600 °C, the content of Fe^{3+} and U(VI) in the slag decreased compared with that at 500 °C. The above results illustrated that roasting could oxidize Fe^{2+} to Fe^{3+}, and Fe^{3+} was used as an excellent oxidant in the reaction to facilitate the increase of the valence state of U(IV), this was conducive to leaching. 500 °C was the optimal temperature for roasting, which was consistent with the previous results.

Table 8. XPS element analysis diagram of tantalum-niobium slag after roasting.

Roasting Temperature (°C)	Fe^{2+} (%)	Fe^{3+} (%)	U(IV) (%)	U(VI) (%)
200 °C	47.89	52.11	62.54	37.46
300 °C	23.01	76.99	52.52	47.48
400 °C	13.13	86.87	38.96	61.04
500 °C	10.57	89.43	33.21	66.79
600 °C	19.11	80.89	45.55	54.45

Figure 8 showed that the raw slag was light gray and the color deepened after roasting with the temperature from 200 °C to 600 °C. As the roasting temperature reached 500 °C, the color of the slag changed to red. The results also indicated the conversion of divalent iron to trivalent iron in the slag, which corresponded to the results in the XPS. Thus, roasting had an oxidizing effect on the iron in the slag.

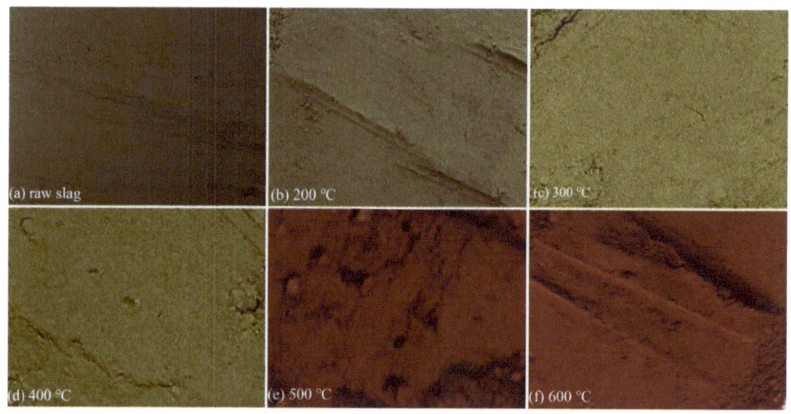

Figure 8. The images of (**a**) raw sample and roasted tantalum-niobium slag at (**b**) 200 °C; (**c**) 300 °C; (**d**) 400 °C; (**e**) 500 °C; (**f**) 600 °C.

The curves of the raw and the roasted slag at five temperatures were type III adsorption isotherms (Figure 9). The N_2 adsorption and desorption curves of slag after roasting at each temperature did not change much at relative pressures less than 0.8, and the adsorption of N_2 by the slag increased significantly at relative pressures of 0.8–1.0. The maximum adsorption of slag was the highest at 500 °C roasting. Thus 500 °C can be chosen as the optimal temperature for roasting, corresponding to the previous results.

There are many narrow pore sizes in the pore channels inside the tantalum-niobium slag (Table 9). In addition, the specific surface area the pore size as well as volume of the slag are relatively small, which is in general agreement with the data in the relevant literature [29,30]. The pore size (9.2171 nm) and pore volume (12.720×10^{-3} cm^3/g) of slag was maximum after roasting at 500 °C, which corresponded to the results of the maximum adsorption of the slag and verified the optimal roasting temperature. Roasting enlarged the pore size and gap of slag and increased the contact area between the leaching agent and the sample particles, allowing the leaching experiment to proceed more fully.

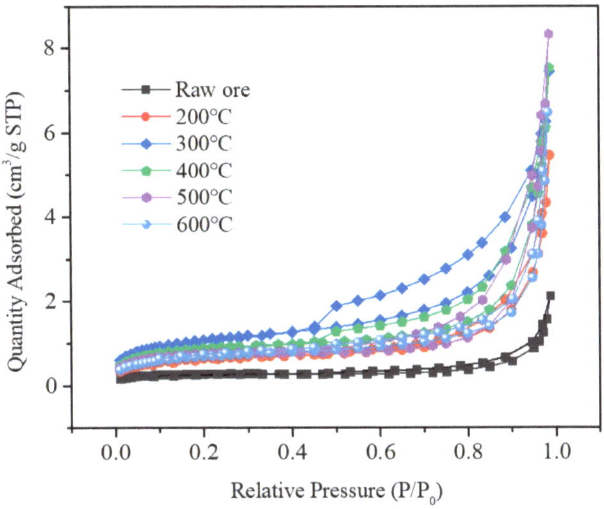

Figure 9. Adsorption and desorption isotherm of tantalum niobium slag to N_2.

Table 9. The tantalum-niobium slag after roasting performance parameters.

Roasting Temperature (°C)	Pore Volume (cm^3/g)	Average Pore Size (nm)	BET Surface Area (m^2/g)
200 °C	8.233×10^{-3}	7.4207	2.2868
300 °C	11.619×10^{-3}	7.3708	3.8298
400 °C	11.489×10^{-3}	7.8396	3.1209
500 °C	12.720×10^{-3}	9.2171	2.6696
600 °C	9.553×10^{-3}	6.7793	2.4359

3.6. Morphological Mechanism Analysis of U

Based on mineralogical characterization and roasting experiments, we can obtain the mechanism of morphological changes of uranium in the slag (Figure 10). In fact, the valence conversion of uranium involves the redox problem [31–33]. In the tantalum-niobium slag experiments, O_2, Fe^{3+}, and Mn^{4+} act as excellent oxidizing agents and influence the reaction atmosphere. These oxidants can promote the conversion of insoluble U(IV) to soluble U(VI), thus facilitating the U leaching in the subsequent stages [34]. However, due to the presence of organic carbon and sulfur in samples, the reduction of organic matter causes the conversion of U(IV) back to U(VI) again, limiting the extraction of U [35,36]. In the roasting experiments, by rising the temperature, the reaction intensity of oxidant is improved, while the content of organic carbon and sulfur is also reduced, reaching of a minimum at 500 °C, which improving the recovery effect. The treatment of samples by means of high-temperature roasting enables the removal of inclusions from the surface of the target elements and the unblocking of pore sizes. Therefore, roasting the slag at 500 °C can promote the leaching of radionuclides. The samples by means of high-temperature roasting removes impurities from the slag surface, strips the dense inclusions surrounding the target elements, and eliminates pore size obstruction. Therefore, calcination of slag at 500 °C can maximize the leaching of radionuclides.

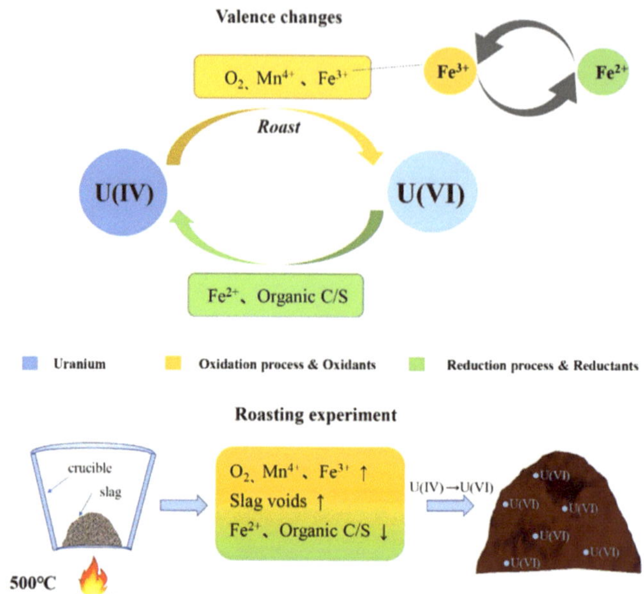

Figure 10. Morphological mechanism analysis of U.

4. Conclusions

The high concentration of H_2SO_4 and HF added during the smelting process destroys the mineral structure of tantalum-niobium ore, making U and Th in slag. The content of U and Th in the raw slag are 2.26×10^3 mg/kg and 7.84×10^3 mg/kg respectively, which have large recovery value. The initial leaching ratio of U and Th with H_2SO_4 as leaching agent is 51.86% and 69.64%. The average pore size of the slag particles is 5.5453 nm. There are many narrow pore channels between the particles, and the specific surface area is small, hindering the full contact between H_2SO_4 and slag. Besides U and Th in the slag, reducing substances such as Fe^{2+}, organics and Mn can also affect the redox atmosphere conditions during leaching.

Considering the possibility of some special inclusions to encapsulate uranium and thorium, the slag is roasted to improve the leaching ratio. Various methods of characterization are used to analyze the mineralogical properties of roasted slag at different temperatures. The leaching ratio of U-Th is 90.84% and 96.62% at the optimum roasting temperature of 500 °C, which are about 39% and 27% higher than original samples. The valence of Fe changes from divalent to trivalent with increasing temperature (Fe^{3+} content increases by 52.04%), and the content of U(VI) reaches 66.79% (increases by 30.62%). The oxidants O_2 and Mn can also promote the conversion of insoluble U(IV) to soluble U(VI). The organic C and S content is reduced after roasting, so this effectively prevented the conversion of U(VI) back to U(IV). The pore size of the slag increases from 5.5453 nm to 9.2171 nm, and the pore volume expands from 3.12×10^{-3} cm^3/g to 12.72×10^{-3} cm^3/g, which obviously facilitates the contact area between H_2SO_4 and samples. Consequently, the roasting method at 500 °C can destroy the surface wrapping structure of radionuclides, reduce the internal density of minerals, and improve uranium-thorium leaching ratio significantly. It is of great practical significance to reduce the radioactive hazard of waste tantalum-niobium slag and to strengthen the sustainable utilization of resources by suitable process improvement techniques.

Author Contributions: M.H.: Ideas, investigation, methodology, data curation and writing original draft.; K.H.: Performing the experiments and data/evidence collection.; X.L.: Performing the experiments and data/evidence collection.; Y.W.: Provision of study materials and reagents.; J.O.: Provision of study materials and reagents.; L.Z.: Provision of study materials and reagents.; Z.L.: Supervision, writing-review & editing, funding acquisition and project administration. All authors have read and agreed to the published version of the manuscript.

Funding: This research was funded by National Nature Science Foundation of China (11875105, 21866006); General Project of Jiangxi Province Key Research and Development Program (20203BBFL63070, 20192BBG70062).

Institutional Review Board Statement: Not applicable.

Informed Consent Statement: Not applicable.

Data Availability Statement: Data is contained within the article. The data presented in this study are available in this published article.

Acknowledgments: The authors thank financial support from National Nature Science Foundation of China (11875105, 21866006); General Project of Jiangxi Province Key Research and Development Program (20203BBFL63070, 20192BBG70062).

Conflicts of Interest: The authors declare that they have no conflict of interest to report regarding the present study.

References

1. Nguyen, T.H.; Lee, M.S. A review on the separation of niobium and tantalum by solvent extraction. *Miner. Process. Extr. Metall. Rev.* **2018**, *40*, 265–277. [CrossRef]
2. Shikika, A.; Sethurajan, M.; Muvundja, F.; Mugumaoderha, M.C.; Gaydardzhiev, S. A review on extractive metallurgy of tantalum and niobium. *Hydrometallurgy* **2020**, *198*, 105496. [CrossRef]
3. Belay, A.N.; Venter, J.A.; Roodt, A. Coordination chemistry: Niobium(V) and tantalum(V) with O-donor ligands. *Acta Crystallogr. Sect. A Found. Adv.* **2017**, *73*, C1163. [CrossRef]
4. Băltatu, S.; Vizureanu, P.; Mareci, D.; Burtan, L.C.; Chiruţă, C.; Trincă, L.C. Effect of Ta on the electrochemical behavior of new TiMoZrTa alloys in artificial physiological solution simulating in vitro inflammatory conditions. *Mater. Corros.* **2016**, *67*, 1314–1320. [CrossRef]
5. Ray, R.; Dutta, B.; Mandal, S.K.; González, A.G.; Pokrovsky, O.S.; Jana, T.K. Bioaccumulation of vanadium (V), niobium (Nb) and tantalum (Ta) in diverse mangroves of the Indian Sundarbans. *Plant Soil* **2020**, *448*, 553–564. [CrossRef]
6. Ibrahim, M.E.; Orabi, A.H.; Falila, N.I.; Ismaiel, D.A.; Salem, H.M. Processing of the mineralized Black Mica for the recovery of uranium, rare earth elements, niobium, and tantalum. *Hydrometallurgy* **2020**, *197*, 105474. [CrossRef]
7. Nete, M.; Purcell, W.; Nel, J.T. Tantalite beneficiation through sequential separation of radioactive elements, iron and titanium by magnetic separation and acid leaching. *Adv. Mat. Res.* **2014**, *1019*, 419–425. [CrossRef]
8. Cui, C.Q.; Wang, B.; Zhao, Y.X.; Wang, Q.; Sun, Z.M. China's regional sustainability assessment on mineral resources: Results from an improved analytic hierarchy process-based normal cloud model. *J. Clean. Prod.* **2019**, *210*, 105–120. [CrossRef]
9. Wu, B.; Shang, H.; Wen, J.K. Sulfuric acid leaching of low-grade refractory tantalum–niobium and associated rare earths minerals in Panxi area of China. *Rare Met.* **2015**, *34*, 202–206. [CrossRef]
10. Jacob-Tatapu, K.J.; Albert, S.; Grinham, A. Sediment arsenic hotspots in an abandoned tailings storage facility, Gold Ridge Mine, Solomon Islands. *Chemosphere* **2020**, *269*, 128756. [CrossRef]
11. Pakhomova, S.; Yakushev, E.; Schaanning, M. Modeling nickel leaching from abandoned mine tailing deposits in Jøssingfjorden. *Water* **2021**, *13*, 967. [CrossRef]
12. Kusin, F.M.; Sulong, N.A.; Affandi, F.N.A.; Molahid, V.L.M.; Jusop, S. Prospect of abandoned metal mining sites from a hydrogeochemical perspective. *Environ. Sci. Pollut. Res.* **2021**, *28*, 2678–2695. [CrossRef]
13. Ding, Y.G.; Wang, J.S.; Wang, G.; Xue, Q.G. Innovative methodology for separating of rare earth and iron from Bayan Obo complex iron ore. *ISIJ Int.* **2012**, *52*, 1772–1777. [CrossRef]
14. Melcher, F.; Graupner, T.; Gäbler, H.E.; Sitnikova, M.; Henjes-Kunst, F.; Oberthür, T.; Gerdes, A.; Dewaele, S. Tantalum–(niobium–tin) mineralisation in African pegmatites and rare metal granites: Constraints from Ta–Nb oxide mineralogy, geochemistry and U–Pb geochronology. *Ore Geol. Rev.* **2015**, *64*, 667–719. [CrossRef]
15. Rodriguez, M.H.; Rosales, G.D.; Pinna, E.G.; Suarez, D.S. Extraction of niobium and tantalum from ferrocolumbite by hydrofluoric acid pressure leaching. *Hydrometallurgy* **2015**, *156*, 17–20. [CrossRef]
16. Zhang, Y.; Hu, Y.; Sun, N.; Liu, R.; Wang, Z.; Wang, L.; Sun, W. Systematic review of feldspar beneficiation and its comprehensive application. *Miner. Eng.* **2018**, *128*, 141–152. [CrossRef]
17. Dai, Y.Y.; Zhong, Z.; Zhong, H.Y. Recovery leaching of niobium from metallurgical residues. *Adv. Mat. Res.* **2011**, *236–238*, 2554–2559. [CrossRef]
18. Li, Z.; Hadioui, M.; Wilkinson, K.J. Conditions affecting the release of thorium and uranium from the tailings of a niobium mine. *Environ. Pollut.* **2019**, *247*, 206–215. [CrossRef]

19. Lin, Y.; Jiao, Y.; Zhao, M.; Wang, G.; Wang, D.; Xiao, W.; Li, H.; Xu, Z.; Jiang, Y. Ecological restoration of wetland polluted by heavy metals in Xiangtan manganese mine area. *Processes* **2021**, *9*, 1702. [CrossRef]
20. Xiang, M.; Li, Y.; Yang, J.; Lei, K.; Li, Y.; Li, F.; Zheng, D.; Fang, X.; Cao, Y. Heavy metal contamination risk assessment and correlation analysis of heavy metal contents in soil and crops. *Environ. Pollut.* **2021**, *278*, 116911. [CrossRef]
21. Soesoo, A.; Vind, J.; Hade, S. Uranium and thorium resources of Estonia. *Minerals* **2020**, *10*, 798. [CrossRef]
22. Nalivaiko, K.; Skripchenko, S.; Titova, S.; Rychkov, V. Characterization and processing of radioactive uranium containing waste sludge by sulfuric acid leaching. *J. Environ. Chem. Eng.* **2022**, *10*, 106972. [CrossRef]
23. Zhang, Y.; Zhang, T.A.; Dreisinger, D.; Lv, C.; Lv, G.; Zhang, W. Recovery of vanadium from calcification roasted-acid leaching tailing by enhanced acid leaching. *J. Hazard. Mater.* **2019**, *369*, 632–641. [CrossRef]
24. Korehi, H.; Blöthe, M.; Sitnikova, M.A.; Dold, B.; Schippers, A. Metal mobilization by iron- and sulfur-oxidizing bacteria in a multiple extreme mine tailings in the Atacama desert, Chile. *Environ. Sci. Technol.* **2013**, *47*, 2189–2196. [CrossRef]
25. Stasik, S.; Wick, L.Y.; Wendt-Potthoff, K. Anaerobic BTEX degradation in oil sands tailings ponds: Impact of labile organic carbon and sulfate-reducing bacteria. *Chemosphere* **2015**, *138*, 133–139. [CrossRef]
26. Lv, S.Y.; Li, M.; Wu, X.Y.; Zhang, X.W.; Hua, Y.L.; Bi, L.; Fang, Q.; Cai, T. A non-polluting method for rapidly purifying uranium-containing wastewater and efficiently recovering uranium through electrochemical mineralization and oxidative roasting. *J. Hazard. Mater.* **2021**, *416*, 125885. [CrossRef]
27. Sun, Y.W.; Zeng, B.Y.; Dai, Y.T.; Liang, X.J.; Zhang, L.J.; Ahmad, R.; Su, X.T. Modification of sludge-based biochar using air roasting-oxidation and its performance in adsorption of uranium(VI) from aqueous solutions. *J. Colloid Interface Sci.* **2022**, *614*, 547–555. [CrossRef] [PubMed]
28. Hu, K.; Liu, Z.R.; Xiu, T.Y.; Zhou, L.M.; Wang, Y. Removal of thorium from aqueous solution by adsorption with Cu3(BTC)2. *J. Radioanal. Nucl. Chem.* **2020**, *326*, 185–192. [CrossRef]
29. Brunauer, S.; Deming, L.S.; Deming, W.E.; Teller, E. On a theory of the van der Waals adsorption of gases. *J. Am. Chem. Soc.* **1940**, *62*, 1723–1732. [CrossRef]
30. Naik, B.; Ghosh, N. A review on chemical methodologies for preparation of mesoporous silica and alumina based materials. *Recent Pat. Nanotechnol.* **2009**, *3*, 213–224. [CrossRef] [PubMed]
31. Boreen, M.A.; Arnold, J. The synthesis and versatile reducing power of low-valent uranium complexes. *Dalton Trans.* **2020**, *49*, 15124–15138. [CrossRef]
32. Puchkova, E.V.; Bogdanov, R.V.; Gieré, R. Redox states of uranium in samples of microlite and monazite. *Am. Mineral.* **2016**, *101*, 1884–1891. [CrossRef]
33. Fu, H.Y.; Zhang, H.; Sui, Y.; Hu, N.; Ding, D.X.; Ye, Y.J.; Li, G.Y.; Wang, Y.D.; Dai, Z.R. Transformation of uranium species in soil during redox oscillations. *Chemosphere* **2018**, *208*, 846–853. [CrossRef]
34. Wang, P.M.; Anderko, A.; Kosinski, J.J.; Springer, R.D.; Lencka, M.M. Modeling speciation and solubility in aqueous systems containing U(IV, VI), Np(IV, V, VI), Pu(III, IV, V, VI), Am(III), and Cm(III). *J. Solut. Chem.* **2017**, *46*, 521–588. [CrossRef]
35. Leventhal, J.S.; Daws, T.A.; Frye, J.S. Organic geochemical analysis of sedimentary organic matter associated with uranium. *Appl. Geochem.* **1986**, *1*, 241–247. [CrossRef]
36. Fuller, A.J.; Leary, P.; Gray, N.D.; Davies, H.S.; Mosselmans, J.F.W.; Cox, F.; Robinson, C.H.; Pittman, J.K.; McCann, C.M.; Muir, M.; et al. Organic complexation of U(VI) in reducing soils at a natural analogue site: Implications for uranium transport. *Chemosphere* **2020**, *254*, 126859. [CrossRef]

toxics

MDPI

Article

Utilization of Waste Amine-Oxime (WAO) Resin to Generate Carbon by Microwave and Its Removal of Pb(II) in Water

Chunlin He [1,2,*], Yun Liu [1,2], Chunhui Zheng [1,2], Yanming Jiang [1,2], Yan Liao [1,2], Jiaxin Huang [1,2], Toyohisa Fujita [1,2], Yuezhou Wei [3,4] and Shaojian Ma [1,2]

1 School of Resources, Environment and Materials, Guangxi University, Nanning 530004, China
2 Guangxi Key Laboratory of Processing for Non-Ferrous Metal and Featured Materials, Guangxi University, Nanning 530004, China
3 School of Nuclear Science and Technology, University of South China, Hengyang 421000, China
4 School of Nuclear Science and Engineering, Shanghai Jiao Tong University, Shanghai 200240, China
* Correspondence: helink1900@126.com

Abstract: Utilising waste amine-oxime (WAO) resin through microwave semi-carbonization, a carbon adsorbent (CA) was obtained to remove Pb(II). After microwave treatment, the pore size of the skeleton structure, three-dimensional porous network, and lamellar pore structure of WAO was improved. The distribution coefficient (Kd) of Pb(II) onto CA is 620 mL/g, and the maximum adsorption capacity of Pb(II) is 82.67 mg/g after 20 min of WAO microwave treatment. The adsorption kinetics and adsorption isotherms conform to the quasi-second-order kinetic equation and Langmuir adsorption isotherm model, respectively. The surface of MT-WAO is negatively charged and the adsorption mechanism is mainly electrostatic interaction. Pb(II) elution in hydrochloric acid solution is more than 98%, and its recovery is high at 318 K and for 1 h.

Keywords: waste amine-oxime; Pb(II); adsorption; adsorbent; microwave

Citation: He, C.; Liu, Y.; Zheng, C.; Jiang, Y.; Liao, Y.; Huang, J.; Fujita, T.; Wei, Y.; Ma, S. Utilization of Waste Amine-Oxime (WAO) Resin to Generate Carbon by Microwave and Its Removal of Pb(II) in Water. *Toxics* **2022**, *10*, 489. https://doi.org/10.3390/toxics10090489

Academic Editor: Xin-Gui Li

Received: 27 July 2022
Accepted: 19 August 2022
Published: 23 August 2022

Publisher's Note: MDPI stays neutral with regard to jurisdictional claims in published maps and institutional affiliations.

Copyright: © 2022 by the authors. Licensee MDPI, Basel, Switzerland. This article is an open access article distributed under the terms and conditions of the Creative Commons Attribution (CC BY) license (https://creativecommons.org/licenses/by/4.0/).

1. Introduction

Lead (Pb) is one of the major heavy metal pollutants in the natural environment, which mainly comes from many industrial wastewaters such as metal mining, metal smelting, storage battery manufacturing, pulp, leather tanning and paper, etc. [1–4]. Once released, Pb can exist in the environment for a long time and is difficult to degrade; if it accumulates through the food chain, it can cause potential toxicity to plants, animals and even humans. Conventional Pb(II) contamination treatments include chemical precipitation, ion exchange, adsorption, membrane filtration, and electrodialysis [5–7]. Adsorption is widely considered a simple and low-cost method to remove metals from water [8–10].

China is the world's key producer of ion exchange resin, which is mainly used in water treatment, hydrometallurgy, biopharmaceutical, food processing, and chemical industries. After prolonged use, pore collapse, shrinkage, reduction in the specific surface area, failure of functional groups, and generation of waste resin occur. For example, in the Guangxi branch of Chalco, amidoxime chelating resins utilized for gallium (Ga) adsorption from Bayer's solution are discarded after reusing 50–60 times a month, and 225 tons of waste resin are produced and shelved each year [11,12]. Besides the Guangxi branch of Chalco, the waste amine-oxime is also produced in Henan, Shandxi, and Guizhou provinces in China, which used amine-oxime in the extraction of Ga. Currently, the central treatment methods for amidoxime chelating resins include stockpiling, landfill, curing and incineration. Incineration is more common, and the discarded resin is sent to solid waste handling companies for incineration at about 2000 RMB/ton [12]. After incineration, many polluting gases, such as CO_2, CO, SO_2, NO and NO_2, are produced, which is not conducive to controlling carbon emissions. In addition, the storage and landfill of waste resin occupy a larger land area, which neither produces economic benefits nor meets the requirements of environmental protection.

Much research has been carried out on the recycling of waste resin. Long et al. [13] obtained a carbon adsorbent by carbonization and activation of the waste polysulfonated cation exchange resin. The activated carbon exhibited adsorption performance for naphthalene. Shi et al. [14] impregnated ferric nitrate into waste resin (D001), and then the waste resin was carbonized and activated to obtain C/Fe composite magnetic adsorption material. Shi et al. [15] used waste polystyrene ion-exchange resin to prepare activated carbon under nitrogen at 450 °C for 30 min and then activated at 800 °C; it exhibited excellent adsorption performance for naphthene and other polycyclic aromatic hydrocarbons. Wei et al. [16–19] used waste ion-exchange resin to prepare activated carbon for CO_2 adsorption, with a specific surface area of 400–900 cm^2/g. Bratek et al. [20–22] studied the carbonization of waste ion exchange resin at 600~850 °C to prepare an activated carbon adsorbent, which exhibited excellent adsorption performance for toluene, oil, copper, and lead ions in wastewater. Gun'ko et al. [23] investigated 5 kinds of waste resins to obtain activated carbon with a specific surface area of 200~600 m^2/g, and they could effectively remove formaldehyde (0.33 cm^3/g). Miura et al. [24] conducted a study on the preparation of porous activated carbon by carbonization of two waste resins at 500~900 °C, and the results showed that adding different metal ions could change the pore size distribution.

In summary, for the recycling of waste resin, many works focus on the preparation of activated carbon by using ion-exchanged resin. Most of the activated carbon consumes large energy, and activated carbon becomes powder, not spherical. The recycling of waste amine oxime resin has not been studied systematically. Amine oxime resin is different from ion exchange resins; it receives strong acid and strong base impact during the extraction of Ga. The internal pores of WAO resin are not developed, which exhibits collapse and shrinkage. In this study, for waste utilization, the waste amine oxime (WAO) resin was treated by microwave semi-carbonization (to improve the pores) to obtain a granular adsorbent to remove Pb(II). After microwave treatment, the resin also retains the sphere. In addition, the effects of microwave treatment time, adsorption time, and co-existing ions in the solution on the adsorption of Pb(II), as well as the effects of time and temperature on elution, were investigated; this work could also provide a reference for other waste resins.

2. Materials and Methods

2.1. Materials

The waste amine oxime (WAO) resin, with a particle diameter of 0.2–0.6 mm, was obtained from the Guangxi branch of Chalco. Reagent grade $Pb(NO_3)_2$ (Aladdin) and deionized water were used in all the experiments. The WAO resin was rinsed with hydrochloric acid (analytical-grade) and ultrapure water to remove the organic and inorganic impurities.

2.2. Characterization Methods

The surface morphology and corresponding elemental distribution were analyzed by scanning electron microscopy with energy dispersive spectroscopy (SEM-EDS, Phenom Pro 800-07334, Eindhoven, The Netherlands). The adsorption mechanism was studied by X-ray photoelectron spectroscopy (XPS, ESCALAB 250XI+, Waltham, MA, USA) equipped with a monochromatized Al-Kα X-ray source and Fourier-transform infrared spectroscopy (FT-IR, THS-108, USA). The C, N, O and H contents of FAO and WAO were estimated using an elemental analyzer (Vario EL cube, Han, Germany). The zeta potential of the suspension (100 mg/L) containing nonmetallic particles of the MT-WAO (200 mesh) was measured by using a high-sensitivity zeta potential analyzer (NanoBrook Omni, New York City, NY, USA) to explore the surface charge as a function of pH. Thermogravimetry and differential thermal analysis (TG-DTA) were carried out in a nitrogen atmosphere with a heating rate of 10 °C/min.

2.3. Microwave Semi-Carbonisation Methods

Due to the collapse and shrinkage, the internal pores of WAO resin are not developed. Therefore, a household microwave oven (EV025LC7-NR, Midea) with a microwave output power of 1000 W was used to improve the pores. Microwave heating is fundamentally different from conventional heating methods, as it uses microwave interactions with materials at molecular or atomic levels to generate heat. Microwave processing involves penetrating radiation, controllable electric field distribution, rapid heating, selective heating of materials through differential absorption and self-limiting reaction. Therefore, 10 g WAO resin was introduced into a quartz cup, the mouth of the cup was completely sealed with paraffin film, and kept in the microwave oven for different times (2–20 min) to obtain a new absorbent (MT-WAO).

2.4. Adsorption Experiments

The static adsorption experiments were performed in a thermostatic water bath to investigate the influence of various experimental parameters on the adsorption capacity. The solution concentration was analyzed using a standard curve method by ICP-AES (Inductively coupled plasma-atomic emission spectrometry, Shimadzu ICPS-7510, Kyoto, Japan). The speed of the stirrer in the water bath oscillator is 140 rpm. The relevant experimental results were calculated using Equations (1) and (2)

$$Q = (C_0 - C) \times (V/m) \quad (1)$$

$$K_d = [(C_0 - Cs)/Cs] \times (V/m) \quad (2)$$

where Q (mg/g) and K_d (mL/g) are the adsorption capacity at the equilibrium state and the distribution coefficient, respectively. C_o (mg/L) and C (mg/L) denote metal ions' initial and equilibrium concentrations, respectively. C_s represents the metal concentration in the aqueous phase after adsorption. V (L) is the volume of the aqueous phase, and m (g) represents the mass of WAO resin.

3. Results

3.1. Characteristics of WAO

The structural features and elements of fresh amine oxime resin (FAO) and WAO are presented in Figure 1. WAO exhibits a compact spherical particle with shrinkage and pore clog (Figure 1d). Compared with FAO, N in WAO is lower, and O is higher (Table 1). C and H contents of the resin changed slightly, while O content increased from 57.86% to 88.29%, and the N content decreased from 23.96% to 8.24%. Based on Figure 1e, the intensity of the N 1 s peak of WAO resin is significantly weakened. In comparison, the intensity of the O 1 s peak is significantly enhanced, consistent with the results shown in Table 1.

Figure 1f shows the peaks of FAO at 1650 cm^{-1}, 1109 cm^{-1} and 935 cm^{-1}, which correspond to tensile vibrations of C=N, C-N and N-O bonds [25]; however, in the case of WAO resin, these peaks shift to higher wavenumbers, and their intensities decrease significantly; furthermore, two new and extremely sharp characteristic peaks appear at 1697 cm^{-1} and 1195 cm^{-1}, corresponding to the tensile vibrations of C=O and C-O bonds in the carboxyl group [26,27], attributed to the oxidation of amine oxime group into carboxyl group [28].

Table 1. C, N, O and H contents in the FAO and WAO.

Element (%)	FAO	WAO
C	46.07	46.04
O	57.86	88.29
N	23.96	8.240
H	3.047	5.373

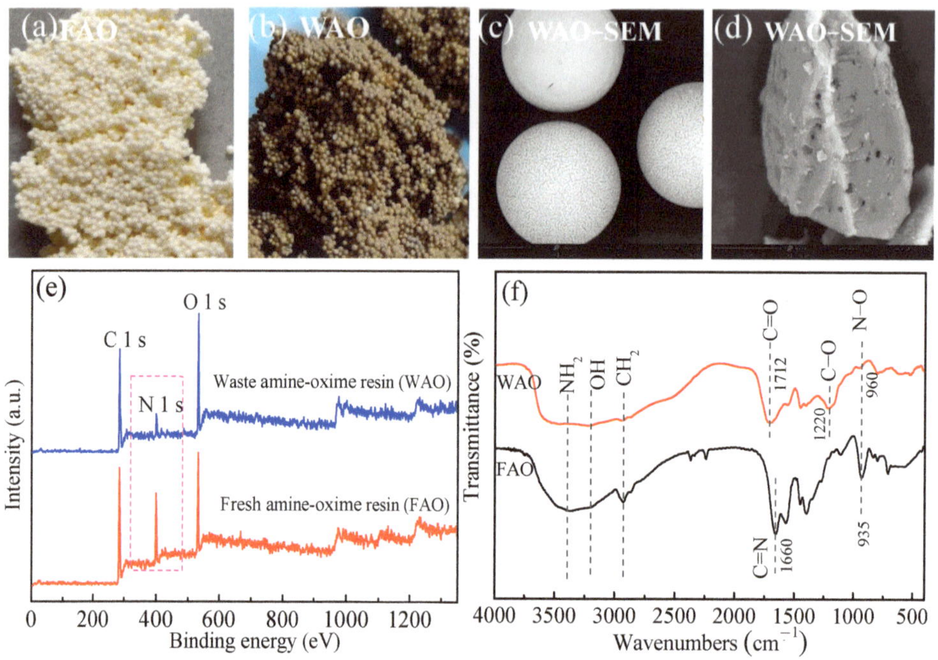

Figure 1. The structural features and elements of fresh and waste amine oxime resin: (**a**) fresh amine-oxime resin (FAO), (**b**) waste amine-oxime resin (WAO), (**c**,**d**) SEM of WAO, (**e**) XPS of WAO and FAO, (**f**) FT-IR of WAO and FAO.

WAO's micropore and mesopore structures were almost completely blocked when the BET and BJH models were used to analyze the specific surface area and pore volume. Therefore, WAO's pore sizes were measured by Mercury Porosimeter (Micrometric Autopore 9620, Sarasota, Florida, USA), as shown in Figure 2a.

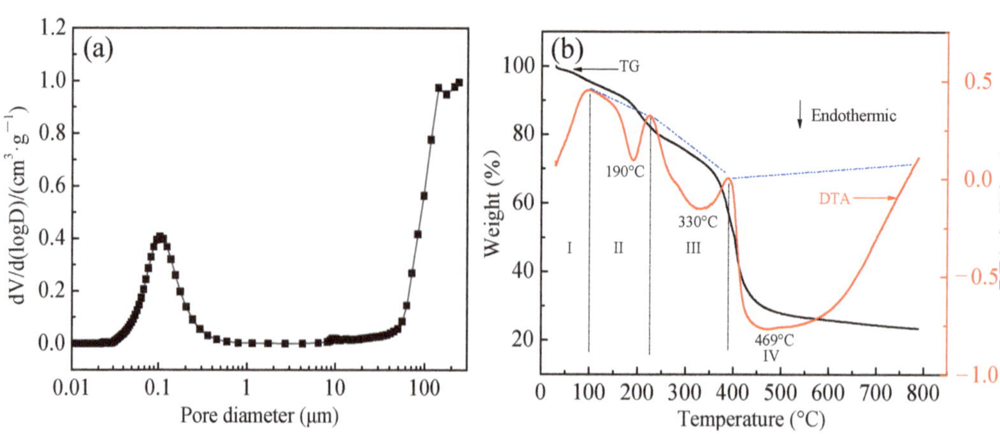

Figure 2. (**a**) Differential pore volume distribution of pores with different diameters in the WAO resin (**b**) TG and DTA curves of WAO resin.

Figure 2a shows that the total porosity of WAO resin is 49.29%, the intra-particle porosity is 34.56%, the interparticle porosity is 14.73%, and the pore volume is 0.65 cm^3/g, and the specific pore surface area is 8.49 m^2/g. The pore size distribution of WAO resin presents bimodal characteristics, and its pore volume and pore-specific surface area concentrate in the macro pores. The pore size of the main peak ranges from 8.1 μm to 96.2 μm, accounting for 10.19% of the total pore volume. The pore size of the small peak ranges from 0.02 μm to 0.3 μm, accounting for 28.63% of the total pore volume. The distribution of the largest pore size ranges from 96.2 to 238.4 μm, with a pore volume ratio of 59.5%; however, it is not in the effective pore size test range (7 nm–100 μm), which may be due to the interference of interparticle voids under low-pressure conditions, resulting in a large pore volume ratio.

The stability of WAO resin was characterized by TG-DTA, as shown in Figure 2b. When the temperature increases to 100 °C (step I), the mass loss is 4.6%. And then, a downward endotherm peak appears at 190 °C in the DTA curve (step II); this may be due to the volatilization of hydrated water. The decomposition at the temperature of 220–390 °C (step III) is attributed to the depolymerization of polystyrene–divinylbenzene structures present in the resin. Finally, the mass of WAO resin rapidly reduced by nearly 24% over the temperature range of 390–600 °C (step IV), possibly due to polystyrene and divinylbenzene structure degradation. Therefore, to obtain certain mechanical strength and spherical shape absorbent, the temperature shall be controlled below 220 °C when a microwave is applied to waste resin.

3.2. Characteristics of MT-WAO

The microstructure of WAO resin before and after microwave treatment was characterized by SEM-EDS, and the obtained images are shown in Figure 3.

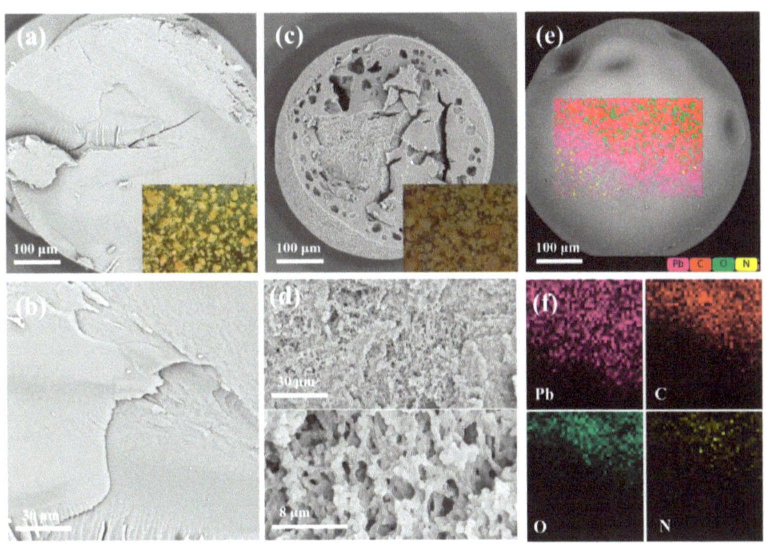

Figure 3. SEM images of (**a**,**b**)WAO resin (**c**,**d**) MT-WAO-resin (**e**,**f**) EDS of MT-WAO resin after adsorbing Pb(II).

No pores could be observed in the WAO resin (Figure 3a, magnified 500 times). The cross-section is rough with uneven holes (Figure 3c) after microwave treatment. When the waste resin was heated by a high-power microwave, the temperature of waste resin was quickly increased in a short time, especially the part inside. Then, the adsorbed water and hydrated water were heated up quickly and became gaseous, and then high pressure occurred inside of waste resin, making pores develop. As shown in Figure 3c, the resin also retains its sphere. MT-WAO resin presents a lamellar stack structure and a connected three-dimensional porous network structure (Figure 3d), indicating that the pore structure of WAO resin was improved after microwave heating; it can be observed from Figure 3e that MT-WAO resin still presents a regular spherical shape after adsorbing Pb(II). The adsorbed Pb(II) is evenly distributed on the surface of MT-WAO resin (Figure 3f).

For MT-WAO resin, the total porosity is 55.37%, the intra-particle porosity is 46.31%, the interparticle porosity is 9.06%, the pore volume is 0.84 cm^3/g, and the pore-specific surface area is 3.89 m^2/g (Figure 4, measurements obtained using mercury porosimeter). The pore size distribution of MT-WAO resin presents bimodal characteristics, and its pore volume and pore-specific surface area are mainly distributed in the macroporous range. The main peak appears in the range of 4.0–57.4 μm, and the total pore volume is 72.3%. The pore size of the small peak ranges from 0.02 μm to 0.3 μm, with a total pore volume of 3.12%. The transitional pore size between the two peaks ranges from 0.3 μm to 4.0 μm, with a total pore volume of 10.25%.

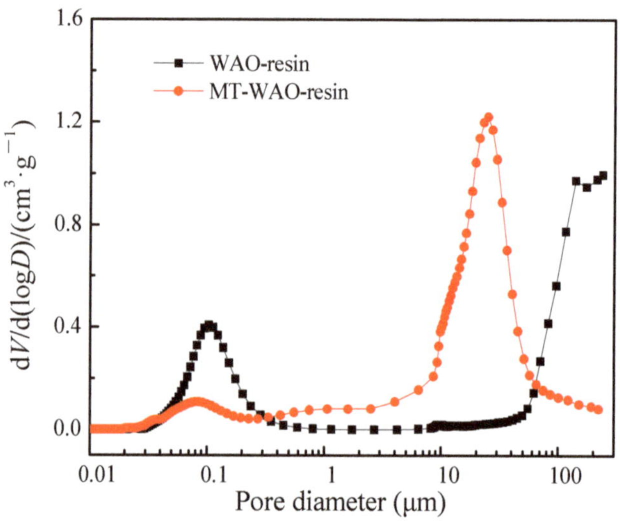

Figure 4. Differential pore volume distribution of pores with different diameters of MT-WAO resin.

By comparing the pore structure of WAO resin, it is found that the pore volume of the pore size of less than 1 μm decreases. Microwave treatment causes the pores (<1 μm) to develop into 4.0–57.4 μm super pores. The total porosity and internal porosity of MT-WAO resin increase slightly, making WAO resin's pore structure more developed.

3.3. *Adsorption*

3.3.1. Effect of Microwave Treatment Time

Microwave heating was applied to treat WAO for different times (2–20 min). Then, a batch of MT-WAO was obtained for the adsorption test. The adsorption test conditions were: pH 4.0, solid-liquid ratio 250 g/mL, initial metal ion concentration 100 mg/L, temperature 298 K, and time 7 h.

As Figure 5 shows, with an increase in the microwave treatment time, the adsorption capacity and efficiency of MT-WAO resin for Pb(II) remain the same at first and then gradually increase. When the microwave treatment time is 20 min, the adsorption capacity and Pb(II) adsorption rate are 24.9 mg/g and 99.7%, respectively. Therefore, the optimal contact time of WAO resin treated by microwave is 20 min which was considered in the subsequent tests.

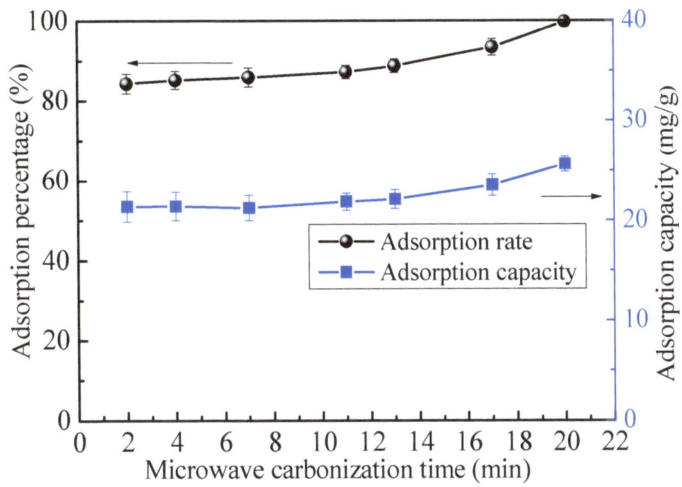

Figure 5. Effect of microwave treatment time on the adsorption of Pb(II).

3.3.2. Effect of pH

The effect of initial pH on adsorption is shown in Figure 6a. The concentration of ions is 100 mg/L. To prevent the formation of hydroxide precipitation, the pH range was controlled at 1.0–4.0, the liquid-solid ratio was 250 mL/g, the temperature was 298 K, and the time was 7 h.

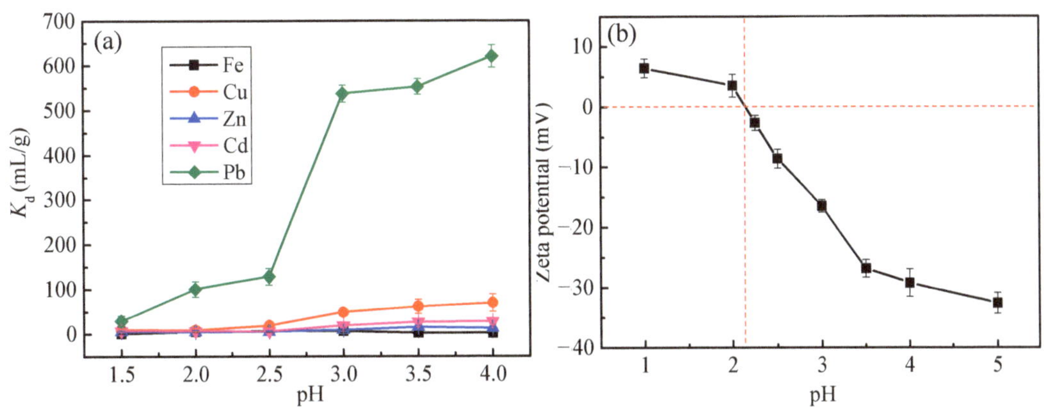

Figure 6. (**a**) Effect of pH on the adsorption of Pb(II) onto MT-WAO resin, (**b**) zeta potential of MT-WAO.

When the pH value is between 1.5 and 2.5, the *Kd* of Cu(II), Zn(II), Cd(II) and Fe(II) metal ions is small, indicating that the ability of MT-WAO resin for adsorption of these metal is weak (Figure 6a). *Kd* of Pb (II) increases with the increase of pH, which means that the ability of MT-WAO resin for Pb(II) adsorption is enhanced as pH increases. When the pH is above 3.0, MT-WAO resin exhibits better adsorption for Pb(II), and the maximum *Kd* of Pb(II) reaches 620 mL/g.

Figure 6b shows that the zeta potential of the MT-WAO decreases as pH increases, and the point of zero charge pH is 2.1. In the pH range 1.0–2.1, the value of the zeta potential is positive, as the pH value exceeds 2.1, the MT-WAO possesses an overall negative surface charge, which makes it to attract the positively charged metal ions in the pores by electrostatic interaction. Based on this, the subsequent tests were carried out at pH 4.0.

3.3.3. Effect of Time and Adsorption Kinetics

The influence of time on the adsorption of Pb(II) was investigated in the range of 5 min to 24 h (Figure 7). The temperature was 298 K, the pH was 4, the liquid-solid ratio was 250 mL/g, and the initial metal ion concentration was 100 mg/L. During 5–300 min of adsorption, the adsorption efficiency of Pb(II) increased with time and remained unchanged after it was extended to 300 min. When the adsorption equilibrium time is 300 min, the equilibrium adsorption capacity of Pb(II) reaches 17.4 mg/g.

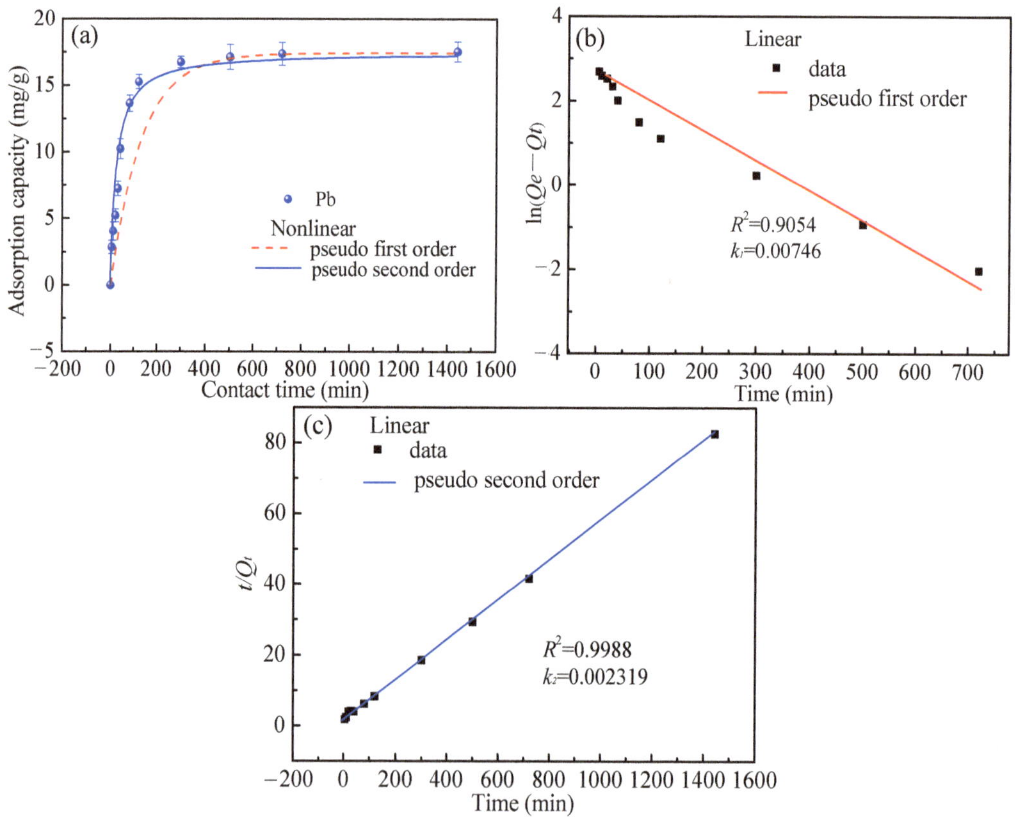

Figure 7. The adsorption kinetics. (**a**) effect of time on adsorption; (**b**) the fitting of pseudo first order; (**c**) the fitting of pseudo second order.

To analyze the reaction kinetics, the pseudo-first-order-kinetic and pseudo-second-order-kinetic [11,29,30] models were applied, as presented in Equations (3) to (6), respectively.

$$\text{Lagergren's pseudo-first-order}: \ Q_t = Q_e(1 - e^{-k_1 t}) \quad (3)$$

$$\text{Linear equation}: \ \ln(Q_e - Q_t) = \ln Q_e - k_1 t \quad (4)$$

$$\text{Ho's pseudo-second-order}: \ Q_t = \frac{Q_e^2 k_2 t}{1 + Q_e k_2 t} \quad (5)$$

$$\text{Linear equation}: \ \frac{t}{Q_t} = \frac{1}{k_2 Q_e^2} + \frac{1}{Q_e} \quad (6)$$

where k_1 (min^{-1}) and k_2 (g/mg/min) are the kinetic rate constants. Q_t and Q_e (mg/g) represent the extent of metal ions adsorbed per unit adsorbent at time t and equilibrium state, respectively.

The detailed parameters for the two models are depicted in (Figure 7b,c); it can be concluded that the adsorption of Pb^{2+} obeys the pseudo-second-order kinetic model owing to the following aspects: the correlation coefficients (R^2) obtained for the pseudo-second-order model are higher and near to 1.

3.3.4. Effect of Temperature and Adsorption Isotherm

The influence of the initial concentration of Pb(II) on the adsorption isotherm was investigated at room temperature (Figure 8a). The test conditions were: the liquid to solid ratio of 300 mL/g, pH 4.0, temperature 298 K, time 7 h, and initial metal ion concentration 50–6000 mg/L. As shown in Figure 8a, the adsorption capacity of MT-WAO resin for Pb(II) increases first; it then remains the same with an increase in the initial concentration of metal ions. When the initial concentration was 1000 mg/L, it reached the adsorption equilibrium, and the actual saturated adsorption capacity was 82.67 mg/g at 318 K.

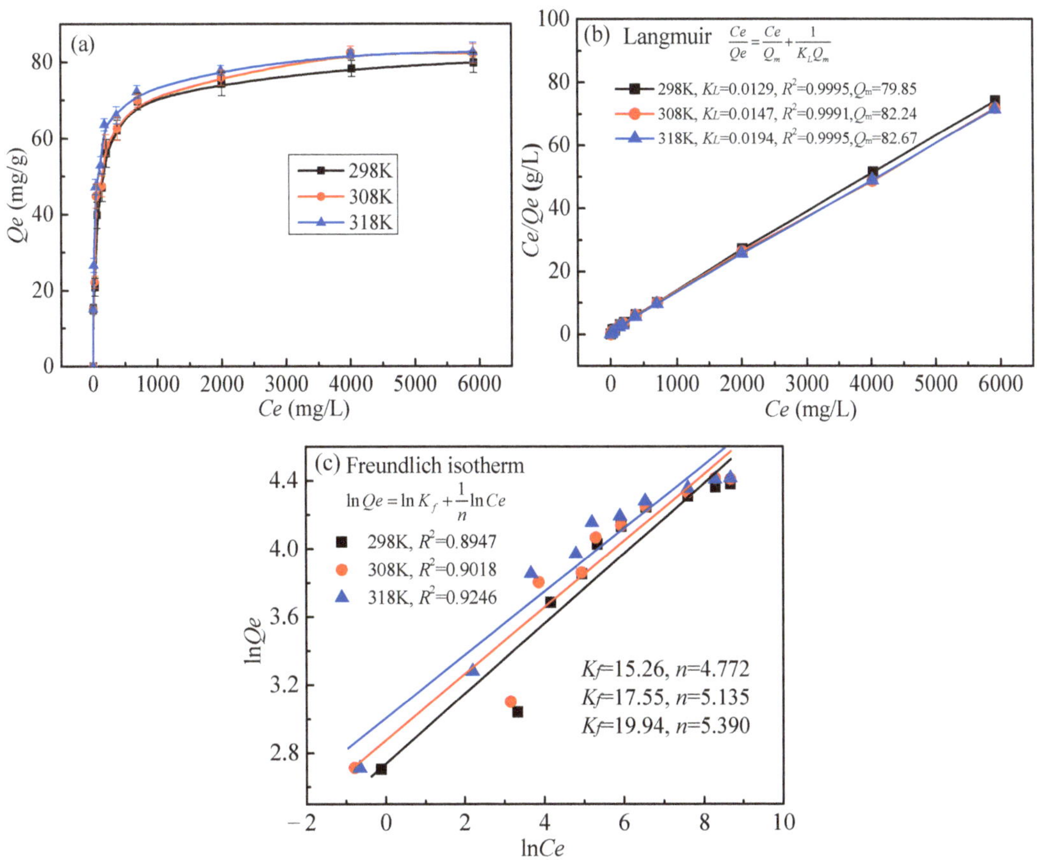

Figure 8. The adsorption isotherms of Pb(II) on the MT-WAO resin: (**a**) the influence of the initial concentration of Pb(II) on the adsorption, (**b**) the fitting of Langmuir model, (**c**) the fitting of Freundlich model.

The adsorption isotherms can be fitted by Langmuir [31] and Freundlich [32] isotherm models expressed as Equations (7) and (8), respectively (Figure 8a,c), and the fitting parameters are shown in Table 2. The Langmuir model (Figure 8b) is more consistent with the adsorption data (R^2 = 0.99), indicating that Pb(II) is adsorbed on MT-WAO resin as a monolayer and uniformly; this suggests that the metal ion adsorption on the waste resin surface is an entirely homogeneous process and follows a monolayer adsorption mechanism. The adsorption capacity of MT-WAO resin for Pb(II) reaches 82.67 mg/g at 318 K. Therefore, MT-WAO resin is an excellent adsorbent and can effectively adsorb Pb(II) from the solution.

$$\text{Langmuir model}: \frac{C_e}{Q_e} = \frac{C_e}{Q_m} + \frac{1}{K_L Q_m} \quad (7)$$

$$\text{Freundlich model}: \ln Q_e = \ln K_f + \frac{1}{n}\ln C_e \quad (8)$$

where Q_e and Q_m represent the equilibrium and maximum adsorption capacities (mg/g), respectively. C_e is the equilibrium metal ion concentration (mg/L), K_L and K_f are the Langmuir and Freundlich constants separately, and $1/n$ is the Freundlich heterogeneity factor.

Table 2. The fitting parameters of the adsorption isotherms for the Langmuir and Freundlich models.

Metal Ion	T(K)	Langmuir Isotherm Model			Freundlich Isotherm Model		
		Q_m (mg/g)	K_L	R^2	K_f	n	R^2
Pb(II)	298	79.85	0.0129	0.9995	15.26	4.772	0.8947
	308	82.24	0.0147	0.9991	17.5	5.135	0.9018
	318	82.67	0.0194	0.9992	19.94	5.390	0.9246

Thermo-dynamic parameters [33] were calculated for this system by using Equations (9) and (10).

$$\text{Langmuir model}: \ln K_d = \frac{\Delta S}{R} - \frac{\Delta H}{RT} \quad (9)$$

$$\text{Freundlich model}: \Delta G = \Delta H - T\Delta S \quad (10)$$

where ΔH, ΔS, and T are the enthalpy, entropy, and temperature in kelvin, respectively, and R (8.314J/(mol·K)) is the gas constant.

The values of enthalpy (ΔH) and entropy (ΔS) were obtained from the slope and intercept of lnK_d vs. $1/T$ plots, which were calculated by a curve-fitting program. The values of the thermodynamic parameters for the sorption of Pb(II) on MT-WAO are given in Table 3. The positive value of enthalpy change ΔH, shows that the adsorption of Pb(II) is endothermic. The numerical value of ΔG decreases with the increase in temperature, indicating that the reaction is spontaneous and more favorable at a higher temperature.

Table 3. Thermodynamic parameters for the adsorption of Pb^{2+} on MT-WAO resin.

Concentration of Pb^{2+} (mg/g)	ΔH (J/mol)	ΔS (J/mol)	ΔG (KJ/mol)		
			298 K	309 K	318 K
50	21.15	94.16	−6.21	−6.42	−6.63
100	54.01	177.48	−52.83	−54.61	−56.38
200	26.07	83.73	−24.93	−25.76	−26.60
300	10.46	25.66	−7.64	−7.89	−8.15
400	10.01	22.55	−6.71	−6.94	−7.16

In this study, the low-cost MT-WAO exhibits good extractive properties for Pb(II) in terms of the adsorption capacity as compared to that of the adsorbents reported previously (Table 4); it reveals that the MT-WAO is a good adsorbent for the treatment of Pb(II) from the waste aqueous.

Table 4. Comparison of Pb^{2+} and Cu^{2+} adsorption performance of the MT-WAO with other adsorbents.

Adsorbents	Adsorption Capacities (mg/g) Pb^{2+}	References
MT-WAO	82.67	This work
Sago waste	109.7	[34]
Aminated polyacrylonitrile fibers	76.1	[35]
Lignocellulosic biomaterial	62.1	[36]
Activated carbon	30.4	[37]
Guanyl-modified cellulose	52.0	[38]
Weak acidic cation resin	58.1	[39]
Pigeon peas hulls	20.8	[40]
Chars from Prosopis Africana	45.3	[41]

3.4. Adsorption Mechanism

WAO and MT-WAO resins before and after the adsorption of Pb(II) were characterized by FT-IR and XPS, as shown in Figures 9 and 10, respectively. The FTIR spectrum of MT-WAO resin, as shown in Figure 9, illustrates that the peak with high intensity can still be observed at wavelengths 1697 cm^{-1} and 1220 cm^{-1}, corresponding to the tensile vibrations of C=O and C−O in the carboxyl group [38]; this indicates that −COOH functional group still exists in MT-WAO resin; however, by comparing the infrared spectra of MT-WAO resin before and after the adsorption of Pb(II), the peaks of MT-WAO resin at 1697 cm^{-1} and 1220 cm^{-1} are not shifted significantly after the adsorption of Pb(II).

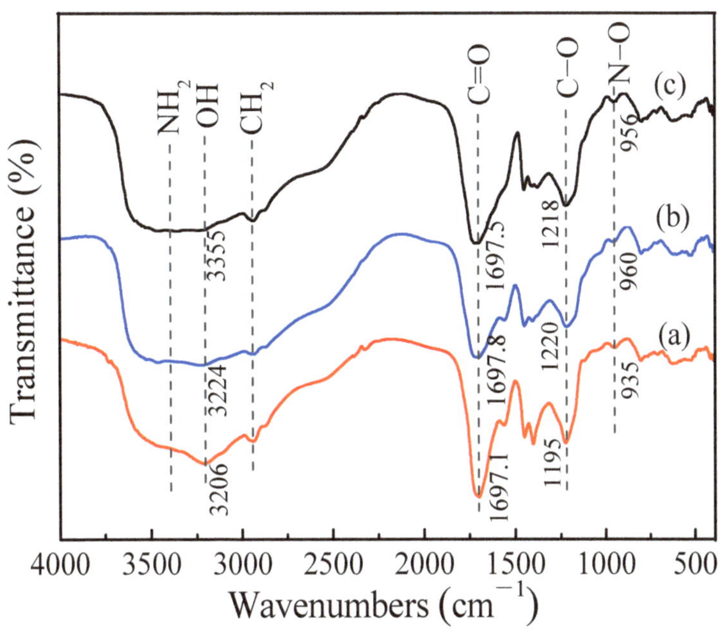

Figure 9. FT-IR spectra of (**a**) WAO and MT-WAO resins (**b**) before and (**c**) after the adsorption of Pb(II).

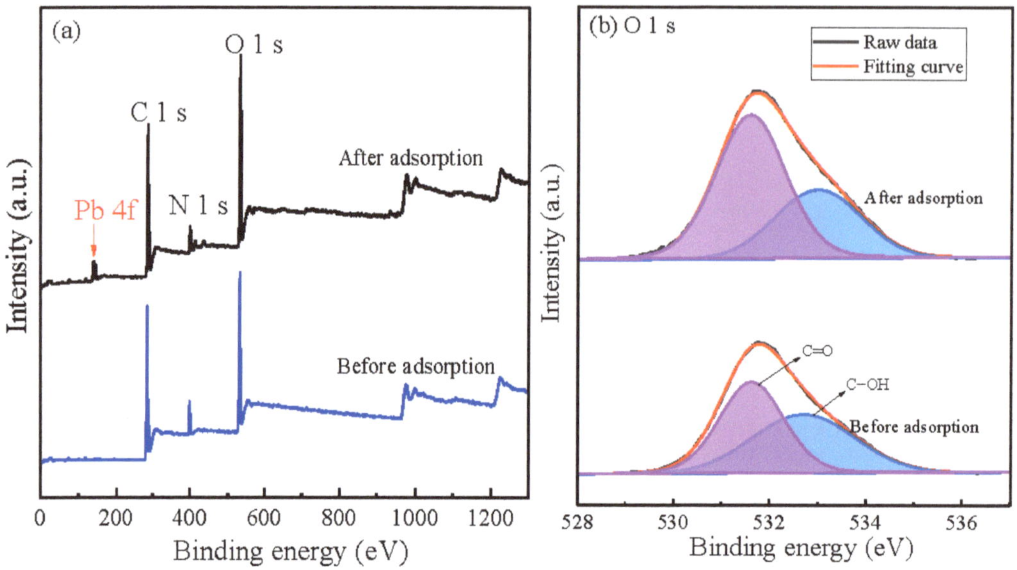

Figure 10. XPS (**a**) survey spectra and (**b**) regions of O 1 s of MT-WAO resin before and after the adsorption of Pb(II).

Figure 10a shows the characteristic peak corresponding to Pb 4f at the binding energy of 138.8 eV in the XPS spectra after the adsorption of Pb(II). Figure 10b exhibits the high-resolution XPS spectra of the O 1 s region. O 1 s peaks of MT-WAO resin before adsorption are fitted into 531.63 eV and 532.69 eV peaks, corresponding to C=O and C-OH. After Pb(II) adsorption, the corresponding binding energies shift to 531.58 eV and 532.99 eV. The characteristic peak C=O and C-OH are almost unchanged, indicating that the adsorption mechanism is not in coordination.

FT-IR shows that the surface oxygen-containing groups of MT-WAO are mainly -OH, =C=O and –COOH; these groups may react with H^+ to form $–OH_2^+$, $=COOH_2^+$ and $–C=OH^+$, resulting in the surface of MT-WAO being positively charged. When pH increases, it will reduce the number of positive ions on the surface of MT-WAO. When pH exceeds the point of zero charge pH is 2.1, the surface of MT-WAO will be negatively charged. Therefore, the adsorption mechanism is electrostatic interaction. The schematic diagram of the possible adsorption mechanism of the MT-WAO is shown in Figure 11.

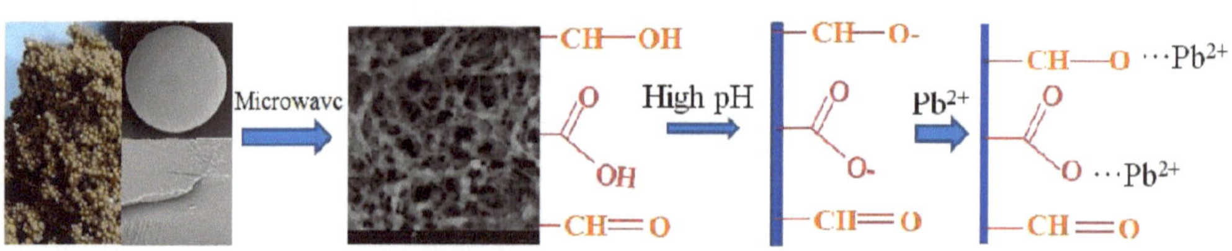

Figure 11. The schematic diagram of the possible adsorption mechanism of the MT-WAO.

3.5. Elution

In order to explore the recovery effect of Pb(II), MT-WAO resin loaded with Pb(II) under optimal adsorption conditions was used as the raw sample for elution and hydrochloric acid solution of 1 mol/L was selected as the desorbing agent. The effects of time and temperature on Pb(II) resolution were investigated, as shown in Figure 11. The desorbed conditions were: 1 mol/L hydrochloric acid solution and the liquid-solid ratio was 50 mL/g.

Figure 12a shows that the elution rate of Pb(II) increases gradually with time, which reaches more than 75% when the time reaches 1 h. The elution time greatly influences the elution of Pb(II). Figure 12b exhibits that the elution rate of Pb(II) increases gradually with temperature. When the temperature increased to 318 K during 1 h, 1 mol/L hydrochloric acid solution was sufficient to completely elute the Pb(II) adsorbed from MT-WAO resin with an elution rate of 98%.

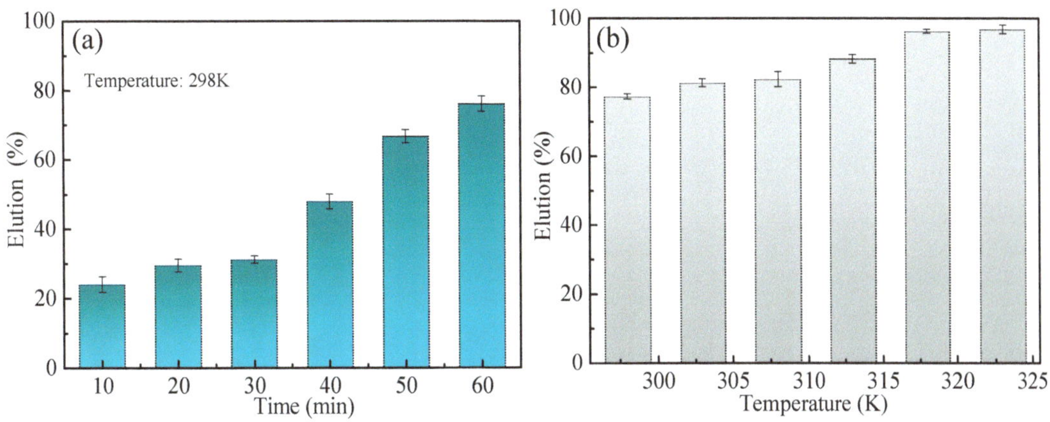

Figure 12. Variation of elution percentage of Pb(II) with (**a**) time and (**b**) temperature.

4. Conclusions

In this investigation, after subjecting WAO-resin to microwave, chaotic pore structure and the original dense structure changed into lamellar stack structure or interconnected three-dimensional porous network structure. WT-WAO resin exhibited good selectivity towards adsorbing Pb(II) (Kd 620 mL/g). The optimal microwave treatment time and optimal pH are 20 min and 4, respectively. The adsorption reached equilibrium at 300 min. The maximum adsorption capacity of Pb(II) reached 82.67 mg/g at 318 K. The adsorption kinetics and adsorption isotherms of WT-WAO resin for Pb(II) conform to the quasi-second-order kinetic equation and Langmuir adsorption isotherm model, respectively. The adsorption behavior of WT-WAO-resin for Pb(II) is homogeneous and adsorbed in a monolayer. The surface of MT-WAO is negatively charged and the adsorption mechanism is mainly electrostatic interaction. When the temperature increased to 318 K during 1 h, 1 mol/L hydrochloric acid solution was sufficient to completely elute the Pb(II) adsorbed from MT-WAO resin with an elution rate of 98%.

Author Contributions: Methodology, C.Z. and C.H.; writing—original draft preparation Y.L. (Yun Liu) and C.H.; writing—review and editing, C.H.; investigation, C.Z., Y.J., J.H. and Y.L. (Yan Liao); conceptualization, S.M.; project administration, C.H., Y.W. and T.F. All authors have read and agreed to the published version of the manuscript.

Funding: This research was funded by the Natural Science Foundation of China, grant number 51804084; Natural Science Foundation of Guangxi Province, grant number 2021GXNSFAA220096; and Guangxi University innovation and entrepreneurship training project, grant number 202110593349; Science and Technology Major Project of Guangxi, grant number AA17204100.

Institutional Review Board Statement: "Not applicable" for studies not involving humans or animals.

Informed Consent Statement: "Not applicable" for studies not involving humans.

Data Availability Statement: Not applicable.

Conflicts of Interest: The authors declare no conflict of interest. The funders had no role in the design of the study; in the collection, analyses, or interpretation of data; in the writing of the manuscript; or in the decision to publish the results.

Abbreviations

WAO	Waste amine-oxime resin
FAO	Fresh amine oxime resin
MT-WAO	Waste amine-oxime resin after microwave pretreatment
FT-IR	Fourier-transform infrared spectroscopy
SEM-EDS	Scanning electron microscopy with energy dispersive spectroscopy
XPS	X-ray photoelectron spectroscopy
ICP-AES	Inductively coupled plasma-atomic emission spectrometry
TG-DTA	Thermogravimetry and differential thermal analysis
K_d	distribution coefficient (Ratio of resin metal concentration to solution metal concentration at equilibrium, $K_d = Qe/Ce$ or $K_d = [(C_0 - Cs)/Cs] \times (V/m)$.

References

1. Lee, J.C.; Hong, H.J.; Chung, K.W.; Kim, A.Y. Separation of platinum, palladium and rhodium from aqueous solutions using ion exchange resin: A review. *Sep. Purif. Technol.* **2020**, *246*, 116896. [CrossRef]
2. Chidiac, C.; Kim, Y.; Lannoy, C. Enhanced Pb(II) removal from water using conductive carbonaceous nanomaterials as bacterial scaffolds: An experimental and modelling approach. *J. Hazard. Mater.* **2022**, *431*, 128516. [CrossRef] [PubMed]
3. Ahmed, Y.M.; Al-Mamun, A.; Al Khatib, M.F.R.; Jameel, A.T.; AlSaadi, M.A.H.A.R. Efficient lead sorption from wastewater by carbon nanofibers. *Environ. Chem. Lett.* **2015**, *13*, 341–346. [CrossRef]
4. Shama, P.; Dutta, D.; Udayan, A.; Kumar, S. Industrial wastewater purification through metal pollution reduction employing microbes and magnetic nanocomposites. *J. Environ. Chem. Eng.* **2021**, *9*, 106673. [CrossRef]
5. Barakat, M.A. New trends in removing heavy metals from industrial wastewater. *J. Arab. Chen.* **2011**, *4*, 361–377. [CrossRef]
6. Hashim, M.A.; Mukhopadhyay, S.; Sahu, J.N.; Gupta, B.S. Remediation technologies for heavy metal contaminated groundwater. *J. Environ. Manag.* **2011**, *92*, 2355–2388. [CrossRef]
7. Qureshi, U.A.; Khatri, Z.; Ahmed, F.; Khatri, M.; Kim, I.-S. Electrospun Zein nanofiber as a green and recyclable adsorbent for the removal of reactive black 5 from the aqueous phase. *ACS Sustain. Chem. Eng.* **2017**, *5*, 4340–4351. [CrossRef]
8. Yang, Y.J.; Yang, Y.N.; He, C.L.; Wei, Y.Z.; Fujita, T.; Wei, Z.W.; Yang, W.C.; Fang, S. The adsorption and desorption behavior and mechanism research of cobalt, nickel and copper in nitrite—Sulfuric acid system. *Sep. Sci. Technol.* **2022**, *57*, 1848–1859. [CrossRef]
9. Zhang, H.; Li, C.M.; Chen, X.J.; Fu, H.; Chen, Y.L.; Ning, S.Y.; Fujita, T.; Wei, Y.Z.; Wang, X.P. Layered ammonium vanadate nanobelt as efficient adsorbents for removal of Sr^{2+} and Cs^{+} from contaminated water. *J. Colloid Interf. Sci.* **2022**, *615*, 110–123. [CrossRef]
10. Meng, J.J.; He, C.L.; Li, Y.J.; Zhou, J.; Li, J.; Zheng, C.H.; Zhao, J.; Fujita, T.; Ning, S.Y.; Wei, Y.Z. Enhanced adsorption and separation of gallium using silica-based P507-TBP/SiO2–P adsorbent from sulfuric acid solution. *Micropor. Mesopor. Mat.* **2021**, *314*, 110859. [CrossRef]
11. Zheng, Q.; He, C.L.; Meng, J.J.; Fujita, T.; Zheng, C.H.; Dai, W.; Wei, Y.Z. Behaviours of adsorption and elution on amidoxime resin for gallium, vanadium and aluminium ions in alkaline aqueous solution. *Solvent. Extr. Ion. Exc.* **2021**, *39*, 373–398. [CrossRef]
12. Zheng, C.H.; He, C.L.; Yang, Y.J.; Fujita, T.; Wang, G.F.; Yang, W.C. Characterization of Waste Amidoxime Chelating Resin and Its Reutilization Performance in Adsorption of Pb(II), Cu(II), Cd(II) and Zn(II) Ions. *Metals* **2022**, *12*, 149. [CrossRef]
13. Long, C.; Lu, J.D.; Li, A.M.; Hu, D.B.; Liu, F.Q.; Zhang, Q.X. Adsorption of naphthalene onto the carbon adsorbent from waste ion exchange resin: Equilibrium and kinetic characteristics. *J. Hazard. Mater.* **2008**, *150*, 656–661. [CrossRef]
14. Shi, Q.Q.; Li, A.M.; Zhou, Q.; Shuang, C.D.; Ma, Y.L. Utilization of waste cation exchange resin to prepare carbon/iron composites for the adsorption of contaminants in water. *J. Ind. Eng. Chem.* **2014**, *20*, 4256–4260. [CrossRef]
15. Shi, Q.Q.; Li, A.M.; Zhu, Z.L.; Liu, B. Adsorption of naphthalene onto a high-surface-area carbon from waste ion exchange resin. *J. Environ. Sci.* **2013**, *25*, 188–194. [CrossRef]
16. Wei, M.Q.; Yu, Q.B.; Duan, W.J.; Hou, L.M.; Liu, K.J.; Qin, Q.; Liu, S.H.; Dai, J.J. Equilibrium and kinetics analysis of CO_2 adsorption on waste ion-exchange resin-based activated carbon. *J. Taiwan Inst. Chem. Eng.* **2017**, *77*, 161–167. [CrossRef]
17. Wei, M.Q.; Yu, Q.B.; Mu, T.T.; Hou, L.M.; Zuo, Z.L.; Peng, J.Y. Preparation and characterization of waste ion-exchange resin-based activated carbon for CO2 capture. *Adsorption* **2016**, *22*, 385–396. [CrossRef]
18. Wei, M.Q.; Yu, Q.B.; Duan, W.J.; Zuo, Z.L.; Hou, L.M.; Dai, J.J. CO_2 adsorption and desorption performance of waste ion-exchange resin-based activated carbon. *Environ. Prog. Sustain. Energy* **2017**, *37*, 703–711. [CrossRef]
19. Wei, M.Q.; Yu, Q.B.; Xie, H.Q.; Zuo, Z.L.; Hou, L.M.; Yang, F. Kinetics studies of CO_2 adsorption and desorption on waste ion-exchange resin-based activated carbon. *Int. J. Hydrogen Energy* **2017**, *42*, 27122–27129. [CrossRef]
20. Bratek, K.; Bratek, W.; Kułażyński, M. Carbon adsorbents from waste ion-exchange resin. *Carbon* **2002**, *40*, 2213–2220. [CrossRef]

21. Bratek, W.; Bratek, K.; Kułażyński, M. Properties and structure of spherical sorbents from waste ion exchange resin. *Fuel Process. Technol.* **2003**, *81*, 87–102. [CrossRef]
22. Bratek, W.; Bratek, K.; Kułażyński, M. The utilization of waste ion exchange resin in environmental protection. *Fuel Process. Technol.* **2002**, *77–78*, 431–436. [CrossRef]
23. Gun'ko, V.M.; Leboda, R.; Skubiszewska-ZieBa, J.; Charmas, B.; Oleszczukc, P. Carbon adsorbents from waste ion-exchange resins. *Carbon* **2005**, *43*, 1143–1150. [CrossRef]
24. Miura, K.; Nakagawa, H.; Watanabe, K. Production of porous carbon from ion exchange resin waste. *Tanso* **1999**, *186*, 25–29. [CrossRef]
25. Lin, W.P.; Fu, R.W.; Lu, Y.; Zeng, H.M. Preparation of amidoxime group-containing chelating fibers and their gold absorption properties. II. A preliminary investigation of the absorption behavior of Au3+ onto chelating fibers containing amidoxime groups. *React. Polym.* **1994**, *22*, 1–8. [CrossRef]
26. Melegy, A. Adsorption of lead (II) and zinc (II) from aqueous solution by bituminous coal. *Geotech. Geol. Eng.* **2010**, *28*, 549–558. [CrossRef]
27. Salman, M.; Athar, M.; Farooq, U.; Rauf, S.; Habiba, U. A new approach to modification of an agro-based raw material for Pb(II) adsorption. *Korean J. Chem. Eng.* **2014**, *31*, 467–474. [CrossRef]
28. Lin, W.P.; Lu, Y.; Zeng, H.M. Extraction of gold from Au(III) ion containing solution by a reactive fiber. *J. Appl. Polym. Sci.* **1993**, *49*, 1635–1638. [CrossRef]
29. Mittal, H.; Ray, S.S.; Okamoto, M. Recent progress on the design and applications of polysaccharide-based graft copolymer hydrogels as adsorbents for wastewater purification. *Macromol. Mater. Eng.* **2016**, *301*, 496–522. [CrossRef]
30. Pandey, S.; Son, N.; Kang, M. Synergistic sorption performance of karaya gum crosslink poly(acrylamide-co-acrylonitrile) @ metal nanoparticle for organic pollutants. *Int. J. Biol. Macromol.* **2022**, *210*, 300–314. [CrossRef]
31. Meng, Q.; Liu, J.; Jiang, Y.; Teng, Q. Branched polyethyleneimine-functionalized polystyrene resin: Preparation and adsorption of Cu^{2+}. *J. Chem. Eng. Data.* **2019**, *64*, 2618–2626. [CrossRef]
32. Kongsuwan, A.; Patnukao, P.; Pavasant, P. Binary component sorption of Cu(II) and Pb(II) with activated carbon from Eucalyptus camaldulensis Dehn bark. *J. Ind. Eng. Chem.* **2009**, *15*, 465–470. [CrossRef]
33. Donat, R.; Akdogan, A.; Erdem, E.; Cetisli, H. Thermodynamics of Pb^{2+} and Ni^{2+} adsorption onto natural bentonite from aqueous solutions—Sciencedirect. *J. Colloid Interface Sci.* **2005**, *286*, 43–52. [CrossRef]
34. Quek, S.Y.; Wase, D.; Forster, C.F. The use of sago waste for the sorption of lead and copper. *Water SA* **1998**, *24*, 251–256. [CrossRef]
35. Deng, S.B.; Bai, R.B.; Chen, J.P. Aminated polyacrylonitrile fibers for lead and copper removal. *Langmuir* **2003**, *19*, 5058–5064. [CrossRef]
36. Dupont, L.; Guillon, E.; Bouanda, J.; Dumonceau, J.; Aplincourt, M. EXAFS and XANES Studies of Retention of Copper and Lead by a Lignocellulosic Biomaterial. *Environ. Sci. Technol.* **2002**, *36*, 5062–5066. [CrossRef] [PubMed]
37. Faur-Brasquet, C.; Reddad, Z.; Kadirvelu, K.; Cloirec, P.L. Modeling the adsorption of metal ions (Cu^{2+}, Ni^{2+}, Pb^{2+}) onto ACCs using surface complexation models. *Appl. Surf. Sci.* **2002**, *196*, 356–365. [CrossRef]
38. Kenawy, I.M.; Hafez, M.; Ismail, M.A.; Hashem, M.A. Adsorption of Cu(II), Cd(II), Hg(II), Pb(II) and Zn(II) from aqueous single metal solutions by guanyl-modified cellulose. *Int. J. Biol. Macromol.* **2017**, *107*, 1538–1549. [CrossRef]
39. Vergili, I.; Gülzada, S.; Kaya, Y.; Özçelep, Z.; Çavuş, S.; Guülten, G.R. Study of the removal of Pb(II) using a weak acidic cation resin: Kinetics, thermodynamics, equilibrium, and breakthrough curves. *Ind. Eng. Chem. Res.* **2013**, *52*, 9227–9238. [CrossRef]
40. Ramana, D.; Reddy, D.; Yu, J.S.; Seshaiah, K. Pigeon peas hulls waste as potential adsorbent for removal of Pb(II) and Ni(II) from water. *Chem. Eng. J.* **2012**, *197*, 24–33. [CrossRef]
41. Elaigwu, S.E.; Rocher, V.; Kyriakou, G.; Greenway, G.M. Removal of Pb^{2+} and Cd^{2+} from aqueous solution using chars from pyrolysis and microwave-assisted hydrothermal carbonization of Prosopis Africana shell. *J. Ind. Eng. Chem.* **2014**, *20*, 3467–3473. [CrossRef]

toxics

MDPI

Article

Ecofriendly Composite as a Promising Material for Highly-Performance Uranium Recovery from Different Solutions

Mohammed F. Hamza [1,2,*], Hanaa A. Abu Khoziem [2], Mahmoud S. Khalafalla [2], Walid M. Abdellah [2], Doaa I. Zaki [2], Khalid Althumayri [3] and Yuezhou Wei [1,4,*]

1 School of Nuclear Science and Technology, University of South China, Hengyang 421001, China
2 Nuclear Materials Authority, POB 530, El-Maadi, Cairo 11728, Egypt
3 Department of Chemistry, College of Science, Taibah University, Al-Madinah Al-Munawarah 30002, Saudi Arabia
4 School of Nuclear Science and Engineering, Shanghai Jiao Tong University, Shanghai 200240, China
* Correspondence: m_fouda21@hotmail.com (M.F.H.); yzwei@sjtu.edu.cn (Y.W.); Tel.: +20-1116681228 (M.F.H.); +86-771-3224990 (Y.W.)

Abstract: The development of new materials based on biopolymers (as renewable resources) is substantial for environmental challenges in the heavy metal and radionuclide ions removal contaminations. Functionalization of chitosan with sulfonic groups was achieved for improving the uranium sorption, not only from slightly acidic leachate, but also for the underground water. The prepared hydrogel based on chitosan was characterized by series of analysis tools for structure elucidation as FTIR spectroscopy, textural properties using nitrogen adsorption method, pH_{PZC} (by pH-drift method), thermogravimetric analysis (TGA), SEM, and SEM-EDX analyses. The sorption was performed toward uranium (VI) ions for adjustment of sorption performances. The optimum sorption was performed at pH 4 (prior to the precipitation pH). The total sorption was achieved within 25 min (relatively fast kinetics) and was fitted by pseudo-first order rate equation (PFORE) and resistance to intraparticle diffusion equation (RIDE). The maximum sorption capacity was around 1.5 mmol U g^{-1}. The sorption isotherms were fitted by Langmuir and Sips equations. Desorption was achieved using 0.3 M HCl solution and the complete desorption was performed in around 15 min of contact. The sorption desorption cycles are relatively stable during 5 cycles with limit decreasing in sorption and desorption properties (around 3 ± 0.2% and 99.8 ± 0.1%, respectively). The sorbent was used for removal of U from acid leachate solution in mining area. The sorbent showed a highly performance for U(VI) removal, which was considered as a tool material for radionuclides removing from aquatic medium.

Keywords: uranium; metal decontamination; hydrogel; uptake kinetics; sorption isotherms; recovery of heavy metal

Citation: Hamza, M.F.; Abu Khoziem, H.A.; Khalafalla, M.S.; Abdellah, W.M.; Zaki, D.I.; Althumayri, K.; Wei, Y. Ecofriendly Composite as a Promising Material for Highly-Performance Uranium Recovery from Different Solutions. *Toxics* **2022**, *10*, 490. https://doi.org/10.3390/toxics10090490

Academic Editor: Xiaojun Luo

Received: 28 July 2022
Accepted: 18 August 2022
Published: 24 August 2022

Publisher's Note: MDPI stays neutral with regard to jurisdictional claims in published maps and institutional affiliations.

Copyright: © 2022 by the authors. Licensee MDPI, Basel, Switzerland. This article is an open access article distributed under the terms and conditions of the Creative Commons Attribution (CC BY) license (https://creativecommons.org/licenses/by/4.0/).

1. Introduction

The hazardous metal removal and the valuable metal recovery are challenging techniques for water purification and metal valorization. Different processes were used either in leaching of metal ions from ores, waste materials or extraction processes from solution. The green leaching using Humic acid (HA), as well as mineral acids, were used for removing of uranium from crude and spent materials [1–4]. For this purpose, several techniques were used including solvent extraction, metal precipitation, and membrane-based techniques, which faced limited use on the concentration level of metal ions and the solution complexity. Sorption is the more specifically used in the low concentration of metal ions. Although there is a large variety of commercial resins, there is a demand for developing special types of sorbents (nano particles size or hydrogel) for the removal of heavy metal ions as well as recovering of valuable/economic metals.

A wide variety of functionalized sorbents were identified for recovering valuable metals and removing hazardous elements from the aquatic medium through high sorption capacity and selectivity, in addition to uses in the diluted solution treatments as wastewater or groundwater treatment. These include (a) combination of hydrogels (i.e., alginate and chitosan [5]), (b) Sulfonic functionalized materials, yeast cells [6–10], (c) Amidoxime groups on synthetic sorbents [11–15] (d) Quaternary ammonium-based sorbents [16,17], (e) bio-based composites [18–22], metal-organic framework [23], or (f) Iminodiacetic acid [24].

Industrial effluents and wastewater treatment through heavy and radionuclide removal is a crucial prospective and main target of researchers. The development of the high-tech industries led to increasing the various pollutants, (heavy elements as lead and cadmium, and the increasing demand of nuclear power increasing the pollution of uranium and thorium, which have had a very negative impact on both environment and human beings). These elements can enter the human body directly through water and food [10,25–28]. The importance of this point is due to the increased necessity from several governments for community protection [29]. Discharging of the mining effluents causes several health problems. For making valorization of such process, by recycling valuable and rare metals that are included in such solution. Several methods were used for removing of contaminants such as solvent extraction [30], precipitation [31,32], membrane separation and electrolytic techniques [33]. This faces a problem in technical and economic uses, especially for diluted solution. Ion exchangers and chelating sorbents are a suitable tools metal recovery, even in the low metal concentration [5,34–38].

In previous decades, nanoparticles modified sorbents received a prominent attention due to the fast kinetic sorption and highly sorption capacities. The natural biobased materials as alginate, chitosan, algal biomass, or agriculture wastes can be designed as micro or nano materials [39–41]. These materials have several functional groups as hydroxyls and amines, which gives reasons for the hydrophilic properties and are ready for further modification through functionalization with other groups to enhance the capacity and improve kinetics.

Chitosan substrate has a good property toward sorption of metal ions. As functionalization was achieved, the sorption properties, as well as selectivity, were changed depending on the type of sorption (as electrostatic attraction or ionic exchange (at acidic pH value) or chelation properties (at slightly acidic or neutral pH values)) [14,42–45]. The grafted groups improve the physical properties of the sorbent through sorption in a broad pH value. In parallel application with metal removal on some prepared sulfonated sorbents, complementary tests were used for the antimicrobial effect. Most of these composite derivatives have been reported as a reactive to the antimicrobial activity [46].

This study compares the sorption properties of U(VI) on functionalized sulfonic groups (derived from 2-Acrylamido-2-methylpropane sulfonic acid (AMPS)) as well as the amine groups (from chitosan and N,N′-Methylenebisacrylamide (MBA; the crosslinker)) of the functionalized composites hydrogel. The prepared sorbent was characterized by textural properties (BET), FTIR spectroscopy, and morphological characterizations (SEM, and SEM-EDX), thermogravimetric analysis (TGA). The sorption properties towards U(VI) were investigated through sorption tests including pH effect, uptake kinetics, sorption isotherms, selectivity (sorption from multi-component solutions), metal desorption and sorbent recycling. The material was investigated for feasibility evaluation through industrial applications, with the composite being tested on real solution bearing complex metal ions.

2. Materials and Methods

2.1. Materials

The chemicals used in this study are of analytical grade. N,N-Methylenebis(acrylamide) (MBA; 99%), potassium persulfate (>99%), chitosan (Medium molecular weight; 75–85% of AD, acetylation degree), sodium hydroxide ($\geq$98%), and Bromoform (99%) contains 60–120 ppm 2-methyl-2-butene as stabilizer, were supplied from Sigma Aldrich (Frankfurter Str. 250, 64293-Darmstadt-Germany). 2-Acrylamido-2-methylpropane sulfonic acid

(AMPS) was supplied from SHANDONG ZHI SHANG CHEMICAL Co., Ltd.-China; Hisense Intelligence Vally, High-tech Zone, Jinan, Shandong.

2.2. Synthesis of Sorbent

2.2.1. Synthesis of the Reference Material (Chitosan Crosslinked)

Dissolving of 2 g of chitosan particles in 30 mL of 7% (*v*/*v*) acetic acid solution was performed in a three nicked flask (equipped with condenser, agitator, and thermostat). Addition of 5 mL (15% *v*/*v*) glutaraldehyde solution dropwise with vigorous stirring. The produced precipitation was continued stirring (with velocity 190 rpm) for 5 h at 75 °C before filtered and washed with acetone and water to remove the unreacted materials and dried at 60 °C for 10 h to yield CH, as shown in Scheme 1.

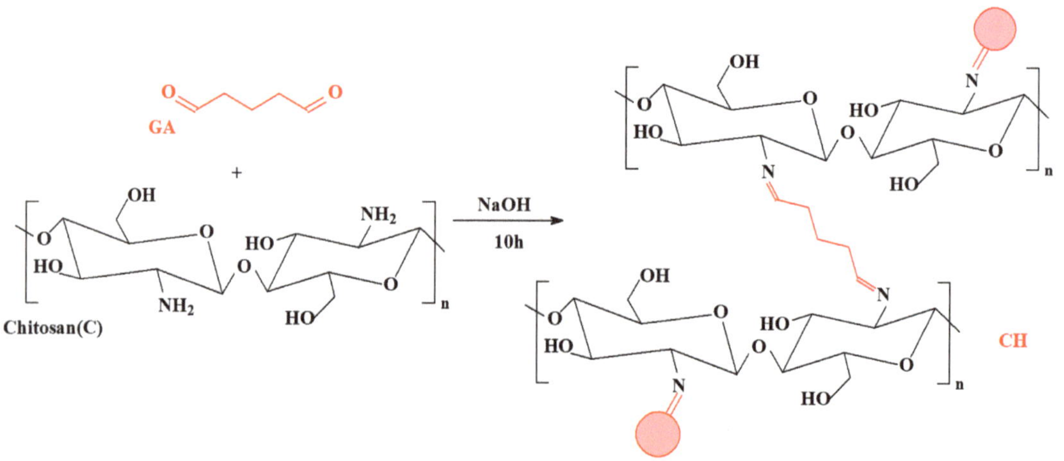

Scheme 1. Synthesis of the CH sorbent (as a reference material).

2.2.2. Synthesis of the Functionalized Chitosan Composite

Two grams of chitosan and 0.2 g of potassium persulfate were dissolved in 7% (*v*/*v*) acetic acid solution (50 mL) in three nicked flask. There as an addition of 2 g of AMPS and 0.1 g of MBA (as a crosslinker) to the solution with vigorous stirring (190 rpm) till dissolve. The flask was equipped with a spiral condenser and heated at 75 °C for 9 h under the vigorous stirring conditions. After cooling, the mixture was poured to a 500 mL solution of NaOH (0.5% (*w*/*v*)) contains 15% *v*/*v* glutaraldehyde solution (for enhancing the stability through further crosslinking effect) for 10 h at room temperature (21 ± 2 °C) with gentle stirring (95 rpm). The precipitated hydrogel was filtered and washed several times with deionized water and acetone then dried at 60 °C for 10 h to yield CH-S as shown in Scheme 2.

2.3. Characterization of Materials

The produced sorbents (CH and CH-S) were subjected to grind and sieving. The sorbents that used in the experiments were obtained with scale up to 10 µm. The (C, S, N and H composition) was measured using elemental analysis through element analyzer; CHNOS; Vario ELIII, Elementar-Analyzer system; GmbH, Sonaustraβe-20354 Hamburg, Germany. Chemical analysis and morphological studies were performed by the scanning electron microscope joint with the energy dispersive X-ray analyzer (SEM and EDX, respectively); XL30-ESEM; Philips, Thermo Fisher Scientific, Hillsboro, OR, USA. FTIR spectroscopy (Fourier-transform infrared) were used from the Mobile IR-Portable; Bruker Optics, Billerica, MA, USA. The textural properties of the synthesized material were carried out through the high-speed surface area analyzer using Nova-e Series, Model-25, Quantachrome, Kingsville,

TX, USA. The thermal analysis for material degradation was carried out by the TGA thermogravimetric analyzer; N5320011-Perkin Elmer, Villebon-sur Yvette, France), using nitrogen atmosphere as medium for the analysis, with 10 °C min^{-1} as a temperature ramp. The pH_{PZC} (zero-charge) was measured by the pH-drift method; a 100 mg of the prepared sorbents (both CH and CH-S) were mixed with 50 mL of series of prepared solution (11 prepared samples in closed bottle) with concentration of 0.1 M NaCl at fixed pH_0 (initial pH), that ranged between 1 and 11. After stirring for around 48 h, the pH_f (final pH) was measured (using pH meter with the specification described below) and the difference in pH values of the initial and final were determined, in which the pH_{PZC} was equal to $pH_0 = pH_f$. The pH of the solution was established through pH/ionometer, S220 Seven compact, Mettler Toledo-China. The collected solution (samples after pH effect, sorption isotherms, uptake kinetics, desorption, selectivity and after treatment with real samples) was firstly filtrated using micromembrane (1.2 µm) before measuring the metal contents.

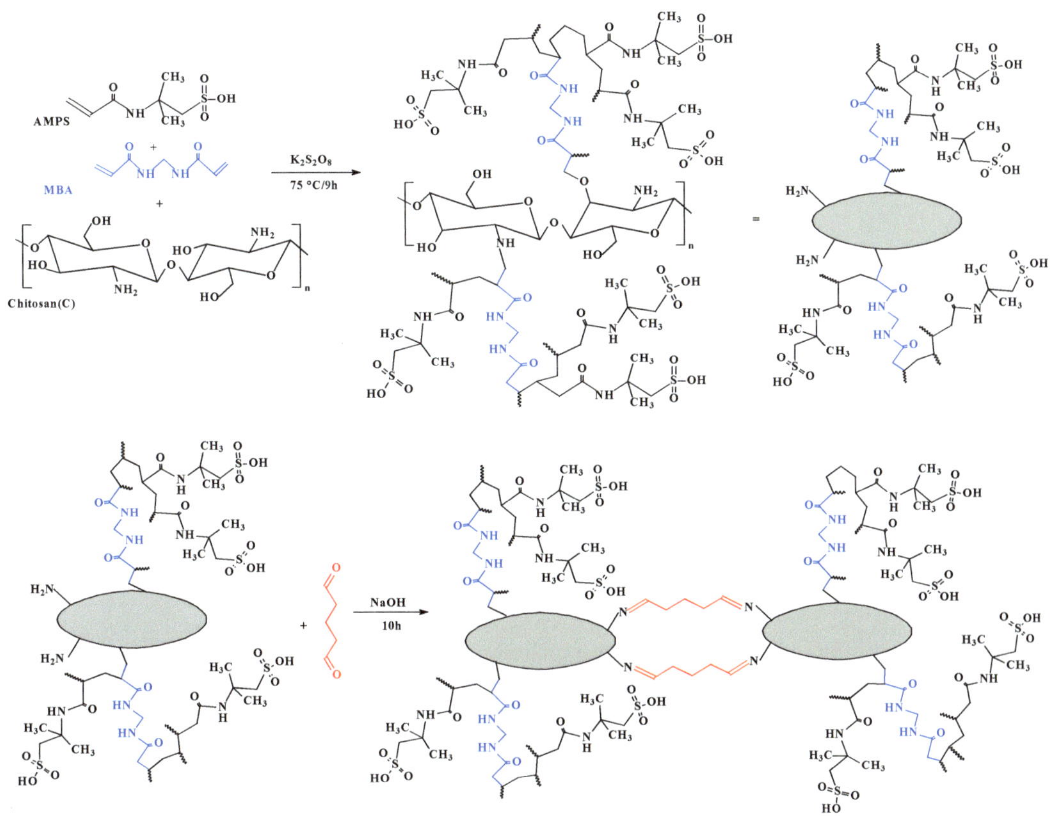

Scheme 2. Synthesis route of the sulfonic hydrogel (CH-S).

2.4. Sorption Procedures

The batch method (agitation technique) was used for carrying out the uranium sorption tests in closed bottles. Fixed volume of the solution (V, L), have a fixed concentration of the metal ions (C_0, mmol L^{-1}) at initial pH_0 value (before sorption), this solution was mixed with a specific amount of the sorbent (m, g) in a specific time (t, hours) at room temperature (21 ± 1 °C).

For the uptake kinetics, the samples collected from this experiment at each specific time were filtrated using filter membrane (with specification 1.2 µm). The experimental conditions were carefully adjusted in the specific experiments. The standard condition is as follow; the sorbent with size below 10 μm were added to solution with initial metal concentration (C_0) around 100 (±5) mg U L^{-1} (corresponding to 0.42 mmol U L^{-1}); at initial pH value (pH_0) 4; with sorbent dose (which described as amount of sorbent in a specific volume of the solution (SD; m/V)): 0.833 g L^{-1} and agitated for 24 h. The pH of the mixture wasn't adjusted through the sorption processes but the pH_{eq} (equilibrium pH) was monitored after sorption experiment.

The sorption isotherms were used with different initial uranyl concentration that varied between 0.04 (±0.02) and 2.00 (±0.05) mmol U L^{-1}. The models used for uptake kinetics and sorption isotherms are listed in Tables S1 and S2. Desorption experiments were investigated using 0.3 M HCl solution, it was tested from the sorbents collected after treatment in the kinetic experiments (uranium-loaded sorbent; around 0.650 g sorbent in each exp.). The condition used in the desorption procedure is as follow; the eluent volume 250 mL (SD: 2.52 (±0.02) g L^{-1}) and the total time of desorption experiment is around 4 h. The recycling experiment was used with the same procedures in which, the water rinsing steps (approximately 30 (±5) mL) were performed between each sorption/desorption run.

The residual concentration (C_{eq}, mmol U L^{-1}) was analyzed using an inductively coupled plasma atomic emission spectrometer (ICP-AES) for synthetic and nature solutions. The uranium content in the diluted solution was measured using ICP-AES; Activa M; Horiba/France. The concentration of uranium ions from ore materials and leachate solution was analyzed using ammonium meta vanadate in oxidimetric titration method [47,48], while the measurements of the rare earth elements (REEs) were performed by colorimetric method using Arsenazo III at λ: 654 nm [49,50] through Shimadzu-160A; Shimadzu-Corporation, Kyoto-Japan. The sorption capacity for each element (q, mmol g^{-1}) was measured by the mass balance equation: $q = (C_0 - C_{eq}) \times V/m$.

2.5. Ore Specification

Gabal (G) El-Sela area is considered as a signifying portion of late Precambrian early Paleozoic Pan Africa Orogeny. It involves two-mica granite as polymetallic intra-granitic vein-type U-deposits. It is known as the one of the most favorable areas in Egypt for hosting U-mineralization [51–55]. G El-Sela is sited in southern part of Egypt between latitudes 22°17′50″–22°18′06″ N and longitudes 36°13′36″–36°14′22″. It is in-between latitudes; (22°16′25″ N) and (22°18′40″ N), while the longitudes are described as (36°12′50″ E) and (36°16′30″ E). It is a topographically low, as well as moderate to slightly high, found as remarkably scattered isolated hills that are separated by vast sand sheets (Figure 1).

Many promising features to locate uranium mineralization were studied, particularly those related to the ENE-WSW shear zone [54,56]. Geology/mineralogy of G. El Sela two mica granite were premeditated by several researchers confirming prevalence of the primary and secondary uranium minerals [54,56,57]. The chemical composition of the collected sample that was used in this study was performed using XRF facilities and reported in Table 1. From chemical composition it was high concentration of U (1158 mg U kg^{-1}), while some valuable elements were found in high concentration as REE (1330 mg U kg^{-1}) and others in low concentration as V, Zr, Hf, Nb and Ta.

2.6. Mineralogical Characteristics of the Studied Sample

Mineralogical characteristics of the studied sample of the G. El Sela were firstly sieved to appropriate liberation size as determined microscopically (0.063–0.5 mm) and then separated into light and heavy fractions by using bromoform (sp. gr. 2.8 g cm^{-3}). Different mineral grains were picked from the obtained heavy fractions and carefully investigated under binocular stereomicroscope. The separated grains were examined by SEM attached by EDX microanalysis unit for minerals identification. The results of EDX analysis reports of notable uranium minerals associated with non-radioactive elements. Accordingly, the studied ore material referred to the poly-mineral associations: autunite, uranophane and apatite in association with hematite, goethite and magnetite minerals.

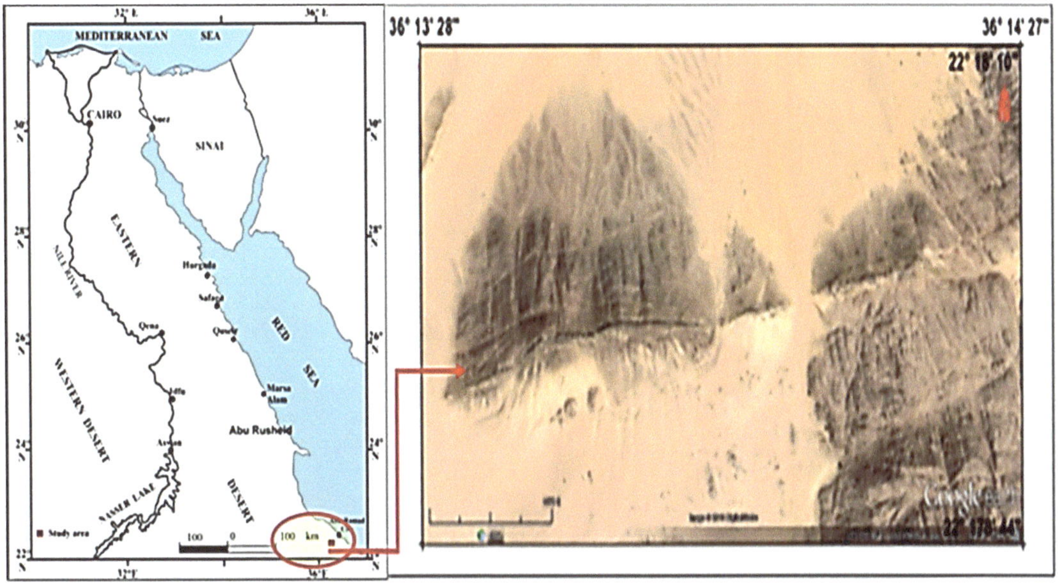

Figure 1. Google image, location map and geological map of El Sela area, South Eastern Desert, Egypt (after Nagar et al., 2016) [58].

Table 1. XRF analysis of the study G. El Sela raw materials.

Major Oxides (%)	Wt. (%)	Trace Elements	ppm
SiO_2	76.59	U	1158
Al_2O_3	8.6	REE	1330
TiO_2	1.03	Th	16
$Fe_2O_3{}^{total}$	3.76	Mn	36
CaO	1.22	V	48
MgO	0.46	Zn	79
Na_2O	0.42	Pb	139
K_2O	1.27	Zr	177
P_2O_5	0.51	Hf	5.4
L.O.I	5.7	Nb	58
Total	99.56	Ta	2.7

2.7. Results of H_2SO_4 Agitation Leaching Process

From the foregoing study, it can be concluded that H_2SO_4 agitation leaching technique is more efficient for U leaching efficiency (89.7%) than other techniques (percolation and pelletization) from the studied G El-Sela at the optimum of the leaching conditions as 15% H_2SO_4 and adding 0.25 M NaCl with a 1/2 solid/liquid ratio at 2 h and 90 °C.

3. Results and Discussion

3.1. Sorbent Characterization

3.1.1. Textural Properties

The functionalization process of the chitosan particles involved increasing in the SSA (specific surface area) through the adsorption/desorption of nitrogen isotherms. The SSA values were increased by functionalization from 5.75 $m^2\ g^{-1}$ (for CH), to 13.7 $m^2\ g^{-1}$ (for CH-S). This is due to the efficient reaction of the MBA and AMPS through crosslinking via C=C (lead to decreasing the size and high network) than that of MBA and chitosan (through C=C and NH_2, respectively). According to this, the porous volume (volume of micropores) is increased as well from 0.0473 $cm^3\ g^{-1}$ (for CH), to 0.0796 $cm^3\ g^{-1}$ (for CH-S). The other criterion was observed through increasing the cumulative volumes (total volume) of pore from 0.051 to 0.059 $cm^3\ g^{-1}$.

3.1.2. Thermogravimetric Analysis

The thermal degradation and the derivative thermogram analyses (DTG) were performed for the functionalized chitosan sorbent (CH-S) as shown in Figure 2. It is characterized by two loss steps. The first stage is achieved at around 160 °C, this is attributed to the loss of the surface water (physically adsorbed) and inside the pores, the wight loss at this stage is 5.88%. The second stage of the loss is noticed between 160 °C and 579.9 °C, the additional loss was 88.22% and this was attributed to different modes for the polymer degradation, decomposition of the amine moieties from chitosan, degradation of sulfonic groups (which reports to be in the range of temp. 200–400 °C [59]), polymer network frame [60], and the char decomposition. It was concluded that the total weight loss of this polymer was close to 94.1%. Similar profiles were observed from Akköz et al. [61] through the use of sulfonated-agriculture waste. Several shoulders were found by DTG, the peaks at 43.6 and 192.7 °C for water loss either from the surface or inside the pores, while at 307.4 °C and 457.1 °C for depolymerization and degradation of the functional groups.

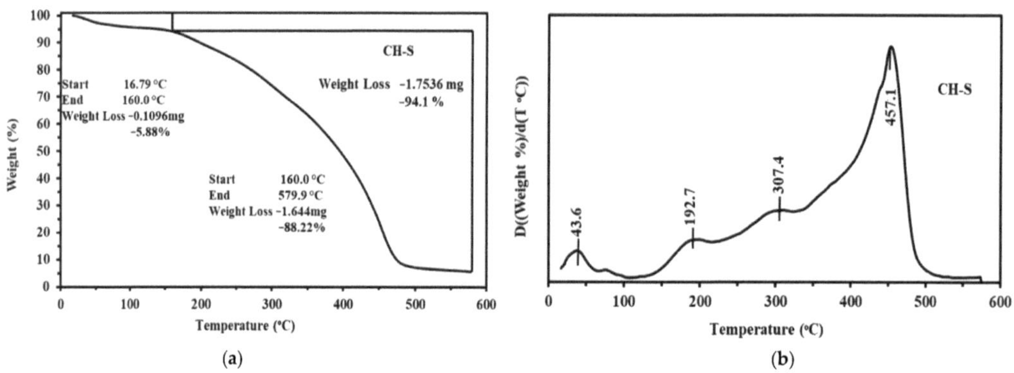

Figure 2. TGA (**a**) and DTG (**b**) for sulfonated chitosan sorbent (CH-S).

3.1.3. FTIR Spectroscopy

Figure 3 shows a comparison study of the crosslinked chitosan (before functionalization) and after functionalization process. The FTIR spectra present in this section represents with full wavenumber ranges, while Figure S1, shows the most important peaks for comparison the shifts and intensity during functionalization or loading and elution processes. The FTIR spectra were performed for sorbent before and after functionalization (CH and CH-S, respectively), after loading and after 5 cycles of sorption desorption for the CH-S. More specifically the -OH, -NH, and $-SO_3H$ groups are identified and reported in Table 2. As sulfonation was performed on chitosan a new series of peaks were identified associated with substantial changes, the C-S, O-S, and SO_3H peaks were identified for the modified composite with increasing the intensity of OH and NH stretching bands (derived from the new substates and the sulfonic groups).

Increasing the intensity of the OH and NH bands as functionalization was performed was attributed to the new grafted groups (from the MBA and the sulphonic group) [62,63]. Disappearing of OH band at 649 cm^{-1} with the new broad band for C-O-S of sulfonic group was observed. Overlapping the N-H bands (which appeared at 1575 cm^{-1} in CH) with that of C=O for CH-S, verifying the changes occurred in the environment of this groups. There was an appearance of new broad band at 1096 cm^{-1} which is related to NCS group [64]. The other interested bands were reported in the Table 2. Most of these bands decreased in the intensity or shifts, especially for OH, NH, S-O, NCS, and C=O as uranyl ion adsorbed, which shared in the binding mechanism. After five cycles of sorption desorption processes, the spectra restored with full intensity of decreased peaks, confirming the high chemical stability and lack of changes in the chemical structure of the sorbent.

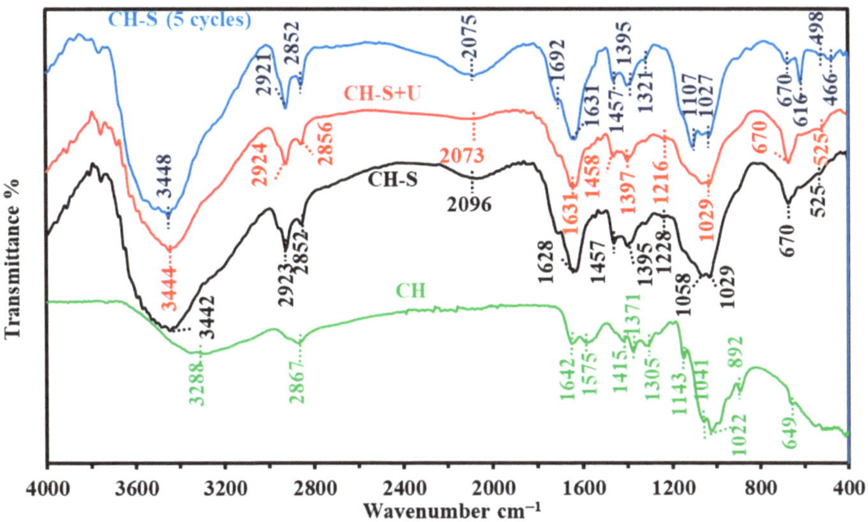

Figure 3. FTIR spectra of CH, CH-S, after loading and after 5 cycles of sorption desorption process (full wavenumber).

Table 2. Assignments peaks in FTIR spectra for CH and CH-S (before and after U(VI) sorption, and after five cycles of sorption and desorption).

Assignment	CH	CH-S	CH-S + U	CH-S (5th Cycle)	Ref.
O-H and N-H (stretching)	3288	3442	3444	3448	[65,66]
C-H (stretching)	2867	2923, 2852	2924, 2856	2921, 2852	[66,67]
-NCS group		2096	2073	2075	[68]
C=O (stretching) and NH of amide	1642, 1575	1628	1631	1692, 1631	[66,69]
$-CH_2$ (bending) and C-N (stretching)	1415	1457	1458	1457	[66,69]
CH_3 (symmetric deformation)	1371, 1305	1395	1397	1395, 1321	[66,67,69]
-N-C- (stretching) and sulfonamide group		1228	1216	1107	[68]
C-O-C (stretching)	1143				
C-O (skeletal stretching) and C-H out-of-plane (bending)	1041, 1022	1058, 1029	1029	1027	[65,66]
C-O- of epoxy ring	892				
C-O-S (stretching) and $-(CH_2)_n-$ rocking	649	670	670	670, 616	[68]
O-H out of plane (bending), C-S and/or C-O-S (stretching).		525	525	498	[70–72]
Polysulfides(S-S stretching)				466	our

3.1.4. Elemental Analysis

The elemental analysis was performed for each sorbent (CH and CH-S), which appeared in Table 3. As sulfonation was performed (by grafting of AMPS) on chitosan surface, with the MBA crosslinker, it followed an increase of N, O and S contents from 3.44, 26.408 and 0 mmol g^{-1} to 4.934, 26.952 and 0.683 mmol g^{-1}, respectively.

Table 3. Elemental analysis of CH and CH-S sorbents.

Sorbent	C	N	H	O	S
CH (%)	45.94	4.83	6.98	42.25	0
CH (mmol g^{-1})	38.248	3.449	69.253	26.408	0
CH-S (%)	41.56	6.91	6.22	43.12	2.19
CH-S (mmol g^{-1})	34.602	4.934	61.712	26.952	0.683

Figure 4a,b, shows the semi-quantitative EDX analysis of the CH and after functionalization (CH-S). It shows a trace of Na element (from the sodium hydroxyl solution) in the final synthesis step. However, this analysis was parallel with the elemental analysis for increasing the N, and O as well as the appearance of S in the final sorbent, which is derived from AMPS and MBA.

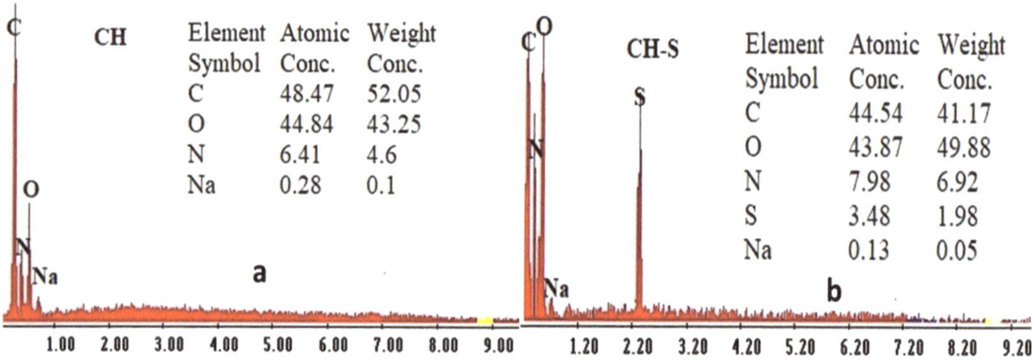

Figure 4. EDX analysis of CH (**a**) and CH-S (**b**).

3.1.5. Surface Charge—pH_{PZC}

Through applying the pH-drift method for determination of the pH_{PZC} values, Figure 5 shows the result of the pH_{pzc} value. It shows a shift based on acid base properties. The pH_{PZC} is strongly affected by functionalization of the sulfonic acid groups. In fact, the chitosan compound enhances the basic characters, which is derived from the amine and hydroxyl groups on the surface. By grafting the sulfonic acid moiety, the acid strength is noticeably changed and shifts to acidic part. The pH_{PZC} decreased from 6.6 to 4.9 for CH and CH-S, respectively. This reflects the effective functionalization of sulfonic groups on the chitosan behavior. On the other words, it means that, the surface of CH remains positively charged at a wide pH range than CH-S and this may cause a repulsion effect (especially at strong acidic pH) with positively charge metal ions. After pH 4.9 the CH-S surface becomes negative, this kind of repulsion is negligible even before this point. The sorbent is partially negatively charged, more specifically on the chelating atoms. Similar results (decreasing in the pH_{pzc} values) were obtained by Akkoz et al. through sulfonation of hawthorn kernel (as agriculture wastes) that decreased from 7 to 3.9 [61], while Urbano and Rivas [73] shows the pH_{PZC} values of 3.4 for the sulfonated groups of polyacrylamide and montmorillonite composites.

3.2. Sorption Studies

3.2.1. pH Effect

Figure 6a compares the loading capacities of both sorbents (before and after modification) toward U (VI) ions. Regardless of the type of sorbent, the sorption capacity of U (VI) was progressively increased as pH raised from 1 to 5 before becoming stabilized at pH around 4. It was increased from 0.08 to 0.23 ± 0.02 mmol U g^{-1} for CH sorbent, and from 0.22 to 0.95 ± 0.05 mmol U g^{-1} for CH-S. From these data, it was shown that as functionalization was performed, the loading capacity enhanced by more than 4 times. Increasing the pH caused partial deprotonation of the functional groups and the electrons on the N, O and S are ready for binding through chelating properties. While repulsion between positively charged uranyl ions and protonated groups at acidic pH values exists, on the other side the sorption properties were improved by increasing pH through decreasing the repulsion effects. The three repeated experiments show the successes in the reproducibility properties of the sorbent. Figure 6b shows the pH deviation of the three experiments, it shows deviation of about 0.3 values. From Figure S2, the uranium was found in different species depending on the pH values. The most interesting points are those that lay below pH 4 and 5 for the maximum sorption, before the precipitation point (>5.2) and the existing of hydrolyzed species. The uranyl oxide (UO_2^{2+}) is the mainly sorbed species that are found in the solution (for our case/C_0, 100 mg L^{-1}, uranyl nitrate as uranium source, at pH 4), which can bound with the deprotonated (or partially deprotonated) functional groups as NH, OH and SO_3H.

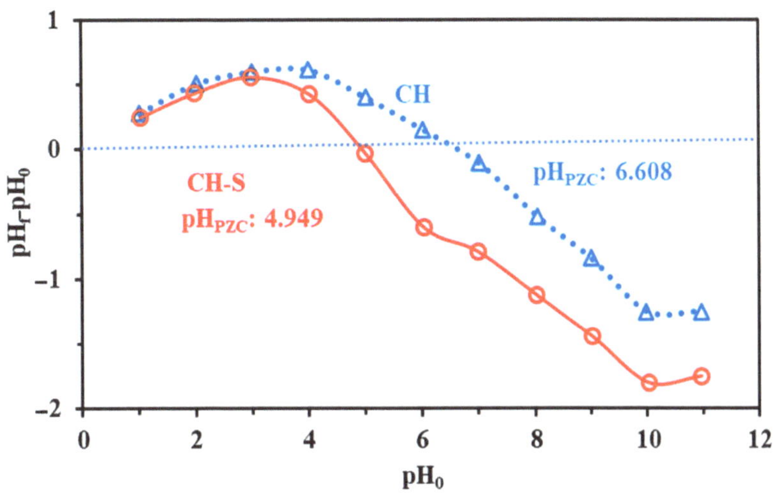

Figure 5. pH_{pzc} of the CH and CH-S sorbents.

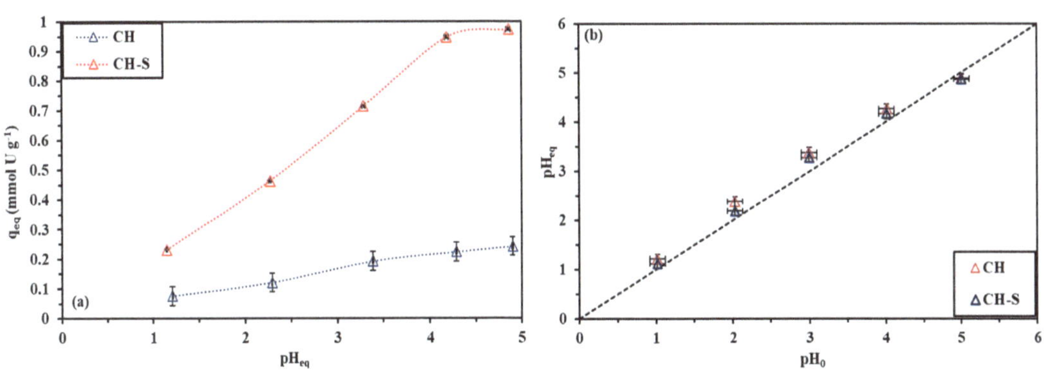

Figure 6. Effect of pH on uranyl sorption (**a**) and pH variation of the three experiments (**b**).

3.2.2. Uptake Kinetics

Under the selected experimental conditions (for the pH, sorbent dose and uranium concentration in the solution), the sorption equilibrium is reached during the first 40 min, while around 90% of the total sorption was attained during 25 min as shown in Figures 7 and S3.

The superimposition of the three experiments (small in deviation error), which emphasizes the reproducibility. The large pores character (see BET surface area) of the sorbent may explain the fast sorption kinetics and the fast mass transfer, which is controlled through the resistances to film diffusion, as well as the intraparticle diffusion properties, so the resistance to intraparticle diffusion equation (RIDE) cannot be neglected for the kinetic models. Moreover, the proper sorption rate, which is modeled by pseudo-first order (PFORE); it is described as physical sorption, or pseudo-second order rate equation, PSORE, by correspond the chemical sorption.

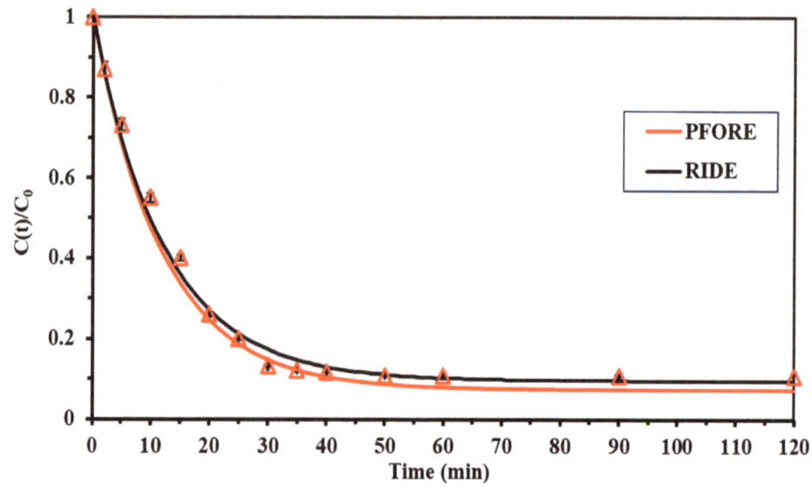

Figure 7. Uptake Kinetic models PFORE and RIDE that fitted the experimental data.

Hubbe et al. [74] concluded that the PSORE in a lot of cases was inferred with the control of mass transfer through resistance to intraparticle diffusion. Simonin [75] was declared the critical impacts of selection of experimental condition, which was responsible for the quality of the models used (i.e., PFORE and PSORE).

Table S1 summarizes the different equations which used in describing of the kinetic profiles. Table 4 reports the experimental values and the parameters for uranium sorption. The values of (R^2) of the determination coefficient and the comparisons of both experimental and calculated of the sorption parameters ($C(t)_{fitted}/C_0$ vs. $C(t)_{exp}/C_0$) are demonstrated that the PFORE and RIDE are more appropriate for kinetic profile modeling. This is also confirmed by the values of AIC, which indicating the lowest (negative) values obtained by these two models.

3.2.3. Sorption Isotherms

Figure 8 shows the plotted of sorption isotherms (with error bars) for U(VI) sorption from the mono component solution at pH_{in} 4 ± 0.1 (pH_{eq}: 4.1 ± 0.1) using CH-S sorbent. Figure S4 reports the Freundlich and Temkin models for fitting the sorption isotherms of U sorption. It was shown that the sorption capacities are progressively increased with concentration before tends to the saturation plateau (which seems to be high for U(VI) 1.54 mmol U g^{-1}).

The sorption of uranium is controlled by the pH of the solution as well as the uranium species (mainly UO_2^{2+}). The fitted equation for the isotherm profiles is the Langmuir equation. This equation is considered as a good fitting model of the experimental data over the entire range of the uranyl concentrations. The Freundlich equation (fails for fitting the sorption experiments either in the initial or saturation sections) is a power-like equation. Sips equation (has a good fit in the whole sections), which includes an adjustable parameter (n). From the data collected from Table 5 concerns with the R^2, variations coefficient of the affinity and AIC values (including the triplicate experiments), the preferred models for fitting the data in Figure 8 is the Langmuir and the Sips equations. The sorption capacity at saturation of the monolayer (i.e., $q_{m,L}$) slightly overestimates the experimental values.

Table 4. The parameters of the three models for U(VI) uptake kinetics for the three experiments.

Model	Parameter	1	2	3
Exp.	q_{eq} (mmol U g^{-1})	0.968	0.966	0.959
PFORE	$q_{eq,1}$ (mmol U g^{-1})	0.978	0.974	0.966
	$k_1 \times 10^2$ (min^{-1})	3.27	3.71	3.28
	R^2	0.995	0.979	0.989
	AIC	−136.6	−146.8	−144.9
PSORE	$q_{eq,2}$ (mmol U g^{-1})	0.802	0.796	0.817
	$k_2 \times 10^2$ (L $mmol^{-1}$ min^{-1})	11.7	9.87	10.68
	R^2	0.797	0.815	0.856
	AIC	−48.3	−39.7	−49.3
RIDE	$D_e \times 10^8$ (m^2 min^{-1})	2.19	1.97	2.05
	R^2	0.992	0.989	0.973
	AIC *	−137	−143.3	−144.5

* Where AIC means Akaike Information Criterion and discusses in Table S2.

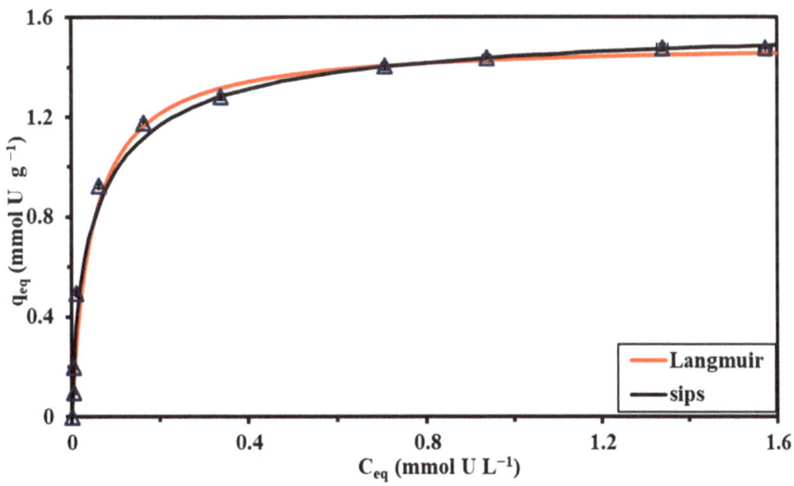

Figure 8. Sorption isotherms models for the experimental data of Langmuir and SIPS equations.

Table 6 reports the comparison of U(VI) sorption of some alternative sorbents. It is hard to establish such a comparison due to different functional groups (the bases used) and the detected solution (the source of uranium bearing solution, initial concentration of the uranium ions in either nature liquids or synthetic solution and the uranium salt used in the synthetic solution) but try to make select the matches condition for our work. The sorption capacity is relatively with high values (1.5 mmol U g^{-1}). Some sorbents were used at pH 6, with high loading capacity (4.48 mmol g^{-1}) using functionalized silica with diethylenetriamine tethered, but there is a risk in our cases of using such condition to avoid precipitation [76]. The other functionalized of mesoporous silica with Phosphonate groups [77] but with lower level of affinity and capacity. Others derived from marine fungus with capacity (about 1.56 mmol U g^{-1}) but exhibited a lower affinity coefficient (≈0.38 L $mmol^{-1}$) [78]. The sorbent represents a promising tool for extraction of uranium with respect the kinetic uptake, the capacity, and the affinity coefficients.

Table 5. Modeling of U(VI) sorption isotherms.

Model	Parameter		1	2	3
Experimental	$q_{m,exp.}$	mmol U g^{-1}	1.543	1.477	1.43
Langmuir	$q_{m,L}$	mmol U g^{-1}	1.572	1.483	1.451
	b_L	L $mmol^{-1}$	1.32	1.29	1.31
	R^2	-	0.993	0.984	0.979
	AIC	-	−144	−147	−169
Freundlich	k_F	$mmol^{1-1/n}$ g^{-1} $L^{1/n}$	0.85	0.79	0.68
	n_F	-	2.1	2.51	35
	R^2	-	0.894	0.911	0.928
	AIC	-	−74.3	−69.3	−58.4
Sips	$q_{m,S}$	mmol U g^{-1}	1.59	1.481	1.447
	b_S	L $mmol^{-1}$	1.17	1.08	1.14
	n_S	-	1.16	1.23	1.27
	R^2	-	0.993	0.989	0.981
	AIC	-	−152	−176	−185
Temkin	$A_T \times 10^{-3}$	L $mmol^{-1}$	22.56	30.2	28.1
	b_T	kJ mol^{-1}	17.4	16.4	15.3
	R^2	-	0.786	0.769	0.832
	AIC	-	−36	−46	−25

Table 6. U(VI) sorption properties of alternative sorbents.

Sorbent	pH	t_{eq}	$q_{m,exp}$	$q_{m,L}$	b_L	Ref.
Functionalized of the activated carbon	5	140	-	0.808	81.4	[79]
Phosphonated mesoporous silica	4	10	1.48	1.15	12.6	[77]
Duolite (ES-467)	3	90	-	0.326	10.9	[80]
Functionalized magnetic with amino and phosphine oxide composite	0.5	180	0.727	0.825	5.47	[81]
Functionalized polyethylene non-woven fabrics	4	720	0.084	0.087	119	[82]
Cyanobacterium	5	60	0.708	0.799	28.6	[83]
Manganese oxide functionalized with amine	4	20	-	0.416	19.0	[84]
Water hyacinth-biochar	6	720	0.571	0.663	407	[85]
201X8 ion exchanger	1.57	120	-	0.282	5.71	[86]
Trimesoyl chloride-melamine-palygorskite composite	6	75	-	1.16	0.95	[87]
Carboxylated chitosan nanotubes aerogels	5	150	1.07	1.29	23.6	[88]
activated carbon-Nano-HAP-alginate	6	480	0.042	0.078	0.95	[89]
Urea-formaldehyde resin	6	180	0.412	0.417	381	[90]
Chitosan/Amidoxime PAN/GO	6	120	1.04	1.24	169	[91]
Carboxymethyl konjac-glucomannan-gellan gum	6	720	0.411	0.428	65.0	[92]
Yarrowia lipolytica-alginate beads	7.5	90	-	0.102	2.86	[93]
Amidoxime-marine fungus	5	120	1.56	1.56	0.38	[78]
TiO_2-alginate	5	1440	0.105	0.132	-	[94]
Gum-ghatti polyacrylamide	6	240	-	1.54	29.8	[95]
Resin of carminic acid	5	120	0.807	0.808	36.3	[96]
Functionalized picolylamine resin	5.3	120	2.10	2.31	164	[97]
Chitosan film	5	600	0.735	0.827	223	[98]
Colloid montmorillonite	6	60	-	0.076	187	[99]
Impregnated bentonite	4	15	-	0.268	40.0	[100]
Dowex 50W X8/Alizarin Red-S	3	30	-	0.512	141	[101]
Acrylic Membrane/Algal fiber	3	60	0.260	0.268	9.05	[102]
Urea/SiO_2	4	90	1.14	1.17	21.3	[18]
Thiourea/SiO_2	4	90	1.02	1.16	5.98	[18]
Silica gel with β-cyclodextrin/	4.5	60	-	0.070	50.3	[103]
silica Functionalized MCM-41	5	40	1.89	2.00	850	[104]
PGG@C	4	60	1.28	1.32	47.1	[105]

Table 6. *Cont.*

Sorbent	pH	t_{eq}	$q_{m,exp}$	$q_{m,L}$	b_L	Ref.
PGG@MC	4	60	1.15	1.22	22.0	[105]
CH-S	4	30	1.53	1.55	1.3	This work

Units: t_{eq}: min; q_m: mmol U g^{-1}; b_L: L $mmol^{-1}$.

3.2.4. Sorption Mechanism

The collected information from the pH_{PZC} (global charge of the functional groups) and the sorption behavior during different pH values (depending on the speciation of uranyl ions), FTIR analysis (shifts and change the peaks resolution, which used in the interaction models; mainly for OH, SO_3H and NH), and the data of the EDX analyses (increasing the ratio of some components as S element during synthesis and sorption of uranyl ions, which appeared clearly). It is possible to conclude a different interaction mode of the reactive functional groups on CH-S sorbent and the uranyl ions. The interaction expected to perform through protons on either sulfonic acid, and hydroxyl groups beside the free electrons available on the amine groups, this interaction be mixed of ion exchange and chelating modes, especially that the sorbent is completely negative, and the optimism sorption pH as shown in Figure 9.

3.2.5. Metal Desorption and Recycling Properties

Desorption of adsorbed uranyl ions and recycling of the CH-S sorbent is a key of challenge in valorization and in term of metal recovery. This is the important step in evaluating of the competitiveness process. From FTIR spectra, it was shown the chemical stability of sorbent during recycling process and restoring the original peaks of the sorbent. The composite material produced from the kinetic experiments applied for desorption using 0.3 M HCl solution, which shows fast kinetics. The maximum time required for complete desorption is around 20 min as shown in Figure 10.

From Table 7 shows stability of sorption and desorption recycling which shows a remarkable stability in both sorption (with limited decreasing; around 3% after five cycles) and the desorption efficiency which remains stable (higher than 99.5%). This means that, the prepared composite is shown a remarkably stable of reuses.

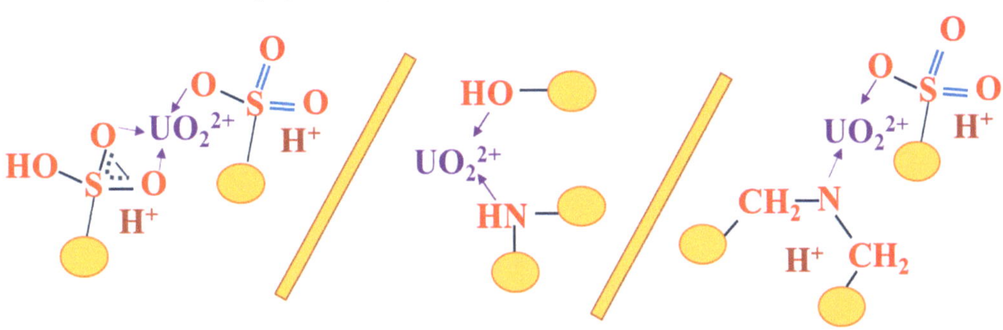

Figure 9. Suggest binding mechanism of functionalized sorbent toward U(VI) ions.

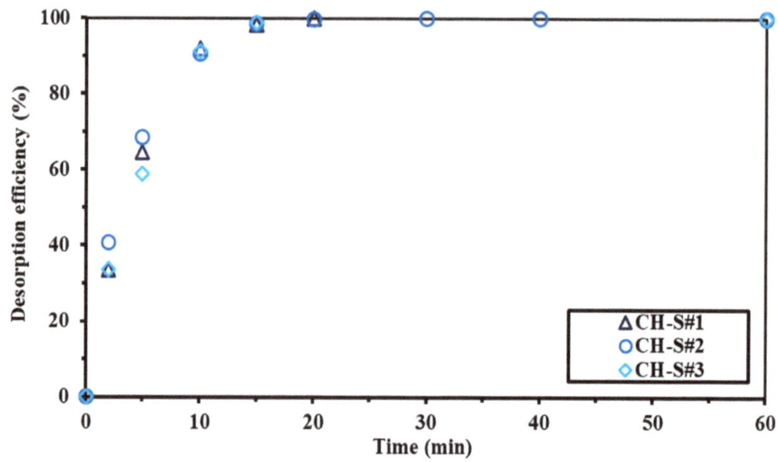

Figure 10. Desorption kinetics of uranium on CH-S sorbent.

Table 7. Sorption and desorption recycling for the sorbent toward uranium (VI) ions.

Cycles	Sorption		Desorption	
	Removal Effeciency (%)	S.D. (Re. %)	Desorption (%)	S.D. (De %)
1	97.6857	0.73187	99.846	0.22026
2	97.003	0.72855	99.893	0.14502
3	96.4471	0.79283	99.989	0.17285
4	96.1154	0.71282	99.672	0.12039
5	95.6361	0. 7312	99.864	0.17931

3.2.6. Results of the Uranium Extraction

Results of the uranium extraction applying the optimum leaching conditions upon 200 g of El Sela highly altered granitic rocks. Agitation leaching technique using sulfuric are more efficient than other techniques (percolation and pelletization). Treatment of the collected ore is sampled with 15% H_2SO_4 in the presence of 0.25 M NaCl with a 1/2 of solid/liquid ratio for 2 h at 90 °C, which yields a solution with uranium leaching efficiency of 89.7%. The produced leaching liquor from 200 g ore sample is about 0.6 L of sulfate solution. The produced solution contains 386 mg U L^{-1} with pH 0.3. The most interesting metal ions concentration is given in Table 8.

Table 8. Chemical composition of the prepared leach liquor at (pH = 0.3).

Constituants	Conc. (mgL^{-1})	Constituants	Conc. (mgL^{-1})
U	386	Al_2O_3	2540
REE	448	Pb	20
Fe	2700	Zr	29

The sorption was performed in batch method at different pH values 1 to 5. As pH increased, the sorption efficiency also increased. The capacity of uranium increased from 0.03, 0.156, 0.20, 0.224 and 0.253 mmol U g^{-1} from pH_{eq} 1.2, 2.3, 3.2, 4.1 to 4.8, respectively. The selectivity coefficient (SC) was increased to pH_{eq} 3.2 then decreased to pH_{eq} 4.8. The $SC_{U/Metal}$ was shown in Figure 11 at the optimum pH_{eq} 3.2, which is around 14.99, 6.67, 29.49, 37.68, 5.22 and 2.70 for Ca, Mg, Fe, Al, Zn, and Pb, respectively. This indicates the efficient uses of the modified chitosan for uranium removal from low concentrate solution.

This can be used for the heavy metal removal of Fe, Zn, and Pb from aquatic medium, which appear to have highly efficient sorption toward these ions.

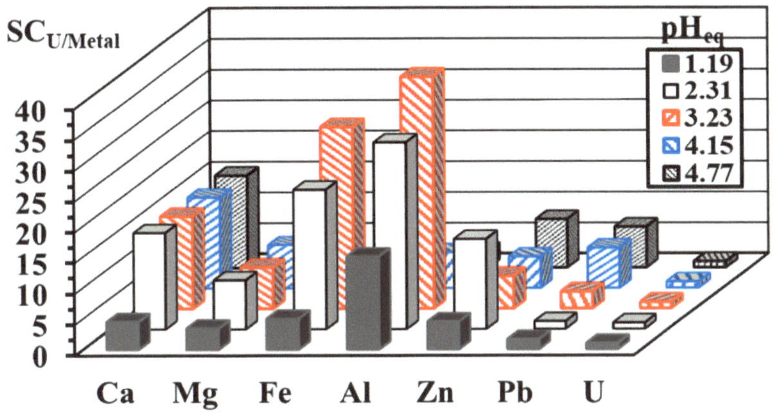

Figure 11. Selectivity studies of uranium sorption from acidic ore leachate.

4. Conclusions

Successive synthesis of novel sorbent based on chitosan particles for high performance uranium recovery was performed through functionalization of sulfonic groups (source; AMPS) on the chitosan surface via redox polymerization. The produced hydrogel was characterized by FTIR spectroscopy, TGA analysis, textural properties (BET surface area) through nitrogen adsorption method, pH_{PZC} (by pH-drift method), SEM, and SEM-EDX analyses. The most favorable pH is around 4 (prior to precipitation limit), the sorption kinetics is relatively fast; it is around 25 min is sufficient for total sorption, and 15 min for the total desorption with more than 99.9% of desorption efficiency. Using 0.3 M HCl solution is efficient eluent. The PFORE and RIDE are the most fitting model for kinetic uptake than the PSORE, while Langmuir and Sips equations for sorption isotherms. The sorbent was applied for uranium recovery from leachate waste solution and is considered as a promising tool for the recovery of uranium from acid leaching in polymetallic solution.

Supplementary Materials: The following supporting information can be downloaded at: https://www.mdpi.com/article/10.3390/toxics10090490/s1, Table S1: Reminder on equations used for modeling uptake kinetics [106,107]; Table S2: Reminder on equations used for modeling sorption isotherms [108–113]; Figure S1: FTIR spectra of most interested vibrational bands for CH, CH-S, after loading and after 5 cycles of sorption desorption process; Figure S2: Uranyl species with different pH values; Figure S3: The PSORE of the uptake kinetics; Figure S4: The Freundlich and Tamkin models for application to fit the sorption isotherms.

Author Contributions: Conceptualization, M.F.H., W.M.A., D.I.Z. and Y.W.; methodology, M.F.H., H.A.A.K. and M.S.K.; software, H.A.A.K., M.S.K., K.A. and Y.W.; validation and formal analysis, M.F.H., W.M.A. and D.I.Z.; investigation, and resources, M.F.H., H.A.A.K., M.S.K., K.A. and Y.W.; data curation, M.F.H., H.A.A.K., M.S.K., D.I.Z. and K.A.; writing—original draft preparation, M.F.H. and D.I.Z.; writing—review and editing, M.F.H., M.S.K. and Y.W.; visualization, H.A.A.K., W.M.A., D.I.Z. and K.A.; project administration, M.F.H. and Y.W.; funding acquisition, M.F.H. and Y.W. All authors have read and agreed to the published version of the manuscript.

Funding: This research received no external funding.

Institutional Review Board Statement: Not applicable.

Informed Consent Statement: Not applicable.

Data Availability Statement: Data available from authors.

Conflicts of Interest: The authors declare no conflict of interest.

References

1. Khalafalla, M.S. Biotechnological recovery of uranium (VI) from Abu Zeneima spent ore residue using green lixiviant. *J. Radioanal. Nucl. Chem.* **2022**, *331*, 2503–2513. [CrossRef]
2. Hamza, M.F. Uranium recovery from concentrated chloride solution produced from direct acid leaching of calcareous shale, Allouga ore materials, southwestern Sinai, Egypt. *J. Radioanal. Nucl. Chem.* **2018**, *315*, 613–626. [CrossRef]
3. Hamza, M.F. Grafting of quaternary ammonium groups for uranium (VI) recovery: Application on natural acidic leaching liquor. *J. Radioanal. Nucl. Chem.* **2019**, *322*, 519–532. [CrossRef]
4. Hamza, M.F.; Sallam, O.R.; Khalafalla, M.S.; Abbas, A.E.A.; Wei, Y. Geological and radioactivity studies accompanied by uranium recovery: Um Bogma Formation, southwestern Sinai, Egypt. *J. Radioanal. Nucl. Chem.* **2020**, *324*, 1039–1051. [CrossRef]
5. Hamza, M.F.; Hamad, N.A.; Hamad, D.M.; Khalafalla, M.S.; Abdel-Rahman, A.A.-H.; Zeid, I.F.; Wei, Y.; Hessien, M.M.; Fouda, A.; Salem, W.M. Synthesis of Eco-Friendly Biopolymer, Alginate-Chitosan Composite to Adsorb the Heavy Metals, Cd (II) and Pb (II) from Contaminated Effluents. *Materials* **2021**, *14*, 2189. [CrossRef] [PubMed]
6. Dousti, Z.; Dolatyari, L.; Yaftian, M.R.; Rostamnia, S. Adsorption of Eu(III), Th(IV), and U(VI) by mesoporous solid materials bearing sulfonic acid and sulfamic acid functionalities. *Sep. Sci. Technol.* **2019**, *54*, 2609–2624. [CrossRef]
7. Taha, M.H. Solid-liquid extraction of uranium from industrial phosphoric acid using macroporous cation exchange resins: MTC1600H, MTS9500, and MTS9570. *Sep. Sci. Technol.* **2020**, *56*, 1562–1578. [CrossRef]
8. Ahmad, M.; Yang, K.; Li, L.; Fan, Y.; Shah, T.; Zhang, Q.; Zhang, B. Modified tubular carbon nanofibers for adsorption of uranium(VI) from water. *ACS Appl. Nano Mater.* **2020**, *3*, 6394–6405. [CrossRef]
9. Hamza, M.F.; Fouda, A.; Wei, Y.; El Aassy, I.E.; Alotaibi, S.H.; Guibal, E.; Mashaal, N.M. Functionalized biobased composite for metal decontamination–Insight on uranium and application to water samples collected from wells in mining areas (Sinai, Egypt). *Chem. Eng. J.* **2022**, *431*, 133967. [CrossRef]
10. Chwastowski, J.; Staroń, P. Influence of Saccharomyces cerevisiae yeast cells immobilized on Cocos nucifera fibers for the adsorption of Pb (II) ions. *Colloids Surf. A Physicochem. Eng. Asp.* **2022**, *632*, 127735. [CrossRef]
11. Ma, F.Q.; Dong, B.R.; Gui, Y.Y.; Cao, M.; Han, L.; Jiao, C.S.; Lv, H.T.; Hou, J.J.; Xue, Y. Adsorption of low-concentration uranyl ion by amidoxime polyacrylonitrile fibers. *Ind. Eng. Chem. Res.* **2018**, *57*, 17384–17393. [CrossRef]
12. Wiechert, A.I.; Liao, W.-P.; Hong, E.; Halbert, C.E.; Yiacoumi, S.; Saito, T.; Tsouris, C. Influence of hydrophilic groups and metal-ion adsorption on polymer-chain conformation of amidoxime-based uranium adsorbents. *J. Colloid Interface Sci.* **2018**, *524*, 399–408. [CrossRef] [PubMed]
13. Wongjaikham, W.; Wongsawaeng, D.; Hosemann, P. Synthesis of amidoxime polymer gel to extract uranium compound from seawater by UV radiation curing. *J. Nucl. Sci. Technol.* **2019**, *56*, 541–552. [CrossRef]
14. Hamza, M.F.; Roux, J.-C.; Guibal, E. Uranium and europium sorption on amidoxime-functionalized magnetic chitosan micro-particles. *Chem. Eng. J.* **2018**, *344*, 124–137. [CrossRef]
15. Wei, Y.; Salih, K.A.; Lu, S.; Hamza, M.F.; Fujita, T.; Vincent, T.; Guibal, E. Amidoxime functionalization of algal/polyethyleneimine beads for the sorption of Sr (II) from aqueous solutions. *Molecules* **2019**, *24*, 3893. [CrossRef]
16. Hamza, M.F.; Mubark, A.E.; Wei, Y.; Vincent, T.; Guibal, E. Quaternization of composite algal/PEI beads for enhanced uranium sorption—application to ore acidic leachate. *Gels* **2020**, *6*, 12. [CrossRef]
17. Abu Khoziem, H.; Khalafalla, M.; Abdellah, W. Green recovery of uranium from Abu Zeneima mineralised carbonaceous shale, West Central Sinai, Egypt. *Int. J. Environ. Anal. Chem.* **2021**, 1–12. [CrossRef]
18. Hamza, M.F.; Wei, Y.; Khalafalla, M.S.; Abed, N.S.; Fouda, A.; Elwakeel, K.Z.; Guibal, E.; Hamad, N.A. U (VI) and Th (IV) recovery using silica beads functionalized with urea-or thiourea-based polymers–Application to ore leachate. *Sci. Total Environ.* **2022**, *821*, 153184. [CrossRef]
19. Hamza, M.F.; Salih, K.A.; Zhou, K.; Wei, Y.; Khoziem, H.A.A.; Alotaibi, S.H.; Guibal, E. Effect of bi-functionalization of algal/polyethyleneimine composite beads on the enhancement of tungstate sorption: Application to metal recovery from ore leachate. *Sep. Purif. Technol.* **2022**, *290*, 120893. [CrossRef]
20. Hamza, M.F.; Mira, H.; Wei, Y.; Aboelenin, S.M.; Guibal, E.; Salem, W.M. Sulfonation of chitosan for enhanced sorption of Li (I) from acidic solutions–Application to metal recovery from waste Li-ion mobile battery. *Chem. Eng. J.* **2022**, *441*, 135941. [CrossRef]
21. Fouda, A.; Hassan, S.E.-D.; Eid, A.M.; Abdel-Rahman, M.A.; Hamza, M.F. Light enhanced the antimicrobial, anticancer, and catalytic activities of selenium nanoparticles fabricated by endophytic fungal strain, Penicillium crustosum EP-1. *Sci. Rep.* **2022**, *12*, 11834. [CrossRef] [PubMed]
22. Hamza, M.F.; Abdel-Rahman, A.A.-H.; Negm, A.S.; Hamad, D.M.; Khalafalla, M.S.; Fouda, A.; Wei, Y.; Amer, H.H.; Alotaibi, S.H.; Goda, A.E.-S. Grafting of Thiazole Derivative on Chitosan Magnetite Nanoparticles for Cadmium Removal—Application for Groundwater Treatment. *Polymers* **2022**, *14*, 1240. [CrossRef] [PubMed]
23. Wu, H.Y.; Chi, F.T.; Zhang, S.; Wen, J.; Xiong, J.; Hu, S. Control of pore chemistry in metal-organic frameworks for selective uranium extraction from seawater. *Microporous Mesoporous Mater.* **2019**, *288*, 109567. [CrossRef]
24. Ang, K.L.; Li, D.; Nikoloski, A.N. The effectiveness of ion exchange resins in separating uranium and thorium from rare earth elements in acidic aqueous sulfate media. Part 2. Chelating resins. *Miner. Eng.* **2018**, *123*, 8–15. [CrossRef]

25. Hamza, M.F. Removal of uranium (VI) from liquid waste of calcareous shale, Allouga, southwestern Sinai, Egypt. *Desalin. Water Treat.* **2015**, *54*, 2530–2540. [CrossRef]
26. Hamza, M.F.; El Aassy, I.E. Solid phase extraction of uranium removal from underground water, Wadi Naseib, Southwestern Sinai, Egypt. *Desalin. Water Treat.* **2014**, *52*, 331–338. [CrossRef]
27. Aly, M.M.; Hamza, M.F. A review: Studies on uranium removal using different techniques. Overview. *J. Dispers. Sci. Technol.* **2013**, *34*, 182–213. [CrossRef]
28. Zahra, M.H.; Hamza, M.F.; El-Habibi, G.; Abdel-Rahman, A.A.-H.; Mira, H.I.; Wei, Y.; Alotaibi, S.H.; Amer, H.H.; Goda, A.E.-S.; Hamad, N.A. Synthesis of a Novel Adsorbent Based on Chitosan Magnetite Nanoparticles for the High Sorption of Cr (VI) ions: A Study of Photocatalysis and Recovery on Tannery Effluents. *Catalysts* **2022**, *12*, 678. [CrossRef]
29. WHO. *Guidelines for Drinking-Water Quality*, 4th ed.; World Health Organization: Geneva, Switzerland, 2011; p. 541.
30. Wilson, A.M.; Bailey, P.J.; Tasker, P.A.; Turkington, J.R.; Grant, R.A.; Love, J.B. Solvent extraction: The coordination chemistry behind extractive metallurgy. *Chem. Soc. Rev.* **2014**, *43*, 123–134. [CrossRef]
31. Hamza, M.F.; El-Aassy, I.E.; Guibal, E. Integrated treatment of tailing material for the selective recovery of uranium, rare earth elements and heavy metals. *Miner. Eng.* **2019**, *133*, 138–148. [CrossRef]
32. Hamza, M.F.; Khalafalla, M.S.; Wei, Y.; Hamad, N.A. Effect of bi-functionalization silica micro beads on uranium adsorption from synthetic and washing pregnant uranyl solutions. *J. Radioanal. Nucl. Chem.* **2021**, *330*, 191–206. [CrossRef]
33. Lai, Y.C.; Lee, W.J.; Huang, K.L.; Wu, C.M. Metal recovery from spent hydrodesulfurization catalysts using a combined acid-leaching and electrolysis process. *J. Hazard. Mater.* **2008**, *154*, 588–594. [CrossRef] [PubMed]
34. Hamza, M.F.; Ahmed, F.Y.; El-Aassy, I.; Fouda, A.; Guibal, E. Groundwater purification in a polymetallic mining area (SW Sinai, Egypt) using functionalized magnetic chitosan particles. *Water Air Soil Pollut.* **2018**, *229*, 1–14.
35. Fouda, A.; Hassan, S.E.-D.; Saied, E.; Azab, M.S. An eco-friendly approach to textile and tannery wastewater treatment using maghemite nanoparticles (γ-Fe_2O_3-NPs) fabricated by Penicillium expansum strain (Kw). *J. Environ. Chem. Eng.* **2021**, *9*, 104693. [CrossRef]
36. Fouda, A.; Hassan, S.E.-D.; Abdel-Rahman, M.A.; Farag, M.M.; Shehal-Deen, A.; Mohamed, A.A.; Alsharif, S.M.; Saied, E.; Moghanim, S.A.; Azab, M.S. Catalytic degradation of wastewater from the textile and tannery industries by green synthesized hematite (α-Fe_2O_3) and magnesium oxide (MgO) nanoparticles. *Curr. Res. Biotechnol.* **2021**, *3*, 29–41. [CrossRef]
37. Hamza, M.F.; Wei, Y.; Althumayri, K.; Fouda, A.; Hamad, N.A. Synthesis and Characterization of Functionalized Chitosan Nanoparticles with Pyrimidine Derivative for Enhancing Ion Sorption and Application for Removal of Contaminants. *Materials* **2022**, *15*, 4676. [CrossRef]
38. Hamza, M.F.; Alotaibi, S.H.; Wei, Y.; Mashaal, N.M. High-Performance Hydrogel Based on Modified Chitosan for Removal of Heavy Metal Ions in Borehole: A Case Study from the Bahariya Oasis, Egypt. *Catalysts* **2022**, *12*, 721. [CrossRef]
39. Hamza, M.F.; Wei, Y.; Benettayeb, A.; Wang, X.; Guibal, E. Efficient removal of uranium, cadmium and mercury from aqueous solutions using grafted hydrazide-micro-magnetite chitosan derivative. *J. Mater. Sci.* **2020**, *55*, 4193–4212. [CrossRef]
40. Hamza, M.F.; Gamal, A.; Hussein, G.; Nagar, M.S.; Abdel-Rahman, A.A.H.; Wei, Y.; Guibal, E. Uranium (VI) and zirconium (IV) sorption on magnetic chitosan derivatives–effect of different functional groups on separation properties. *J. Chem. Technol. Biotechnol.* **2019**, *94*, 3866–3882. [CrossRef]
41. Hamza, M.F.; Wei, Y.; Mira, H.I.; Abdel-Rahman, A.A.H.; Guibal, E. Synthesis and adsorption characteristics of grafted hydrazinyl amine magnetite-chitosan for Ni(II) and Pb(II) recovery. *Chem. Eng. J.* **2019**, *362*, 310–324. [CrossRef]
42. Haggag, E.S.A.; Khalafalla, M.S.; Masoud, A.M. Leaching kinetics of uranium, rare earth elements and copper using tartaric acid from El Allouga ore material, Southwestern Sinai, Egypt. *Int. J. Environ. Anal. Chem.* **2021**, 1–16. [CrossRef]
43. Abdellah, W.M.; Khalafalla, M.S.; Abu Khoziem, H.A.; El Hussain, O.M. Physical and chemical processes of Abu Rusheid cataclastic rocks for recovering niobium, zirconium and uranium compounds. *Physicochem. Probl. Miner. Process.* **2021**, *57*, 137–152. [CrossRef]
44. El Sheikh, R.; El Sheikh, E.M.; Khalafalla, M.S.; Mahfouz, L.I.; El-Gabry, M.M.; Gouda, A.A. Simultaneous preconcentration and determination of zirconium in environmental samples using ultrasound-assisted ionic liquid based dispersive liquid-liquid microextraction combined with spectrophotometry. *Int. J. Environ. Anal. Chem.* **2021**, 1–14. [CrossRef]
45. Khoziem, A.; Hanaa, A. Processing of Abu Dob mineralized pegmatites, Central Eastern Desert, Egypt: A study on the kinetics of dissolution process and extraction of some valuable metals. *J. Radioanal. Nucl. Chem.* **2022**, *331*, 937–951. [CrossRef]
46. Yeswanth, S.; Sekhar, K.C.; Chaudhary, A.; Sarma, P. Anti-microbial and Anti-biofilm activity of a novel Dibenzyl (benzo d thiazol-2-yl (hydroxy) methyl) phosphonate by inducing protease expression in Staphylococcus aureus. *Med. Chem. Res.* **2018**, *27*, 785–795. [CrossRef]
47. Davies, W.; Gray, U. A rapid and specific titrimetric method for the precise determination of uranium using iron (II) sulphate as reductant. *Talanta* **1964**, *11*, 1203–1211. [CrossRef]
48. Mathew, K.; Bürger, S.; Vogt, S.; Mason, P.; Morales-Arteaga, M.; Narayanan, U. Uranium assay determination using Davies and Gray titration: An overview and implementation of GUM for uncertainty evaluation. *J. Radioanal. Nucl. Chem.* **2009**, *282*, 939–944. [CrossRef]
49. Marczenko, Z.; Balcerzak, M. Chapter 39—Rare-earth elements. In *Analytical Spectroscopy Library*; Marczenko, Z., Balcerzak, M., Eds.; Elsevier: Amsterdam, The Netherlands, 2000; Volume 10, pp. 341–349.

50. Marczenko, Z.; Balcerzak, M. Chapter 54—Uranium. In *Analytical Spectroscopy Library*; Marczenko, Z., Balcerzak, M., Eds.; Elsevier: Amsterdam, The Netherlands, 2000; Volume 10, pp. 446–455.
51. Ibrahim, M.; Zalata, A.; Assaf, H.; Ibrahim, I.; Rashed, M. El Sella Shear Zone, South Eastern Desert, Egypt. Example of vein type uranium deposit. In Proceedings of the 9th International Mining, Petroleum, and Metallurgical Engering Conference, Cairo University, Giza, Egypt, 21–24 February 2005; pp. 41–55.
52. Ibrahim, T.; Amer, T.; Ali, K.; Omar, S. Uranium potentiality and its extraction from El Sela shear zone, south Eastern Desert Egypt. *Sci. Fac. Sci. Minufia Univ.* **2007**, *21*, 1–18.
53. Ali, K.G. Structural control of El Sela granites and associated uranium deposits, Southern Eastern Desert, Egypt. *Arab. J. Geosci.* **2013**, *6*, 1753–1767. [CrossRef]
54. Gaafar, I.M.; Aboelkhair, H.M.; Bayoumi, M.B. Integration of gamma-ray spectrometric and aster data for uranium exploration in Qash Amer-El-Sela area, Southeastern Desert, Egypt. *Nucl. Sci. Sci. J.* **2017**, *6*, 17–33. [CrossRef]
55. Gawad, A.E.A.; Orabi, A.H.; Bayoumi, M.M. Uranium evaluation and its recovery from microgranite dike at G. El Sela area, South Eastern Desert, Egypt. *Arab. J. Geosci.* **2015**, *8*, 4565–4580. [CrossRef]
56. Karim, M.A.; Gafaar, I.; El-Halim, A.; Hanfi, M.; El-Dine, N.W. Natural radioactivity and radiological implications of granite rocks, El-Sela area, Southeastern Desert, Egypt. *J. Radioanal. Nucl. Chem.* **2021**, *330*, 707–720. [CrossRef]
57. Gaafar, I.; Cuney, M.; Gawad, A.A. Mineral chemistry of two-mica granite rare metals: Impact of geophysics on the distribution of uranium mineralization at El Sela shear zone, Egypt. *Open J. Geol.* **2014**, *4*, 137. [CrossRef]
58. Shahin, H.; Bahige, M. Column Percolation Leaching of Uranium from El-Sela Area, South Eastern Desert, Egypt. *J. Chem.* **2016**, *5*, 32–41.
59. Fan, G.; Liao, C.; Fang, T.; Luo, S.; Song, G. Amberlyst 15 as a new and reusable catalyst for the conversion of cellulose into cellulose acetate. *Carbohydr. Polym.* **2014**, *112*, 203–209. [CrossRef]
60. Zhan, W.; Xu, C.H.; Qian, G.F.; Huang, G.H.; Tang, X.Z.; Lin, B.F. Adsorption of Cu(II), Zn(II), and Pb(II) from aqueous single and binary metal solutions by regenerated cellulose and sodium alginate chemically modified with polyethyleneimine. *RSC Adv.* **2018**, *8*, 18723–18733. [CrossRef]
61. Akkoz, Y.; Coskun, R.; Delibas, A. Preparation and characterization of sulphonated bio-adsorbent from waste hawthorn kernel for dye (MB) removal. *J. Mol. Liq.* **2019**, *287*, 11. [CrossRef]
62. Hamza, M.F.; Abdel-Rahman, A.A.H. Extraction studies of some hazardous metal ions using magnetic peptide resins. *J. Dispers. Sci. Technol.* **2015**, *36*, 411–422. [CrossRef]
63. Hamza, M.F.; Aly, M.M.; Abdel-Rahman, A.A.H.; Ramadan, S.; Raslan, H.; Wang, S.; Vincent, T.; Guibal, E. Functionalization of magnetic chitosan particles for the sorption of U(VI), Cu(II) and Zn(II)—Hydrazide derivative of glycine-grafted chitosan. *Materials* **2017**, *10*, 539. [CrossRef]
64. Coates, J. *Interpretation of Infrared Spectra, a Practical Approach*; John Wiley & Sons Ltd.: Chichester, UK, 2000.
65. Duarte, M.L.; Ferreira, M.C.; Marvao, M.R.; Rocha, J. An optimised method to determine the degree of acetylation of chitin and chitosan by FTIR spectroscopy. *Int. J. Biol. Macromol.* **2002**, *31*, 1–8. [CrossRef]
66. Lawrie, G.; Keen, I.; Drew, B.; Chandler-Temple, A.; Rintoul, L.; Fredericks, P.; Grondahl, L. Interactions between alginate and chitosan biopolymers characterized using FTIR and XPS. *Biomacromolecules* **2007**, *8*, 2533–2541. [CrossRef] [PubMed]
67. Coates, J. Interpretation of Infrared Spectra, A Practical Approach. In *Encyclopedia of Analytical Chemistry*; John Wiley & Sons, Ltd.: New York, NY, USA, 2006; pp. 1–23.
68. Imran, M.; Sajwan, M.; Alsuwayt, B.; Asif, M. Synthesis, characterization and anticoagulant activity of chitosan derivatives. *Saudi Pharm. J.* **2020**, *28*, 25–32. [CrossRef] [PubMed]
69. Corazzari, I.; Nistico, R.; Turci, F.; Faga, M.G.; Franzoso, F.; Tabasso, S.; Magnacca, G. Advanced physico-chemical characterization of chitosan by means of TGA coupled on-line with FTIR and GCMS: Thermal degradation and water adsorption capacity. *Polym. Degrad. Stabil.* **2015**, *112*, 1–9. [CrossRef]
70. Xiang, Y.; Yang, M.; Guo, Z.; Cui, Z. Alternatively chitosan sulfate blending membrane as methanol-blocking polymer electrolyte membrane for direct methanol fuel cell. *J. Membr. Sci.* **2009**, *337*, 318–323. [CrossRef]
71. Caetano, C.S.; Caiado, M.; Farinha, J.; Fonseca, I.M.; Ramos, A.M.; Vital, J.; Castanheiro, J.E. Esterification of free fatty acids over chitosan with sulfonic acid groups. *Chem. Eng. J.* **2013**, *230*, 567–572. [CrossRef]
72. Hamza, M.F.; Salih, K.A.M.; Abdel-Rahman, A.A.H.; Zayed, Y.E.; Wei, Y.; Liang, J.; Guibal, E. Sulfonic-functionalized algal/PEI beads for scandium, cerium and holmium sorption from aqueous solutions (synthetic and industrial samples). *Chem. Eng. J.* **2021**, *403*, 126399. [CrossRef]
73. Urbano, B.; Rivas, B.L. Poly(sodium 4-styrene sulfonate) and poly(2-acrylamido glycolic acid) polymer–clay ion exchange resins with enhanced mechanical properties and metal ion retention. *Polym. Int.* **2012**, *61*, 23–29. [CrossRef]
74. Hubbe, M.A.; Azizian, S.; Douven, S. Implications of apparent pseudo-second-order adsorption kinetics onto cellulosic materials: A review. *BioResources* **2019**, *14*, 45. [CrossRef]
75. Simonin, J.-P. On the comparison of pseudo-first order and pseudo-second order rate laws in the modeling of adsorption kinetics. *Chem. Eng. J.* **2016**, *300*, 254–263. [CrossRef]
76. Amesh, P.; Venkatesan, K.A.; Suneesh, A.S.; Samanta, N. Diethylenetriamine tethered mesoporous silica for the sequestration of uranium from aqueous solution and seawater. *J. Environ. Chem. Eng.* **2020**, *8*, 103995. [CrossRef]

77. Giannakoudakis, D.A.; Anastopoulos, I.; Barczak, M.; Alphantoniou, E.; Terpilowski, K.; Mohammadi, E.; Shams, M.; Coy, E.; Bakandritsos, A.; Katsoyiannis, I.A.; et al. Enhanced uranium removal from acidic wastewater by phosphonate-functionalized ordered mesoporous silica: Surface chemistry matters the most. *J. Hazard. Mater.* **2021**, *413*, 125279. [CrossRef]
78. He, D.X.; Tan, N.; Luo, X.M.; Yang, X.C.; Ji, K.; Han, J.W.; Chen, C.; Liu, Y.Q. Preparation, uranium (VI) absorption and reuseability of marine fungus mycelium modified by the bis-amidoxime-based groups. *Radiochim. Acta* **2020**, *108*, 37–49. [CrossRef]
79. Nezhad, M.M.; Semnani, A.; Tavakkoli, N.; Shirani, M. Efficient removal and recovery of uranium from industrial radioactive wastewaters using functionalized activated carbon powder derived from zirconium carbide process waste. *Environ. Sci. Pollut. Res.* **2021**, *28*, 57073–57089. [CrossRef] [PubMed]
80. Zidan, I.H.; Cheira, M.F.; Bakry, A.R.; Atia, B.M. Potentiality of uranium recovery from G.Gattar leach liquor using Duolite ES-467 chelating resin: Kinetic, thermodynamic and isotherm features. *Int. J. Environ. Anal. Chem.* **2020**, *102*, 2102–2124. [CrossRef]
81. Zhang, S.; Yuan, D.; Zhang, Q.; Wang, Y.; Liu, Y.; Zhao, J.; Chen, B. Highly efficient removal of uranium from highly acidic media achieved using a phosphine oxide and amino functionalized superparamagnetic composite polymer adsorbent. *J. Mater. Chem. A* **2020**, *8*, 10925–10934. [CrossRef]
82. Zhang, M.; Yuan, M.; Zhang, M.; Wang, M.; Chen, J.; Li, R.; Qiu, L.; Feng, X.; Hu, J.; Wu, G. Efficient removal of uranium from diluted aqueous solution with hydroxypyridone functionalized polyethylene nonwoven fabrics. *Radiat. Phys. Chem.* **2020**, *171*, 108742. [CrossRef]
83. Yuan, Y.; Liu, N.; Dai, Y.; Wang, B.; Liu, Y.; Chen, C.; Huang, D. Effective biosorption of uranium from aqueous solution by cyanobacterium *Anabaena Flos-aquae*. *Environ. Sci. Pollut. Res.* **2020**, *27*, 44306–44313. [CrossRef]
84. Yousef, L.A.; Bakry, A.R.; Ahmad, A.A. Uranium(VI) recovery from acidic leach liquor using manganese oxide coated zeolite (MOCZ) modified with amine. *J. Radioanal. Nucl. Chem.* **2020**, *324*, 409–421. [CrossRef]
85. Xu, Z.; Xing, Y.; Ren, A.; Ma, D.; Li, Y.; Hu, S. Study on adsorption properties of water hyacinth-derived biochar for uranium (VI). *J. Radioanal. Nucl. Chem.* **2020**, *324*, 1317–1327. [CrossRef]
86. Wen, Z.; Huang, K.; Niu, Y.; Yao, Y.; Wang, S.; Cao, Z.; Zhong, H. Kinetic study of ultrasonic-assisted uranium adsorption by anion exchange resin. *Colloids Surf. A* **2020**, *585*, 124021. [CrossRef]
87. Tuzen, M.; Saleh, T.A.; Sari, A. Naeemullah Interfacial polymerization of trimesoyl chloride with melamine and palygorskite for efficient uranium ions ultra-removal. *Chem. Eng. Res. Des.* **2020**, *159*, 353–361. [CrossRef]
88. Tang, X.; Zhou, L.; Le, Z.; Wang, Y.; Liu, Z.; Huang, G.; Adesina, A.A. Preparation of porous chitosan/carboxylated carbon nanotube composite aerogels for the efficient removal of uranium(VI) from aqueous solution. *Int. J. Biol. Macromol.* **2020**, *160*, 1000–1008. [CrossRef] [PubMed]
89. Saha, S.; Basu, H.; Rout, S.; Pimple, M.V.; Singhal, R.K. Nano-hydroxyapatite coated activated carbon impregnated alginate: A new hybrid sorbent for uranium removal from potable water. *J. Environ. Chem. Eng.* **2020**, *8*, 103999. [CrossRef]
90. Ma, D.; Wei, J.; Zhao, Y.; Chen, Y.; Tang, S. The removal of uranium using novel temperature sensitive urea-formaldehyde resin: Adsorption and fast regeneration. *Sci. Total Environ.* **2020**, *735*, 139399. [CrossRef]
91. Lu, W.; Dai, Z.; Li, L.; Liu, J.; Wang, S.; Yang, H.; Cao, C.; Liu, L.; Chen, T.; Zhu, B.; et al. Preparation of composite hydrogel (PCG) and its adsorption performance for uranium(VI). *J. Mol. Liq.* **2020**, *303*, 112604. [CrossRef]
92. Liang, L.; Lin, X.; Liu, Y.; Sun, S.; Chu, H.; Chen, Y.; Liu, D.; Luo, X.; Zhang, J.; Shang, R. Carboxymethyl konjac glucomannan mechanically reinforcing gellan gum microspheres for uranium removal. *Int. J. Biol. Macromol.* **2020**, *145*, 535–546. [CrossRef]
93. Kolhe, N.; Zinjarde, S.; Acharya, C. Removal of uranium by immobilized biomass of a tropical marine yeast Yarrowia lipolytica. *J. Environ. Radioact.* **2020**, *223–224*, 106419. [CrossRef]
94. Basu, H.; Pimple, M.V.; Saha, S.; Patel, A.; Dansena, C.; Singhal, R.K. TiO_2 microsphere impregnated alginate: A novel hybrid sorbent for uranium removal from aquatic bodies. *New J. Chem.* **2020**, *44*, 3950–3960. [CrossRef]
95. Shelar-Lohar, G.; Joshi, S. Comparative study of uranium and thorium metal ion adsorption by gum ghatti grafted poly(acrylamide) copolymer composites. *RSC Adv.* **2019**, *9*, 41326–41335. [CrossRef]
96. Rahmani-Sani, A.; Hosseini-Bandegharaei, A.; Hosseini, S.H.; Kharghani, K.; Zarei, H.; Rastegar, A. Kinetic, equilibrium and thermodynamic studies on sorption of uranium and thorium from aqueous solutions by a selective impregnated resin containing carminic acid. *J. Hazard. Mater.* **2015**, *286*, 152–163. [CrossRef]
97. Liu, S.; Yang, Y.; Liu, T.; Wu, W. Recovery of uranium(VI) from aqueous solution by 2-picolylamine functionalized polystyrene-co-maleic anhydride) resin. *J. Colloid Interface Sci.* **2017**, *497*, 385–392. [CrossRef] [PubMed]
98. Wang, Y.; Li, Y.; Li, L.; Kong, F.; Lin, S.; Wang, Z.; Li, W. Preparation of three-dimensional fiber-network chitosan films for the efficient treatment of uranium-contaminated effluents. *Water Sci. Technol.* **2020**, *81*, 52–61. [CrossRef] [PubMed]
99. Yu, S.; Ma, J.; Shi, Y.; Du, Z.; Zhao, Y.; Tuo, X.; Leng, Y. Uranium(VI) adsorption on montmorillonite colloid. *J. Radioanal. Nucl. Chem.* **2020**, *324*, 541–549. [CrossRef]
100. Salah, B.A.; Gaber, M.S.; Kandil, A.H.T. The removal of uranium and thorium from their aqueous solutions by 8-hydroxyquinoline immobilized bentonite. *Minerals* **2019**, *9*, 626. [CrossRef]
101. Yousef, L.A.; Ahmad, A.A.; Bakry, A.R. Separation of uranium ions from acetate medium by Dowex50WX8/Alizarin Red-S and its application on granitic samples, South Um Tawat, Eastern Desert. *Int. J. Environ. Anal. Chem.* **2020**, *8*, 667–681. [CrossRef]
102. Orabi, A.H.; Abdelhamid, A.E.-S.; Salem, H.M.; Ismaiel, D.A. New adsorptive composite membrane from recycled acrylic fibers and Sargassum dentifolium marine algae for uranium and thorium removal from liquid waste solution. *J. Radioanal. Nucl. Chem.* **2020**, *326*, 1233–1247. [CrossRef]

103. Liu, H.-J.; Jing, P.-F.; Liu, X.-Y.; Du, K.-J.; Sun, Y.-K. Synthesis of beta-cyclodextrin functionalized silica gel and its application for adsorption of uranium(VI). *J. Radioanal. Nucl. Chem.* **2016**, *310*, 263–270. [CrossRef]
104. Bayramoglu, G.; Arica, M.Y. MCM-41 silica particles grafted with polyacrylonitrile: Modification in to amidoxime and carboxyl groups for enhanced uranium removal from aqueous medium. *Microporous Mesoporous Mater.* **2016**, *226*, 117–124. [CrossRef]
105. Hamza, M.F.; Fouda, A.; Elwakeel, K.Z.; Wei, Y.; Guibal, E.; Hamad, N.A. Phosphorylation of guar gum/magnetite/chitosan nanocomposites for uranium (VI) sorption and antibacterial applications. *Molecules* **2021**, *26*, 1920. [CrossRef]
106. Ho, Y.S.; McKay, G. Pseudo-second order model for sorption processes. *Process Biochem.* **1999**, *34*, 451–465. [CrossRef]
107. Crank, J. *The Mathematics of Diffusion*, 2nd ed.; Oxford University Press: Oxford, UK, 1975; p. 414.
108. Langmuir, I. The adsorption of gases on plane surfaces of glass, mica and platinum. *J. Amer. Chem. Soc.* **1918**, *40*, 1361–1402.
109. Freundlich, H.M.F. Uber die adsorption in lasungen. *Z. Phys. Chem.* **1906**, *57*, 385–470.
110. Tien, C. *Adsorption Calculations and Modeling*; Butterworth-Heinemann: Newton, MA, USA, 1994; p. 243.
111. Kegl, T.; Kosak, A.; Lobnik, A.; Novak, Z.; Kralj, A.K.; Ban, I. Adsorption of rare earth metals from wastewater by nano-materials: A review. *J. Hazard. Mater.* **2020**, *386*, 121632. [CrossRef]
112. Puccia, V.; Avena, M.J. On the use of the Dubinin-Radushkevich equation to distinguish between physical and chemical ad-sorption at the solid-water interface. *Colloid Interface Sci. Commun.* **2021**, *41*, 100376.
113. Falyouna, O.; Eljamal, O.; Maamoun, I.; Tahara, A.; Sugihara, Y. Magnetic zeolite synthesis for efficient removal of cesium in a lab-scale continuous treatment system. *J. Colloid Interface Sci.* **2020**, *571*, 66–79. [CrossRef]

toxics

Article

Chitosan-Modified Biochar and Unmodified Biochar for Methyl Orange: Adsorption Characteristics and Mechanism Exploration

Nguyen Xuan Loc [1,†], Phan Thi Thanh Tuyen [1], Le Chi Mai [2] and Do Thi My Phuong [2,*,†]

1 Department of Environmental Sciences, College of the Environment and Natural Resources, Can Tho University, Can Tho 900000, Vietnam
2 Department of Environmental Engineering, College of the Environment and Natural Resources, Can Tho University, Can Tho 900000, Vietnam
* Correspondence: dtmphuong@ctu.edu.vn; Tel.: +84-91-918-8834
† These authors contributed equally to this work.

Abstract: In this study, shrimp shell-derived chitosan (CS) and rice husk-derived biochar (RHB) were produced; CS and RHB were then used to synthesize chitosan-modified biochar (CSBC) hydrogel beads. N_2 adsorption (77 K), SEM-EDX and FT-IR techniques were used to evaluate the physicochemical properties of the adsorbents. A batch experiment was conducted to test the methyl orange (MO) adsorption performance of RHB and CSBC. The results showed that the MO adsorption process was strongly pH-dependent. The kinetics were well described by the pseudo-second-order and intra-particle diffusion models, assuming the chemisorption and intraparticle diffusion mechanisms govern the adsorption process. Homogeneous adsorption for MO on the surface of RHB and CSBC was also assumed since the isotherm data showed the best-fit to the Langmuir model. Under the experimental conditions of initial pH 3, dosage 0.2 g, contact time 240 min and temperature 298 K, the maximum adsorption capacity of CSBC and RHB for MO dye adsorption was 38.75 mg.g^{-1} and 31.63 mg.g^{-1}, respectively. This result demonstrated that biochar had better performance after modification with chitosan, which provided more functional groups (i.e., $-NH_2$ and $-OH$ groups) for enhanced electrostatic interactions and complexation between MO and CSBC. Overall, CSBC is an effective adsorbent for the removal of MO from aqueous solution.

Keywords: adsorption; biochar; chitosan; modification; rice husk; methyl orange

Citation: Loc, N.X.; Tuyen, P.T.T.; Mai, L.C.; Phuong, D.T.M. Chitosan-Modified Biochar and Unmodified Biochar for Methyl Orange: Adsorption Characteristics and Mechanism Exploration. *Toxics* **2022**, *10*, 500. https://doi.org/10.3390/toxics10090500

Academic Editors: Yuezhou Wei, Yan Wu and Xinpeng Wang

Received: 2 August 2022
Accepted: 24 August 2022
Published: 27 August 2022

Publisher's Note: MDPI stays neutral with regard to jurisdictional claims in published maps and institutional affiliations.

Copyright: © 2022 by the authors. Licensee MDPI, Basel, Switzerland. This article is an open access article distributed under the terms and conditions of the Creative Commons Attribution (CC BY) license (https://creativecommons.org/licenses/by/4.0/).

1. Introduction

Dye, a complex and soluble organic compound, is commonly used to impart color to fibrous materials. Worldwide, more than 10,000 forms of dyes and an estimated 700,000 tons are produced every year [1,2]. Particularly in the textile industry, roughly 200,000 tons of dyes are annually expelled to effluents during production and application [1]. The high-colored effluents are highly toxic to the environment and humans if they discharge into water bodies without treatment. Most complex synthetic dyes can persist as stable aquatic pollutants, impede the photosynthetic activity of algae, cause depletion of dissolved oxygen in water bodies and thereby pose a threat to aquatic species. They are also categorized as possibly toxic, mutagenic and carcinogenic to humans [3]. Methyl orange (MO) is a common water-soluble and typical azo anionic dye (commonly known as a pH indicator) that is one of the most common dyes used in the textile industry [4,5]. MO contains aromatic and –N = N– groups in its molecules, which are highly toxic, carcinogenic, teratogenic and harmful to the environment and organisms [6]. Thus, MO is identified as a potential organic pollutant that requires proper treatment before its discharge into natural water bodies.

Various physical and chemical biological methods have been employed for the removal of dyes from textile effluents, such as the advanced oxidation process [6], membrane filtration [7], electrochemical oxidation [8], ozonation [9] and coagulation/flocculation-based methods [10]. However, these techniques have their own limitations in terms of complicated design, low removal performance and high operating and maintenance costs. Adsorption is a widely accepted technology due to its high effectiveness, easy operation, inexpensive cost and the possibility of scaling-up to an industrial scale from a laboratory or a pilot scale [11]. Nevertheless, the efficiency of an adsorption process is largely dependent on the adsorbent used in the process [12]. A variety of adsorbents have been developed and used for dye removal, including activated carbon [13], zeolite [14], biochar [15], clay [16], metal–organic framework [17], natural materials [18], agricultural wastes [19] and synthesized products [20]. Biochar-based materials have been widely endorsed as effective and eco-friendly adsorbents for the removal of a range of heavy metals and organic contaminants thanks to their high surface functional groups, high porous structure and large surface area [21,22]. A study by Bussan et al. (2016) used biochar to reduce mercury methylation rates in aquatic sediments [23]. Nevertheless, the heterogeneous nature, high pH and negatively charged surface of biochar make biochar an inconsistent adsorbent [24]. To address this issue, a number of physical, mechanical and chemical modification processes of biochar have been developed in recent years [25]. Many studies have shown that the incorporation of foreign materials (such as zeolite, silica, polymers and metal oxides) into the biochar matrix will enhance functional groups, recalcitrance potential and adsorption efficiencies of the precursor biochar [26,27]. For instance, Fang et al. (2020) used Mg/Ca-modified biochar from peanut shells and achieved a high amount of phosphate (129.79 $mg.g^{-1}$) to be removed from the acid-extract of incinerated sewage sludge ash [28]. In our previous study, a higher adsorption (41.59 $mg.g^{-1}$) of Safranin O was recorded using rice straw biochar modified with iron oxide [29].

Chitosan is a biodegradable and renewable polymer with amazing characteristics of abundantly availability, cost-effectiveness, good biocompatibility and biodegradation, nontoxic nature and a broad range of applications [30]. This polymer is derived from a nitrogenous polysaccharide called chitin, the second most abundant natural polysaccharide, which is extracted from the exoskeleton of invertebrates such as insects or crustaceans. Chitosan is a linear copolymer of β-1,4-linked 2-amino-2-deoxy-β-D-glucose (deacetylated D-glucosamine) and N-acetyl-D-glucosamine units. In general, chitosan possesses two essential functional groups, including (1) free amino (-NH_2) groups (existing in the D-glucosamine units) and (2) free hydroxyl (-OH) groups (attaching to both the N-acetyl-D-glucosamine units and D-glucosamine units). These groups can take on a positive charge (NH_3^+ or H_3O^+) which creates a good chelating ligand capable of binding to metal ions [31] or electrostatic attraction to dye anions [32]. Therefore, chitosan has also received much attention as a promising adsorbent to remove metal ions [33,34] or ionic dyes [35,36] from aqueous solution.

There is increasing interest in modifying biochar with foreign materials such as chitosan since this modification benefits from the combined advantages provided by biochar (high porous network and large surface area) and chitosan (high chemical affinity) [37]. Furthermore, chitosan can be utilized as an organic glue to attach pollutants on the surface of biochar and thus, the adsorption capacities of biochar modified by chitosan to various pollutants are generally improved [38]. For example, chitosan-coated onto biochar surfaces showed an adsorption capacity for phloridzin (28.96 $mg.g^{-1}$), nearly double the unmodified biochar derived from apple branches (15.93 $mg.g^{-1}$) [39]. Additionally, the modification of chitosan on biochar surface has been shown to have greater adsorption capacities for Cd(II) and Pb(II) ions than unmodified biochar [40,41].

The adsorption applications of chitosan-modified biochar hydrogel beads to heavy metal ions have been explored in several studies [42–44]. For dye ions, chitosan-modified biochar hydrogel beads used for Malachite Green and Rhodamine B removal have also been reported [45]. However, the effectiveness of chitosan-modified biochar hydrogel beads for methyl orange removal from contaminated aqueous media has yet to be explored. Therefore, in this study, biochar was prepared from rice husk, chitosan was derived from shrimp shell and then rice husk biochar (RHB) was modified with chitosan (CS) to form chitosan-modified biochar (CSBC) hydrogel beads. The physicochemical characteristics of the synthesized RHB and CSBC were first evaluated using different techniques and measurements, including SEM, EDX, FTIR techniques, N_2 adsorption (77 K) and pH_{pzc} measurements. Then, their adsorption capacities toward methyl orange from aqueous solutions were tested using batch experiments, where the pH solution, the mass of the adsorbent, the contact time and the initial MO concentration were varied. Finally, the adsorption mechanism was explored using various isotherm and kinetic models.

2. Materials and Methods

2.1. Chemicals

All chemicals, including acetic acid (CH_3COOH), hydrochloric acid (HCl) and sodium hydroxide (NaOH), were provided by Merck (Germany). Methyl orange (MO) was supplied by Sigma-Aldrich.

2.2. Chitosan, Biochar and Chitosan-Modified Biochar Production

To synthesize chitosan-modified biochar, shrimp shell-derived chitosan and rice husk-derived biochar were firstly produced, and then biochar was modified with chitosan to form chitosan-modified biochar hydrogel beads following the diagram displayed in Figure 1. For the extraction of chitosan from shrimp shells, raw material *Penaeus monodon* shrimp shell waste was collected from a local seafood company in the Mekong Delta and the procedure of Radwan et al. (2012) was followed [46]. For the pyrolysis of biochar from rice husk, raw rice husk (*OM5451* rice variety) was also collected from a local rice-milling factory in the Mekong Delta and the procedure provided in ref. [29] was followed. The solid collected at the end of this step was identified as rice husk biochar (RHB). Finally, chitosan-modified biochar adsorptive beads were prepared following the protocol summarized by Dewage et al. (2018) [43]. The hydrogel beads collected at the end of this step were identified as chitosan-modified biochar (CSBC).

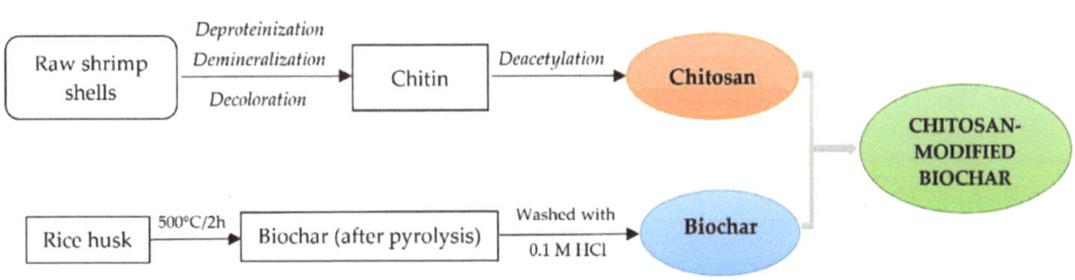

Figure 1. Schematic diagram of preparation process of chitosan-modified biochar.

2.3. Characterization of Adsorbents

The micrographs and elemental compositions of the RHB and CSBC adsorbents were evaluated by scanning electron microscopy with energy dispersive X-ray spectroscopy (SEM-EDX Hitachi S4800, Hitachi Ltd., Tokyo, Japan), while their functional groups were analyzed using Fourier transform infrared spectroscopy (FTIR-PerkinElmer Spectrum 10.5.2). The textural characterization of RHB and CSBC was obtained from the nitrogen adsorption–desorption isotherms at 77 K in a Nova Station A (Quantachrome Instruments version 11.0, Miami, FL, USA). The point of zero charge (pH_{pzc}) was examined for the pH range of 2.0 to 12.0 using 0.1 M NaCl [25].

2.4. Sorption Batch Trials

The batch adsorption experiments were used to evaluate the effect of solution pH (3–10), adsorbent dose (0.1–0.5 g), initial MO concentration (10–200 $mg.g^{-1}$) and adsorption time (1–720 min). The adsorption experiment was performed using #15 mL conical centrifuge tubes, in which a fixed amount of RHB or CSBC was added to a 10 mL MO solution. The mixtures were then agitated in a rotary shaker (HS 250 Basic, IKA Labortechnik) at an agitation speed of 120 rpm and under room conditions (25 ± 2 °C) for a fixed time. After filtering with Whatman No. 6 filter paper, UV-Vis spectroscopy (Shimazdu UV-1900, Kyoto, Japan) was used to analyze the residual MO in the remaining solution.

The amount of MO adsorbed (q_e, mg.g^{-1}) was determined according to the following Equation (1):

$$q_e = \frac{C_0 - C_e}{m} V, \quad (1)$$

where C_0 (mg.L^{-1}) is the initial MO concentration; C_e (mg.L^{-1}) is the MO concentration at equilibrium; m (g) is the mass of adsorbent; V (mL) is the volume of the solution.

All the experiments in this study were conducted in triplicate (n = 3), and data were expressed as the mean ± standard deviation.

2.4.1. Kinetic Modelling

Pseudo-first-order, pseudo-second-order and intraparticle diffusion models were used to analyze the kinetic data of MO on RHB and CSBC. Their equations are shown in Equations (2)–(4):

$$\text{Pseudo-first-order: } q_t = q_e(1 - exp^{-k1t}), \quad (2)$$

where q_e (mg.g^{-1}) is the adsorption capacity at equilibrium; q_t (mg.g^{-1}) is the adsorption capacity at a given t time; k_1 (1.min^{-1}) is the constant rate.

$$\text{Pseudo-second-order}: \frac{dq_t}{dt} = k_2(q_e - q_t)^2, \quad (3)$$

where k_2 (g.mg^{-1}.min^{-1}) is the constant rate.

$$\text{Intraparticle diffusion: } q_t = k_{id}t^{1/2} + C, \quad (4)$$

where C is the intercept; k_{id} (mg.g^{-1}.min$^{-1/2}$) is the intraparticle diffusion rate constant.

2.4.2. Isotherm Modelling

The equilibrium adsorption data were simulated using the Langmuir and Freundlich isotherm models. Their equations are shown in Equations (5) and (6):

$$\text{Langmuir}: q_e = \frac{q_m K_L C_e}{1 + K_L C_e}, \quad (5)$$

where q_e (mg.g^{-1}) is the adsorption capacity at equilibrium; q_m (mg.g^{-1}) is the theoretical maximum adsorption capacity; C_e (mg.L^{-1}) is the equilibrium concentration of the adsorbate; K_L (L.mg^{-1}) is the Langmuir adsorption constant.

$$\text{Freundlich: } q_e = K_F\, C_e^{1/n}, \quad (6)$$

where K_F ((mg.kg^{-1})/(mg.L^{-1})^n) is the Freundlich adsorption constant; $1/n$ represents the intensity of adsorption (unitless).

3. Results

3.1. Physicochemical Properties of RHB and CSBC

The textural properties such as surface area, average pore size and pore volume were estimated from well-known calculation methods such as the Brunauer–Emmett–Teller (BET) and the Barrett– Joyner–Halenda (BJH) models. The nitrogen adsorption–desorption curves on RHB and CSBC are shown in Figure 2A, while the pore size distributions of RHB and CSBC estimated by the BJH method from desorption branches are shown in Figure 2B. In depth, at a low relative pressure $P/P_0 \leq 0.1$, the isotherms of RHB and CSBC exhibit the typical type I characteristic, where nitrogen adsorption increases linearly with the increase of relative pressure, indicating the existence of micropores. At medium relative pressures $0.5 < P/P_0 < 0.9$, an approximately closed hysteresis loop is formed and the isotherms of RHB and CSBC exhibit the type IV characteristic, demonstrating the existence of mesopores [28]. The desorption curves of RHB and CSBC are almost consistent with their adsorption curves, with differences existing at mid to high P/P_0. According to the classification by the International Union of Pure and Applied Chemistry (IUPAC), the nitrogen adsorption–desorption isotherms of RHB and CSBC can be classified as being of type IV shapes and type H4 hysteresis loops, which indicates the existence of slit-shaped pores, attributed to mesoporosity [47]. This result can be further confirmed by the corresponding pore size distribution, as displayed in Figure 2B. The prepared RHB and CSBC possessed a mesoporous surface with a mean pore radius of approximately 2.26 nm and 2.34 nm, respectively.

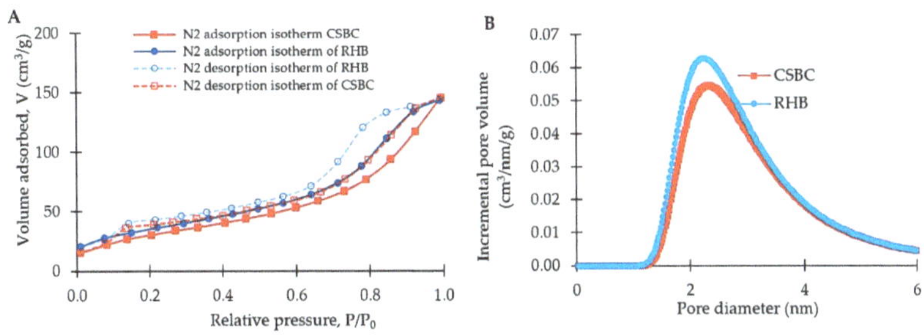

Figure 2. Nitrogen adsorption–desorption isotherms at 77 K (**A**) and pore size distribution (**B**) of RHB and CSBC.

It is obvious that the RHB had a higher adsorbed amount of nitrogen than the CSBC, implying that the RHB possesses a higher specific surface area and pore volume than CSBC. As a matter of fact, the specific surface area calculated using the BET equation was 115.59 $m^2.g^{-1}$ for RHB, slightly higher than that of CSBC with 107.97 $m^2.g^{-1}$. Chitosan-modified biochar produced from other feedstocks also exhibit lower surface areas than their unmodified biochar [37,43,45]. This is probably the reason for the blockage of partial pores by the incorporation of chitosan in the biochar matrix or the changes in the chemical compositions of the biochar surface so that nitrogen has less affinity for surface adsorption [37].

The functional groups on the RHB and CSBC were detected through FT-IR analysis. Figure 3 show that the FT-IR spectra of CSBC were somewhat different from those of RHB. In particular, the FT-IR spectra of RHB in this study were similar to other high-temperature biochar derived from rice husk [48,49], which possess several types of functional groups, including O–H (3437 cm^{-1}),—COOH (1606 cm^{-1}), C–H aliphatic (1369 cm^{-1}), C—O—C (1103 cm^{-1}) and Si-O (465 cm^{-1}) groups [50]. Meanwhile, characteristic bands for CSBC were observed in the infrared spectrum around 3849 cm^{-1} and 3423 cm^{-1}, attributed to amine N–H symmetrical vibration and an H-bonded O–H group. In addition to these functional groups, the FT-IR spectrum of CSBC showed other typical infrared spectrums for chitosan, for example, sp^3-hybridized C–H stretching vibrations of carbohydrate ring appeared at peak 2868 cm^{-1} [51], the amide II band (the N–H stretching vibrations) appeared at peak 1613 cm^{-1} [43] and the amide III band (the C–N or N–H stretching vibrations) appeared at peak 1378 cm^{-1} [52,53].

The SEM morphological structures of CSBC and RHB are shown in Figure 4A. It is obvious that the surface morphology of precursor biochar significantly changed after modification with chitosan. The external morphology of CSBC generally possesses a highly irregular surface compared to the well-defined cylindrical pores of the RHB surface. These results were similar to other observations [40,54].

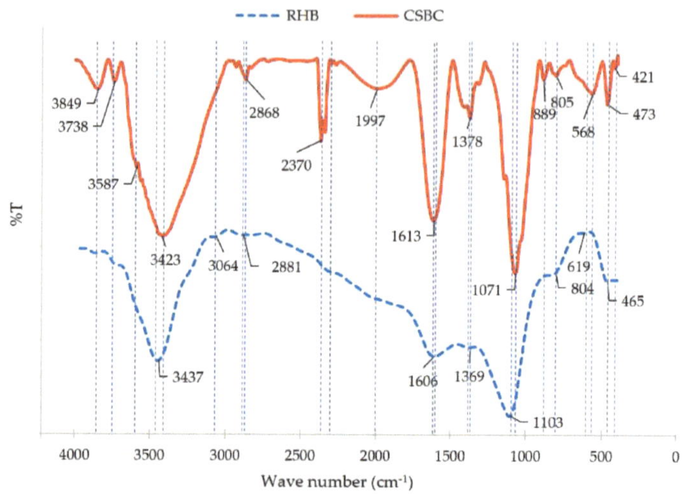

Figure 3. The FTIR analysis of RHB and CSBC.

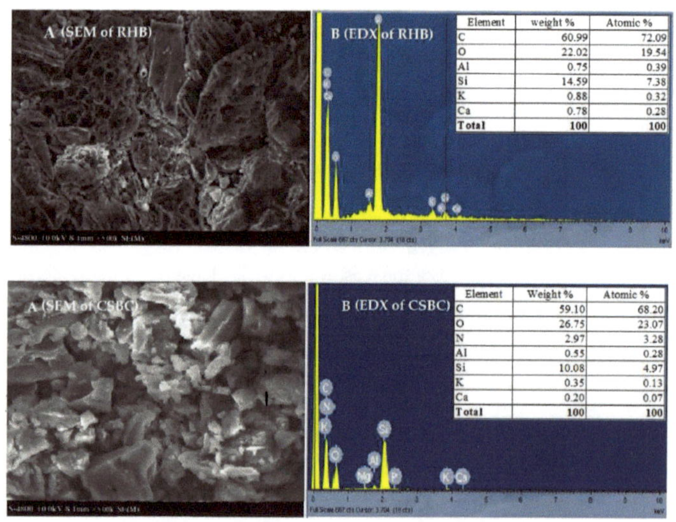

Element	weight %	Atomic %
C	60.99	72.09
O	22.02	19.54
Al	0.75	0.39
Si	14.59	7.38
K	0.88	0.32
Ca	0.78	0.28
Total	100	100

Element	Weight %	Atomic %
C	59.10	68.20
O	26.75	23.07
N	2.97	3.28
Al	0.55	0.28
Si	10.08	4.97
K	0.35	0.13
Ca	0.20	0.07
Total	100	100

Figure 4. SEM-EDX analysis of RHB and CSBC; (**Figure A upper**): SEM of RHB; (**Figure B upper**): EDX of RHB; (**Figure A lower**): SEM of CSBC; (**Figure B lower**): EDX of CSBC.

In the EDX results, both CSBC and RHB mainly consisted of C (59–61%), O (22–27%) and Si (10–15%) (Figure 4B). Significantly, the introduction of $-NH_2$ groups into the chitosan-modified biochar hydrogel beads was confirmed by the appearance of a Nitrogen (N) peak (accounted for 2.97% atomic weight) after chitosan coating. The tapioca peel biochar modified with chitosan also reported similar findings [45].

From the FTIR and SEM-EDX results, it can be safely concluded that chitosan was successfully bonded with the biochar matrix. Based on the physical properties of CSBC and RHB, a feasibility test for MO removal was conducted to compare the performance between CSBC and RHB. The results are described in the following section.

3.2. pH-Dependent Mechanism

The effect of the solution pH on the adsorption capacity of MO by CSBC and RHB was performed in the range of 3–10 at room temperature (298 K), adsorbate dosage of 0.2 g, contact time of 240 min and initial CSBC/RHB concentration of 50 $mg.L^{-1}$. Figure 5A show that MO adsorption onto CSBC and RHB was highly pH-dependent, where the amount of MO adsorbed into CSBC and RHB decreased with increasing solution pH. More specifically, as the initial pH increased from 3 to 10, the amount of MO adsorbed substantially decreased from 16.13 $mg.g^{-1}$ to 8.35 $mg.g^{-1}$, from 15.50 $mg.g^{-1}$ to 7.30 $mg.g^{-1}$ for CSBC and RHB, respectively.

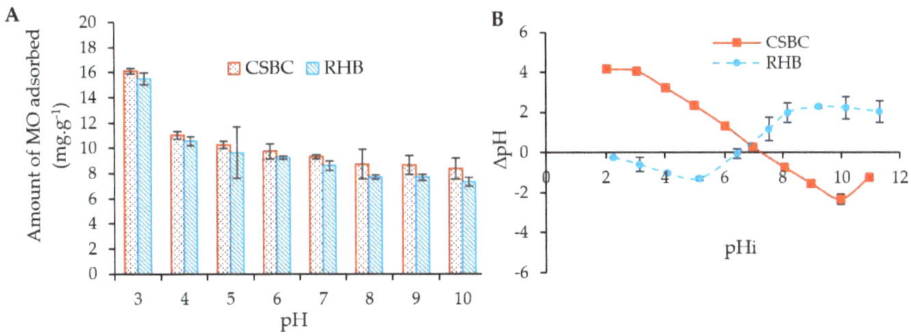

Figure 5. Effect of pH on MO adsorption (**A**) and pH_{pzc} of CSBC and RHB (**B**).

The pH effect could be explained using the point of zero charge (pH_{pzc}), the pH at which the surface of an adsorbent carries an equal amount of positive and negative charge. In comparison, the point of zero charge of CSBC (7.26) was slightly higher than the point of zero charge of RHB (6.48), as depicted in Figure 5B. The increase in the point of zero charge of CSBC is probably due to the addition of the basic groups (i.e., $-NH_2$ and –OH groups) after chitosan was incorporated into the biochar matrix [43]. This finding was similar to other chitosan-modified biochar [40].

It is well recognized that when $pH < pH_{pzc}$ (i.e., high H^+ activity in solution), the RHB surface charges are more positively charged due to the protonation of the hydroxyl –OH group. In this case, a stronger electrostatic attraction between hydrogen (H^+) with MO^- could occur; consequently, the amount of MO absorbed by RHB in an acidic environment was increased. For the CSBC case, in addition to the protonation of hydroxyl –OH groups at $pH < pH_{pzc}$, the amine $-NH_2$ group in the chitosan was also able to accept a proton from a hydronium ion ($-NH_2 + H_3O^+ \rightleftarrows NH_3^+ + H_2O$) [37]. As a result, electrostatic attraction between NH_3^+ and MO^- can easily occur; thus, CSBC generally shows better adsorption capacity for MO compared to RHB under the same adsorption conditions (Figure 5A).

Furthermore, the pK_a of MO was 3.46, and the MO molecule existed as a negatively charged species until pH 3.46. Afterwards, MO existed as neutral, and in high pH conditions, as slightly positive charged species. When pH > 4, the MO adsorption decreased with the increase of pH, which was partially caused by the competition between OH^- excess in the solution and anionic ions of MO. In brief, the pH-dependency of the employed materials can be explained by the electrostatic attraction between the negatively charged MO^- (solution $pH > pK_a$) and the negatively charged surface of adsorbents (solution $pH > pH_{pzc}$).

3.3. Adsorption Isotherms

The obtained optimum results used for kinetic and isotherm studies are shown in Figures S1–S3 (Supplementary Material). The optimum conditions for MO adsorption by both adsorbents were determined to be pH~3, a MO concentration of 50 mg/L, an adsorbent dosage of 0.2 g, adsorption time of 240 min and conditions at room temperature (298 K).

The nonlinear fitting curves of the Langmuir and Freundlich models are presented in Figure 6, while their isotherm parameters obtained from the adsorption experiments are tabulated in Table 1.

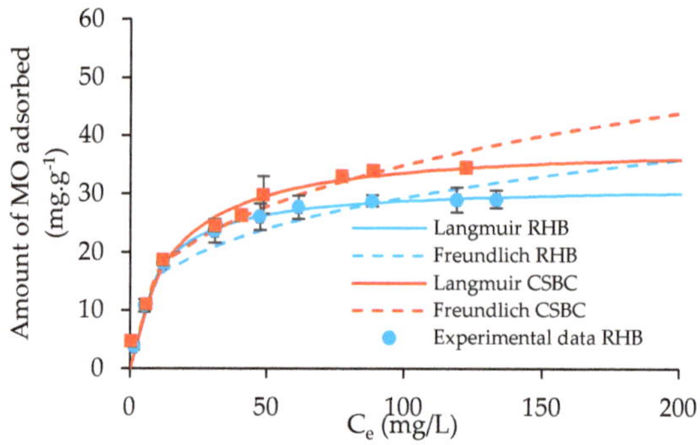

Figure 6. Fitting of Langmuir and Freundlich isotherms for MO adsorption on RHB and CSBC.

Table 1. Fitting results of Langmuir and Freundlich isotherms for MO adsorption by RHB and CSBC.

Model	Isotherm Parameter	Adsorbent	
		RHB	**CSBC**
Langmuir	q_m (mg.g^{-1})	31.63	38.75
	K_L (L.mg^{-1})	0.10	0.17
	R^2	0.998	0.980
Freundlich	1/n	3.45	3.02
	K_F ((mg.kg^{-1})/(mg.L^{-1})^n)	7.73	7.89
	R^2	0.916	0.974

It is obvious that the experimental data of CSBC and RHB for MO adsorption has a trend to stimulate both Langmuir and Freundlich isotherm models, with the determination coefficient R^2 of these models ranging from 0.92 to 0.99. However, a more in-depth comparison would reveal that the Langmuir model could fit the MO experimental data better, with R^2 values close to 1 (0.998 for RHB and 0.980 for CSBC). The fitting of Langmuir suggested homogeneous adsorption for MO on the surface of RHB and CSBC. In addition, the relatively higher Langmuir affinity constant K_L value of CSBC (K_L = 0.17 L.mg^{-1}) than that of RHB (K_L = 0.10 L.mg^{-1}) could be because of the stronger affinity of CSBC toward MO in aqueous solution.

Table 1 also report the maximum adsorption capacity of CSBC and RHB adsorbents obtained from the Langmuir model under the optimum adsorption conditions at room temperature (25 ± 2 °C), pH~3, time 240 min and adsorbent dosage of 0.2 g. It is recorded that the maximum adsorption capacity for MO of CSBC was estimated as 38.75 mg.g^{-1}, comparatively higher than that of RHB with 31.63 mg.g^{-1}.

3.4. Adsorption Kinetics

Three well-known theoretical kinetic models, including pseudo-first-order (PFO), pseudo-second-order (PSO) and intraparticle diffusion (IPD), were employed, and their derived coefficient of determination R^2 was used to obtain the best-fitting kinetic model for the experimental data. The experimental data (q_t, mg.g^{-1}) of RHB and CSBC toward MO fitted with the PFO and PSO kinetic models are displayed in Figure 7A, while the fitting for the IPD kinetic model is depicted in Figure 7B. The calculated kinetic parameters for MO adsorption onto RHB and CSBC are summarized in Table 2.

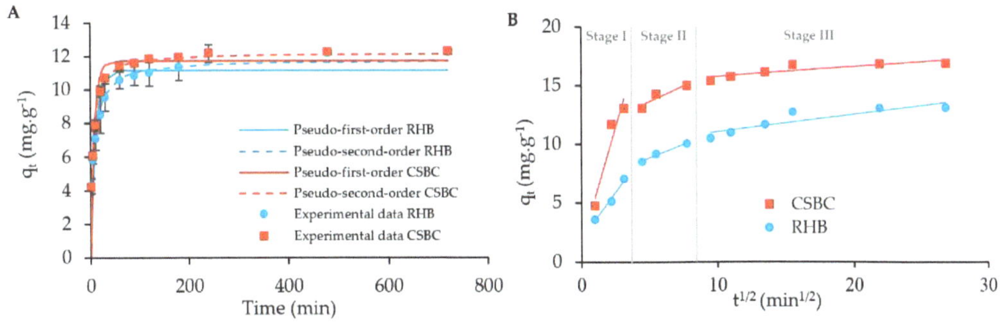

Figure 7. Fitting of PFO and PSO kinetic (**A**) and IPD (**B**) models for MO adsorption on CSBC and RHB.

Table 2. Fitting results of different kinetics for MO Adsorption by RHB and CSBC.

Adsorbent	$q_{e,exp}$ (mg.g^{-1})	Kinetic Model					
		Pseudo-First-Order			*Pseudo-Second-Order*		
		$q_{e,cal}$ (mg.g^{-1})	k_1 (1.min^{-1})	R^2	$q_{e,cal}$ (mg.g^{-1})	k_2 (g.mg^{-1}.min^{-1})	R^2
CSBC	12.20	11.72	0.13	0.864	12.02	0.02	0.942
RHB	12.16	11.16	0.10	0.832	11.81	0.01	0.930

	Intra-particle diffusion								
	Stage I: t = 1–10 min			Stage II: t = 20–60 min			Stage III: t > 60 min		
	C_1 mg.g^{-1}	k_{id1} mg.g^{-1}.min$^{-1/2}$	R^2	C_2 mg.g^{-1}	k_{id2} mg.g^{-1}.min$^{-1/2}$	R^2	C_3 mg.g^{-1}	k_{id3} mg.g^{-1}.min$^{-1/2}$	R^2
CSBC	1.47	3.94	0.927	10.95	0.54	0.926	14.94	0.08	0.858
RHB	1.92	1.58	0.970	6.59	0.46	0.974	9.63	0.14	0.891

According to Table 2, the PSO kinetic model was better fitted to the experimental data than the PFO model for both tested materials, where the R^2 values obtained by the PSO kinetic model were greater than the R^2 values from the PFO kinetic. In depth, the R^2 value of the PFO kinetic equation was 0.86 for CSBC and 0.83 for RHB, while that of the PSO kinetic model was 0.94 for CSBC and 0.93 for RHB. In addition, the adsorption capacity values ($q_{e,cal}$) calculated by PSO kinetics (11.81 mg.g^{-1} for RHB, 12.02 mg.g^{-1} for CSBC) showed closer values to those obtained by experiments ($q_{e,exp}$) than PSO kinetics (12.16 mg.g^{-1} for RHB, 12.20 mg.g^{-1} for CSBC), revealing that the obtained data is well-fitted by the PSO kinetic model. The PSO model comes as an indication that chemisorption might be the rate-limiting step in the adsorption process of methyl orange. A good fit to the PSO kinetic model was also observed in the other study on malachite green and Rhodamine B adsorption by chitosan–tapioca peel biochar [45].

The experimental kinetic data were further analyzed with the IPD model, in which a linear plot of q_t vs. $t^{1/2}$ was used to obtain K_d and C constants. Figure 7B demonstrate that the adsorption process of MO by RHB and CSBC was not a straight line throughout the whole experiment duration, involving three different kinetic stages: the first external film diffusion stage (stage I), the second gradual adsorption stage (stage II) and the third equilibrium adsorption stage (stage III). The R^2 values obtained from the three stages were found between 0.86 and 0.97, indicating that the IPD mechanism could play an important role in the MO adsorption process but not the sole rate-limiting step (C_1, C_2, $C_3 \neq 0$) [29]. In fact, the obtained diffusion rates $k_{id1} > k_{id2} > k_{id3}$ indicated that external film diffusion could play an important role in the rate-limiting step during diffusion [55]. A comparison of the external diffusion rate constants k_{id1} between the CSBC (3.94) and RHB (1.58) in

Table 2 demonstrate that faster external diffusion occurred in CSBC than in RHB, which could be attributed to the higher swelling behavior of chitosan hydrogel beads [55].

3.5. Mechanism Exploration and Comparison Assessment

In light of the above kinetic analysis, it is apparent that the adsorption process of MO by RHB and CSBC hydrogel beads followed both PSO and IPD kinetic models, advising that both chemisorption and diffusion mechanisms were the rate-controlling steps over the whole adsorption process.

In the chemisorption mechanism of RHB, the adsorption of MO was carried out mainly by the electrostatic interactions between the positively charged surface (i.e., H^+) of RHB and the negatively charged (i.e., SO_3^-) of MO. In the chemisorption mechanism of CSBC, the adsorption mechanism predicted would be the contribution of individual material (biochar and chitosan) in chitosan-modified biochar combinations. As demonstrated previously, the modification process by chitosan successfully incorporated the $-NH_2$ and –OH groups into the biochar matrix. The incorporation of these new functional groups is helpful in stimulating the formation of complexes between the MO molecules and CSBC material (illustrated in Figure 8). Thus, in addition to the electrostatic interaction mechanism, the surface complexation mechanism may account for the contribution of chitosan to the CSBC, thereby enhancing the adsorption efficiency of the CSBC material to 1.3 times as compared to RHB (38.75 mg.g^{-1} for CSBC and 31.63 mg.g^{-1} for RHB).

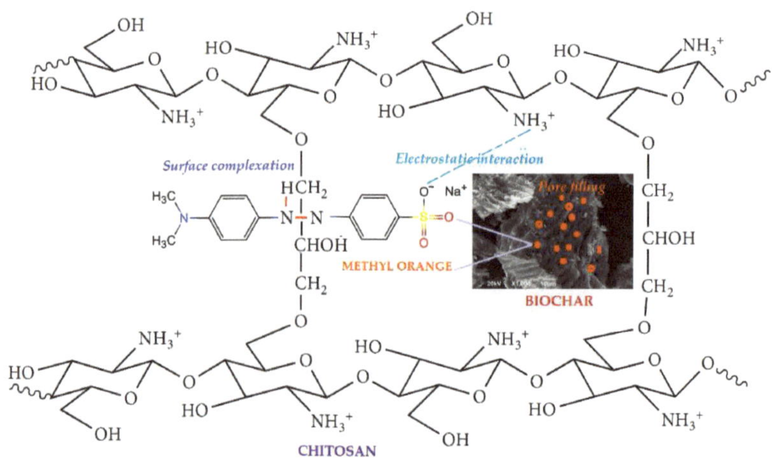

Figure 8. Possible chemisorption and diffusion mechanisms of MO onto CSBC hydrogel beads.

Thus, surface complexation and electrostatic interaction mechanisms may account for the chemisorption of MO ions by CSBC, thereby enhancing the adsorption efficiency of the CSBC material. As found in the present study, the adsorption capacity for MO of CSBC (38.75 mg.g^{-1}) was close to 1.3 times greater than BC (31.63 mg.g^{-1}). Several studies that were conducted on chitosan-modified different raw materials of biochar, such as peels of pomelo [56], apple branches [39], pine wood [43] and corncob [57], have shown that the use of chitosan-modified biochar also contributed to the overall adsorption performance, as listed in Table 3. Although the adsorption capacity of CSBC does not seem to be significantly improved regarding RHB as compared with other studies (Table 3), it is true that no safe comparisons can be made here because the feedstock used, the pyrolytic temperature, the experimental conditions and target substrates are totally different.

Table 3. Effect on removal capacity by chitosan coated on biochar matrix.

Feedstock	Pyrolytic Temperature (°C)	Surface Area ($m^2\ g^{-1}$)	Experimental Conditions	Contaminants	Effect on Removal Capacity	Ref.
Peels of pomelo	450	Not available	pH~3; T = 298 K, C_0 = 10 ÷ 50 (mg/L)	Ciprofloxacin	Near 11 times higher than unmodified biochar ($q_{max,CSBC}$ = 36.72 mg.g^{-1})	[56]
Apple branches	300	S_{BC} = 13.3; S_{CSBC} = 23.3	pH~9; T = 298 K, C_0 = 10 ÷ 500 (mg/L)	Phloridzin	Near 2 times higher than unmodified biochar ($q_{max,CSBC}$ = 28.96 mg.g^{-1})	[39]
Pinewood	425	S_{BC} = 10.5; S_{CSBC} = 7.13	pH~5; T = 318 K, C_0 = 50 ÷ 350 (mg/L)	Pb(II)	Near 3 times higher than unmodified biochar ($q_{max,CSBC}$ = 134 mg.g^{-1})	[43]
Corncob	400	S_{BC} = 301.9; S_{CSBC} = 631.5	pH~12; T = 298 K, C_0 = 10 ÷ 100 (mg/L)	Methylene Blue	Near 2 times higher than unmodified biochar ($q_{max,CSBC}$ = 499.8 mg.g^{-1})	[57]
Rice husk	500	S_{BC} = 115.59; S_{CSBC} = 107.97	pH~3; T = 298 K, C_0 = 10 ÷ 200 (mg/L)	Methyl Orange	Near 1.3 times higher than unmodified biochar ($q_{max,CSBC}$ = 38.75 mg.g^{-1})	This study

4. Conclusions

Chitosan-modified biochar hydrogel beads were successfully synthesized in this study. The batch experimental study demonstrated that chitosan beads coated onto biochar matrix could enhance the adsorption capacities of methyl orange molecules from the aqueous solution since chitosan modification helps increase the number of functional groups (i.e., $-NH_2$ and $-OH$ groups). The theoretical maximum adsorption capacities of the chitosan-modified biochar were close to 1.3 times greater than unmodified biochar (38.75 mg.g^{-1} for CSBC and 31.63 mg.g^{-1} for RHB), which were obtained at room temperature (25 ± 2 °C), pH~3, time 240 min and adsorbent dosage 0.2 g. The experimental data were well described by the Langmuir isotherm, with R^2 values close to 1 (0.998 for RHB and 0.980 for CSBC), suggesting homogeneous adsorption for MO on the surface of RHB and CSBC. The kinetic data for MO adsorption was well-fitted to both pseudo-second-order and intra-particle diffusion kinetic models, indicating that the chemisorption mechanism and intra-particle diffusion mechanism may govern the adsorption process of methyl orange molecules by both RHB and CSBC adsorbents. Briefly, the developed chitosan-modified biochar hydrogel beads can be utilized as an effective adsorbent for the removal of methyl orange molecules from aqueous solution.

Supplementary Materials: The following supporting information can be downloaded at: https://www.mdpi.com/article/10.3390/toxics10090500/s1, Figure S1: The effects of RHB and CSBC dosage on MO adsorption; Figure S2: The effects of initial MO concentration on adsorption by RHB and CSBC; Figure S3: The effects of contact time on MO adsorption onto RHB and CSBC.

Author Contributions: Conceptualization, N.X.L. and D.T.M.P.; methodology, N.X.L.; conducted physicochemical analysis, D.T.M.P.; conducted experiments, P.T.T.T.; conducted experiment, L.C.M.; data analysis, N.X.L.; writing—original draft preparation, D.T.M.P.; writing—review and editing, N.X.L. All authors have read and agreed to the published version of the manuscript.

Funding: This research received no external funding.

Institutional Review Board Statement: Not applicable.

Informed Consent Statement: Not applicable.

Data Availability Statement: Not applicable.

Acknowledgments: The authors gratefully acknowledge Nguyen Ngoc Quynh Lam, Bui Thi Nhu Y and Nguyen Thi Thanh Thao for their assistance with the experiment.

Conflicts of Interest: The authors declare no conflict of interest.

References

1. Ogugbue, C.J.; Sawidis, T. Bioremediation and detoxification of synthetic wastewater containing triarylmethane dyes by Aeromonas hydrophila isolated from industrial effluent. *Biotechnol. Res. Int.* **2011**, *2011*, 967925. [CrossRef] [PubMed]
2. Robinson, T.; McMullan, G.; Marchant, R.; Nigam, P. Remediation of dyes in textile effluent: A critical review on current treatment technologies with a proposed alternative. *Biores. Technol.* **2001**, *77*, 247–255. [CrossRef]
3. Lellis, B.; Fávaro-Polonio, C.Z.; Pamphile, J.A.; Polonio, J.C. Effects of textile dyes on health and the environment and bioremediation potential of living organisms. *Biotechnol. Res. Innov.* **2019**, *3*, 275–290. [CrossRef]
4. Malviya, A.; Jaspal, D.; Sharma, P.; Dubey, A. Isothermal mathematical modeling for decolorizing water-A comparative approach. *Sustain. Environ. Res.* **2015**, *25*, 53–58.
5. Iwuozor, K.O.; Ighalo, J.O.; Emenike, E.C.; Ogunfowora, L.A.; Igwegbe, C.A. Adsorption of methyl orange: A review on adsorbent performance. *Curr. Res. Green Sustain. Chem.* **2021**, *4*, 100179. [CrossRef]
6. Soares, P.A.; Silva, T.F.; Manenti, D.R.; Souza, S.; Boaventura, R.A.; Vilar, V.J. Insights into real cotton-textile dyeing wastewater treatment using solar advanced oxidation processes. *Environ. Sci. Pollut. Res.* **2014**, *21*, 932–945. [CrossRef]
7. Marszałek, J.; Żyłła, R. Recovery of Water from Textile Dyeing Using Membrane Filtration Processes. *Processes* **2021**, *9*, 1833. [CrossRef]
8. Wang, J.; Yao, J.; Wang, L.; Xue, Q.; Hu, Z.; Pan, B. Multivariate optimization of the pulse electrochemical oxidation for treating recalcitrant dye wastewater. *Sep. Purif. Technol.* **2020**, *230*, 115851. [CrossRef]
9. Wang, J.; Chen, H. Catalytic ozonation for water and wastewater treatment: Recent advances and perspective. *Sci. Total Environ.* **2020**, *704*, 135249. [CrossRef]
10. Gadekar, M.R.; Ahammed, M.M. Use of water treatment residuals for colour removal from real textile dye wastewater. *Appli. Water Sci.* **2020**, *10*, 160. [CrossRef]
11. Alguacil, F.J.; Lopez, F.A. Adsorption Processes in the Removal of Organic Dyes from Wastewaters: Very Recent Developments. In *Promising Techniques for Wastewater Treatment and Water Quality Assessment*; IntechOpen: London, UK, 2021; pp. 17–32.
12. Rápó, E.; Tonk, S. Factors affecting synthetic dye adsorption; desorption studies: A review of results from the last five years (2017–2021). *Molecules* **2021**, *26*, 5419. [CrossRef] [PubMed]
13. Kheddo, A.; Rhyman, L.; Elzagheid, M.I.; Jeetah, P.; Ramasami, P. Adsorption of synthetic dyed wastewater using activated carbon from rice husk. *SN Appl. Sci.* **2020**, *2*, 1–14. [CrossRef]
14. Hammood, Z.A.; Chyad, T.F.; Al-Saedi, R. Adsorption performance of dyes over zeolite for textile wastewater treatment. *Ecol. Chem. Eng.* **2021**, *28*, 329–337. [CrossRef]
15. Sutar, S.; Patil, P.; Jadhav, J. Recent advances in biochar technology for textile dyes wastewater remediation: A review. *Environ. Res.* **2022**, *209*, 112841. [CrossRef] [PubMed]
16. Kausar, A.; Iqbal, M.; Javed, A.; Aftab, K.; Bhatti, H.N.; Nouren, S. Dyes adsorption using clay and modified clay: A review. *J. Mol. Liq.* **2018**, *256*, 395–407. [CrossRef]
17. Uddin, M.J.; Ampiaw, R.E.; Lee, W. Adsorptive removal of dyes from wastewater using a metal-organic framework: A review. *Chemosphere* **2021**, *284*, 131314. [CrossRef]
18. Kishor, B. A Review on Removal of Colour and Dye By Using Different Natural Materials. *Int. J. Adv. Sci.* **2005**, *4238*, 5060–5062.
19. Bharathi, K.; Ramesh, S. Removal of dyes using agricultural waste as low-cost adsorbents: A review. *Appl. Water Sci.* **2013**, *3*, 773–790. [CrossRef]
20. Pai, S.; Kini, M.S.; Mythili, R.; Selvaraj, R. Adsorptive removal of AB113 dye using green synthesized hydroxyapatite/magnetite nanocomposite. *Environ. Res.* **2022**, *210*, 112951. [CrossRef]
21. Huang, Q.; Song, S.; Chen, Z.; Hu, B.; Chen, J.; Wang, X. Biochar-based materials and their applications in removal of organic contaminants from wastewater: State-of-the-art review. *Biochar* **2019**, *1*, 45–73. [CrossRef]
22. Tan, X.; Liu, Y.; Zeng, G.; Wang, X.; Hu, X.; Gu, Y.; Yang, Z. Application of biochar for the removal of pollutants from aqueous solutions. *Chemosphere* **2015**, *125*, 70–85. [CrossRef] [PubMed]
23. Bussan, D.D.; Sessums, R.F.; Cizdziel, J.V. Activated carbon and biochar reduce mercury methylation potentials in aquatic sediments. *Bull. Environ. Contam. Toxicol.* **2016**, *96*, 536–539. [CrossRef]
24. Yao, Y.; Gao, B.; Chen, J.; Yang, L. Engineered biochar reclaiming phosphate from aqueous solutions: Mechanisms and potential application as a slow-release fertilizer. *Environ. Sci. Technol.* **2013**, *47*, 8700–8708. [CrossRef] [PubMed]
25. Al-Wabel, M.I.; Ahmad, M.; Usman, A.R.; Al-Farraj, A.S. Designing chitosan based magnetic beads with conocarpus waste-derived biochar for efficient sulfathiazole removal from contaminated water. *Saudi J. Biol. Sci.* **2021**, *28*, 6218–6229. [CrossRef] [PubMed]
26. Ahmad, M.; Ahmad, M.; Usman, A.R.; Al-Faraj, A.S.; Abduljabbar, A.S.; Al-Wabel, M.I. Biochar composites with nano zerovalent iron and eggshell powder for nitrate removal from aqueous solution with coexisting chloride ions. *Environ. Sci. Pol. Res.* **2018**, *25*, 25757–25771. [CrossRef] [PubMed]
27. Chen, Y.; Shi, J.; Du, Q.; Zhang, H.; Cui, Y. Antibiotic removal by agricultural waste biochars with different forms of iron oxide. *RSC Adv.* **2019**, *9*, 14143–14153. [CrossRef]

28. Fang, L.; Li, J.-S.; Donatello, S.; Cheeseman, C.; Poon, C.S.; Tsang, D.C. Use of Mg/Ca modified biochars to take up phosphorus from acid-extract of incinerated sewage sludge ash (ISSA) for fertilizer application. *J. Clean. Prod.* **2020**, *244*, 118853. [CrossRef]
29. Phuong, D.T.M.; Loc, N.X. Rice Straw Biochar and Magnetic Rice Straw Biochar for Safranin O Adsorption from Aqueous Solution. *Water* **2022**, *14*, 186. [CrossRef]
30. Kurita, K. Chitin and chitosan: Functional biopolymers from marine crustaceans. *Mar. Biotechnol.* **2006**, *8*, 203–226. [CrossRef]
31. Amalraj, A.; Jude, S.; Gopi, S. Polymer blends, composites and nanocomposites from chitin and chitosan; manufacturing, characterization and applications. In *Handbook of Chitin and Chitosan*; Elsevier: Amsterdam, The Netherlands, 2020; pp. 1–46.
32. Dotto, G.L.; Vieira, M.L.; Pinto, L.A. Kinetics and mechanism of tartrazine adsorption onto chitin and chitosan. *Ind. Eng. Chem. Res.* **2012**, *51*, 6862–6868. [CrossRef]
33. Zargar, V.; Asghari, M.; Dashti, A. A review on chitin and chitosan polymers: Structure, chemistry, solubility, derivatives, and applications. *ChemBioEng Rev.* **2015**, *2*, 204–226. [CrossRef]
34. Perumal, S.; Atchudan, R.; Yoon, D.H.; Joo, J.; Cheong, I.W. Spherical chitosan/gelatin hydrogel particles for removal of multiple heavy metal ions from wastewater. *Ind. Eng. Chem. Res.* **2019**, *58*, 9900–9907. [CrossRef]
35. Le, H.Q.; Sekiguchi, Y.; Ardiyanta, D.; Shimoyama, Y. CO_2-activated adsorption: A new approach to dye removal by chitosan hydrogel. *ACS Omega* **2018**, *3*, 14103–14110. [CrossRef] [PubMed]
36. Crini, G.; Torri, G.; Lichtfouse, E.; Kyzas, G.Z.; Wilson, L.D.; Morin-Crini, N. Dye removal by biosorption using cross-linked chitosan-based hydrogels. *Environ. Chem. Lett.* **2019**, *17*, 1645–1666. [CrossRef]
37. Zhou, Y.; Gao, B.; Zimmerman, A.R.; Fang, J.; Sun, Y.; Cao, X. Sorption of heavy metals on chitosan-modified biochars and its biological effects. *Chem. Engin. J.* **2013**, *231*, 512–518. [CrossRef]
38. Lyu, H.; Xu, S.; Liu, Y.; Zhang, W.; Duan, Q.; Zhu, M.; Tang, J. Effect of biochar on the emission of greenhouse gas in farmland. In *Biochar in Agriculture for Achieving Sustainable Development Goals*; Elsevier: Amsterdam, The Netherlands, 2022; pp. 251–262.
39. Ma, Z.; Du, W.; Yan, Z.; Chen, X.; Wang, Y.; Mao, Z. Removal of phloridzin by chitosan-modified biochar prepared from apple branches. *Anal. Lett.* **2021**, *54*, 903–918. [CrossRef]
40. Huang, A.; Bai, W.; Yang, S.; Wang, Z.; Wu, N.; Zhang, Y.; Ji, N.; Li, D. Adsorption Characteristics of Chitosan-Modified Bamboo Biochar in Cd (II) Contaminated Water. *J. Chem.* **2022**, *2022*, 6303252. [CrossRef]
41. Tan, Y.; Wan, X.; Ni, X.; Wang, L.; Zhou, T.; Sun, H.; Wang, N.; Yin, X. Efficient removal of Cd (II) from aqueous solution by chitosan modified kiwi branch biochar. *Chemosphere* **2022**, *289*, 133251. [CrossRef]
42. Song, J.; Messele, S.A.; Meng, L.; Huang, Z.; El-Din, M.G. Adsorption of metals from oil sands process water (OSPW) under natural pH by sludge-based Biochar/Chitosan composite. *Water Res.* **2021**, *194*, 116930. [CrossRef]
43. Dewage, N.B.; Fowler, R.E.; Pittman, C.U.; Mohan, D.; Mlsna, T. Lead (Pb^{2+}) sorptive removal using chitosan-modified biochar: Batch and fixed-bed studies. *RSC Adv.* **2018**, *8*, 25368–25377. [CrossRef]
44. Zhang, L.; Tang, S.; He, F.; Liu, Y.; Mao, W.; Guan, Y. Highly efficient and selective capture of heavy metals by poly (acrylic acid) grafted chitosan and biochar composite for wastewater treatment. *Chem. Eng. J.* **2019**, *378*, 122215. [CrossRef]
45. Vigneshwaran, S.; Sirajudheen, P.; Nikitha, M.; Ramkumar, K.; Meenakshi, S. Facile synthesis of sulfur-doped chitosan/biochar derived from tapioca peel for the removal of organic dyes: Isotherm, kinetics and mechanisms. *J. Mol. Liq.* **2021**, *326*, 115303. [CrossRef]
46. Radwan, M.A.; Farrag, S.A.; Abu-Elamayem, M.M.; Ahmed, N.S. Extraction, characterization, and nematicidal activity of chitin and chitosan derived from shrimp shell wastes. *Biol. Fertil. Soils* **2012**, *48*, 463–468. [CrossRef]
47. Dwivedi, P.; Narvi, S.S.; Tewari, R.P.; Tiwary, D. *Nanobiotechnology for Safe Bioactive Nanobiomaterials*; CRC Press: Boca Raton, FL, USA, 2022.
48. Jindo, K.; Mizumoto, H.; Sawada, Y.; Sanchez-Monedero, M.; Sonoki, T. Physical and chemical characterizations of biochars derived from different agricultural residues. *Biogeosci. Discuss.* **2014**, *11*, 11727–11746. [CrossRef]
49. Palniandy, L.K.; Yoon, L.W.; Wong, W.Y.; Yong, S.-T.; Pang, M.M. Application of biochar derived from different types of biomass and treatment methods as a fuel source for direct carbon fuel cells. *Energies* **2019**, *12*, 2477. [CrossRef]
50. Ahmed, M.B.; Zhou, J.L.; Ngo, H.H.; Guo, W.; Chen, M. Progress in the preparation and application of modified biochar for improved contaminant removal from water and wastewater. *Biores. Technol.* **2016**, *214*, 836–851. [CrossRef]
51. Fouad, D.; Bachra, Y.; Ayoub, G.; Ouaket, A.; Bennamara, A.; Knouzi, N.; Berrada, M. A Novel Drug Delivery System Based on Nanoparticles of Magnetite Fe_3O_4 Embedded in an Auto Cross-Linked Chitosan. In *Chitin and Chitosan-Physicochemical Properties and Industrial Applications*; IntechOpen: London, UK, 2020; pp. 3–26.
52. Xiang, J.; Lin, Q.; Yao, X.; Yin, G. Removal of Cd from aqueous solution by chitosan coated MgO-biochar and its in-situ remediation of Cd-contaminated soil. *Environ. Res.* **2021**, *195*, 110650. [CrossRef]
53. Zhang, Z.; He, S.; Zhang, Y.; Zhang, K.; Wang, J.; Jing, R.; Yang, X.; Hu, Z.; Lin, X.; Li, Y. Spectroscopic investigation of Cu^{2+}, Pb^{2+} and Cd^{2+} adsorption behaviors by chitosan-coated argillaceous limestone: Competition and mechanisms. *Environ. Pol.* **2019**, *254*, 112938. [CrossRef]
54. Liu, S.; Huang, B.; Chai, L.; Liu, Y.; Zeng, G.; Wang, X.; Zeng, W.; Shang, M.; Deng, J.; Zhou, Z. Enhancement of As (V) adsorption from aqueous solution by a magnetic chitosan/biochar composite. *RSC Adv.* **2017**, *7*, 10891–10900. [CrossRef]

55. Crini, G. Recent developments in polysaccharide-based materials used as adsorbents in wastewater treatment. *Prog. Polym. Sci.* **2005**, *30*, 38–70. [CrossRef]
56. Afzal, M.Z.; Sun, X.-F.; Liu, J.; Song, C.; Wang, S.-G.; Javed, A. Enhancement of ciprofloxacin sorption on chitosan/biochar hydrogel beads. *Sci. Total Environ.* **2018**, *639*, 560–569. [CrossRef] [PubMed]
57. Liu, X.-J.; Li, M.-F.; Ma, J.-F.; Bian, J.; Peng, F. Chitosan crosslinked composite based on corncob lignin biochar to adsorb methylene blue: Kinetics, isotherm, and thermodynamics. *Colloids Surf. A Physicochem. Eng. Asp.* **2022**, *642*, 128621. [CrossRef]

toxics

MDPI

Article

Performance and Mechanism of As(III/V) Removal from Aqueous Solution by Fe_3O_4-Sunflower Straw Biochar

Yuling Zhao [1], Hao Shi [1], Xin Tang [1], Daihong Kuang [2], Jinlong Zhou [3,*] and Fangyuan Yang [2,*]

1 College of Resources and Environment, Xinjiang Agricultural University, Urumqi 830052, China
2 College of Mathematics and Physics, Xinjiang Agricultural University, Urumqi 830052, China
3 College of Hydraulic and Civil Engineering, Xinjiang Agricultural University, Urumqi 830052, China
* Correspondence: zjzhoujl@163.com (J.Z.); yangfy@xjau.edu.cn (F.Y.)

Abstract: Humans and ecosystems are severely damaged by the existence of As(III/V) in the aquatic environment. Herein, an advanced Fe_3O_4@SFBC (Fe_3O_4-sunflower straw biochar) adsorbent was fabricated by co-precipitation method with sunflower straw biochar (SFBC) prepared at different calcination temperatures and different SFBC/Fe mass ratios as templates. The optimal pH for As(III/V) removal was investigated, and Fe_3O_4@SFBC shows removal efficiency of 86.43% and 95.94% for As(III) and As(V), respectively, at pH 6 and 4. The adsorption effect of calcining and casting the biochar-bound Fe_3O_4 obtained at different temperatures and different SFBC/Fe mass ratios were analyzed by batch experiments. The results show that when the SFBC biochar is calcined at 450 °C with an SFBC/Fe mass ratio of 1:5, the adsorption of As(III) and As(V) reaches the maximum, which are 121.347 and 188.753 mg/g, respectively. Fe_3O_4@SFBC morphology, structure, surface functional groups, magnetic moment, and internal morphology were observed by XRD, FTIR, SEM, TEM, and VSM under optimal working conditions. The material shows a small particle size in the range of 12–14 nm with better magnetic properties (54.52 emu/g), which is suitable for arsenic removal. The adsorption mechanism of As(III/V) by Fe_3O_4@SFBC indicates the presence of chemisorption, electrostatic, and complexation. Finally, the material was used for five consecutive cycles of adsorption–desorption experiments, and no significant decrease in removal efficiency was observed. Therefore, the new adsorbent Fe_3O_4@SFBC can be efficiently used for arsenic removal in the aqueous system.

Keywords: As(III/V); sunflower straw biochar; Fe_3O_4; magnetic composite; adsorption

Citation: Zhao, Y.; Shi, H.; Tang, X.; Kuang, D.; Zhou, J.; Yang, F. Performance and Mechanism of As(III/V) Removal from Aqueous Solution by Fe_3O_4-Sunflower Straw Biochar. *Toxics* **2022**, *10*, 534. https://doi.org/10.3390/toxics10090534

Academic Editors: Yuezhou Wei, Yan Wu and Xinpeng Wang

Received: 17 August 2022
Accepted: 9 September 2022
Published: 11 September 2022

Publisher's Note: MDPI stays neutral with regard to jurisdictional claims in published maps and institutional affiliations.

Copyright: © 2022 by the authors. Licensee MDPI, Basel, Switzerland. This article is an open access article distributed under the terms and conditions of the Creative Commons Attribution (CC BY) license (https://creativecommons.org/licenses/by/4.0/).

1. Introduction

Arsenic is commonly present in the aquatic environment. Arsenic-contaminated water bodies affect populations in most parts of the world, including Mexico, Argentina, China, India, and Bangladesh [1,2]. A total of 2.5 million people living in specific areas of China have greater than 50 μg/L arsenic in drinking water [3]. Prolonged exposure to arsenic in water may cause an increased risk of skin cancer, lung cancer, and esophageal cancer in humans [4]. Hence, the World Health Organization established a 10 μg/L upper limit in drinking water for total arsenic [5]. Therefore, it is important to prepare effective wastewater treatment methods to reduce the health risks associated with arsenic water pollution.

Among these methods, membrane separation [6], ion exchange [7], and bioremediation [8], adsorption are potentially effective for arsenic removal due to their lower cost, highly efficient, convenient, and eco-friendly processes [9]. Until now, multiple adsorbents, including activated carbon [10], clay mineral materials [11], metal oxides/hydroxides [12], natural organic compounds [13], and nanocomposites [14], have been investigated as media for arsenic removal.

In recent years, nanocomposite adsorbents have received increasing attention due to their specific high surface area and excellent reactivity via abundant adsorption sites [15].

There is no doubt that high adsorption and selectivity are key points for materials with adsorption potential. According to Awual et al. [16,17], composite materials are of increasing interest for arsenic removal based on their specific functionality and surface area. Among them, Fe_3O_4 nanomaterials with superparamagnetic characteristics make the separation of adsorbents easy from the aqueous phase under the influence of an applied magnetic field [18]. In addition, these materials exhibit good heavy metal adsorption properties [19]. However, nanoscale Fe_3O_4 materials are highly susceptible to oxidation to non-magnetic materials and loss of dispersibility due to their high chemical activity [20,21]. Ways to overcome these defects and structural modifications are needed to prevent aggregation.

Biochar is a porous, carbonaceous material obtained by the pyrolysis of biomass waste under anaerobic/limited oxygen conditions [22]. New biochar-based adsorbents can be used as environmentally friendly carriers for metal oxide dispersion. Thus far, hydroxides or oxides of metals such as Pb, Cu, Cd, Zn, Fe, Bi Al, Mn, Ce, La, etc., have been fabricated as As(V) adsorbents [23,24]. An amount of 90 mg/g was the highest As(V) adsorption capacity determined by Wen et al. [25], using a nano-biochar composite of magnetically active tea waste. Iron oxide nanoneedles were deposited on a biochar surface made of cotton fiber which increased the adsorption capacities of As(V) and As(III) to 93.94 and 70.22 mg/g, respectively. Furthermore, Wang et al. [26] used pine biochar as zero-valent iron-loaded particles with 124.5 mg/g adsorption capacity for As(V).

Sunflower straw is often used for preparing livestock feed and composting fertilizers; as an agricultural by-product, its yield is high, and its cost is low. Additionally, the adsorption capacity of using it in combination with Fe_3O_4 as a nano sorbent is yet to be investigated, and its possible removal mechanism is not clear. Therefore, the aim of this work was mainly to prepare a cleaner and recyclable Fe_3O_4-biochar composite (Fe_3O_4@SFBC)using sunflower straw (SFBC) for efficient As(III/V) removal from water. The effects of different calcination temperatures (350, 450, 550, and 650 °C) and Fe_3O_4 materials prepared with different SFBC/Fe ratios (1:2, 1:3, 1:4, 1:5, and 1:6) on the As(III/V) adsorption, respectively, were studied. Adsorption studies with varying pH, initial concentration, and reaction time were also conducted to obtain the optimal adsorption conditions, while the adsorption isotherms and kinetics were studied to explore the adsorption process. In addition, adsorption–desorption experiments were conducted to measure the performance of the material for As(III/V) adsorption in terms of low cost and ease of application.

2. Materials and Methods

2.1. Materials

Sunflower stalks were harvested through local fields in Bayannur, in Inner Mongolia, China. Before use, stalks were rinsed twice with tap water, washed three times using deionized water to eliminate impurities, dried in air for 24 h, and then dried thoroughly in an oven at 80 °C. The dried sunflower stalks were crushed, sieved, and sealed for later use. Two iron salts ($FeCl_3 \cdot 6H_2O$ and $FeSO_4 \cdot 7H_2O$) and ammonia water ($NH_3 \cdot H_2O$) were purchased from Tianjin Zhiyuan Chemicals. Deionized water was used for solution preparation. Analytical grade reagents were used without any further purification.

2.2. Sunflower Straw Pretreatment

The crushed sunflower straws were positioned in a 30 mL covered porcelain crucible and heated at 350, 450, 550, and 650 °C under anaerobic conditions. The heating rate was 5°/min for 2 h. After calcination, the materials were washed twice with deionized water–anhydrous ethanol by centrifugation to remove fine particulate matter and water-soluble organic residues. Finally, the materials were oven dried at 80 °C. Synthesized materials were labeled as SFBC 350, SFBC 450, SFBC 550, and SFBC 650.

2.3. Synthesis of Fe_3O_4@SFBC Magnetic Nanoparticles

Different Fe_3O_4@SFBC nano absorbents material were prepared by co-precipitation method using SFBC 450 with an SFBC/Fe ratio of 1:5. First, 5.40 g and 2.78 g of $FeCl_3 \cdot 6H_2O$

and $FeSO_4 \cdot 7H_2O$ were taken and 100 mL deionized water added to dissolve these salts by ultrasound. Then, 1.62 g of SFBC 450 was poured into a beaker containing 10 mL of deionized water, and both liquids were mixed in a double-necked flask and de-oxygenated with a vacuum pump. The water bath was then heated to 90 °C, and 10 mL of $NH_3 \cdot H_2O$ was injected and stirred for 2 h, allowed to precipitate, followed by cooling at room temperature. Centrifugation was performed for 5 min at 9000 rpm, and the precipitate was collected and washed twice using an alternating step of deionized water–anhydrous ethanol. The final product of Fe_3O_4@SFBC was obtained after drying under a vacuum for 12 h at 60 °C.

In addition, the material solutions prepared by different SFBC/Fe mass ratios (1:2, 1:3, 1:4, and 1:6) and biomass char at different calcination temperatures (350, 550, and 650 °C) were used to prepare different composites.

2.4. Characterization

The mineralogy of Fe_3O_4@SFBC was characterized using XRD (Bruker D2, Karlsruhe, Germany) with Cu Kα radiation in the range of 2θ = 5~80°. Morphology of biochar and nanomaterials was observed by using scanning electron microscopy (ZEISS GeminiSEM 300, Oberkochen, Germany) and transmission electron microscopy (JEM2100F, Akishima-shi, Japan). FTIR spectra of nanoparticles were measured using the KBr compression method from 400 to 4000 cm^{-1} with an infrared photo spectrometer (Scientific Nicolet iS5, Waltham, MA, USA). The magnetic behavior of the particles was studied using a vibrating sample magnetometer (MPMS-XL-7, San Diego, CA, USA) in the varying magnetic field strengths of $\pm$2000 magnetic moments. The type and valence of the sample elements were determined with an X-ray electron spectrometer (ESCALAB 250XI, Waltham, MA, USA).

2.5. Adsorption Experiments

As(III/V) removal by Fe_3O_4@SFBC nanocomposites were investigated and optimized. Batch adsorption tests were conducted in the laboratory to optimize parameters for arsenic adsorption, such as pH, adsorption isotherm, and adsorption kinetics. An amount of 0.1 g adsorbent Fe_3O_4@SFBC was added to beakers containing 100 mL solution of As(III) and As(V) of 5 mg/L concentration, respectively. The solution was shaken for 1 h at room temperature under continuous stirring. The adsorbent was removed from the solution using an external magnet.

The parameters studied include solution pH, adsorption time, and initial As(III/V) concentration. In pH experiments, 0.1 mol/L of NaOH or HCl was used to adjust the pH. Generally, adsorption was performed in the pH range of 3–11. A total of 0.1 g of Fe_3O_4@SFBC nanoparticles were added to 100 mL of 5 mg/L As(III/V) solution, mixed, and stirred for some time. The samples were filtered by a 0.45 um membrane filter to collect filtrate and analyzed by an atomic fluorescence spectrophotometer (PF6-3, Beijing Purkinje General Instrument Co., Ltd., Beijing, China).

Before testing the samples using the atomic fluorescence spectrophotometer, the arsenic standard stock solution was used to configure the arsenic standard solution and draw the standard curve. After pretreatment, the sample enters the atomic fluorescence instrument through the sensor and generates arsine gas under the reducing acidic potassium borohydride conditions. The hydride forms ground state atoms in the argon–hydrogen flame, and its ground state atoms are excited by the light emitted from the arsenic lamp to produce atomic fluorescence, and the intensity of atomic fluorescence is proportional to the element to be measured in the test solution under the standard curve.

The adsorption amount (q_t, mg/g) of arsenic at the sampling time (min) and the adsorbent removal efficiency (R, %) were calculated by Equations (1) and (2), respectively.

$$q_t = \frac{(C_0 - C_t)}{m} V \quad (1)$$

$$R(\%) = \frac{(C_0 - C_t)}{C_0} \times 100\% \quad (2)$$

where V is the volume of arsenic solution (L), C_0 is an arsenic concentration at an initial time and C_t at time t (mg/L), and magnetic adsorbent mass (g) is denoted by m.

2.6. Desorption and Regeneration Texts

The adsorption–desorption cycle for each adsorbent was studied for five cycles. In each cycle, 5 mg/L of As(III) or As(V) solution was mixed with 1 g/L of adsorbents for 1 h on a magnetic stirrer, and desorption of As was carried out by mixing with 200 mL NaOH [27] with a concentration 1 mol/L for 30 min. Washing of the adsorbent was carried out by 0.01 mol/L HCl solution until the solution pH turned neutral. Finally, the Fe_3O_4@SFBC nanoparticles were dried in an oven for 2 h at 60 °C under a vacuum. Nanoparticles were collected with an external magnet for the next cycling experiments.

3. Results

3.1. XRD and FTIR Spectroscopy Analysis

In order to evaluate the crystalline structure, an X-ray diffraction analysis of the synthesized Fe_3O_4@SFBC was carried out. Figure 1a shows the XRD patterns of Fe_3O_4@SFBC nanoparticles synthesized at different pyrolysis temperatures (SFBC/Fe mass ratio of 1:5), and Figure 1b reveals the XRD patterns of Fe_3O_4@SFBC nanoparticles produced at different SFBC/Fe mass ratios (biochar calcination temperature of 450 °C). The distinctive peaks of all samples were concentrated in the 2θ range of 30.4°, 35.6°, 43.3°, 53.7°, 57.4°, and 62.9°; agreeing with (400), (422), (220), (311), (511), and (440) crystal planes; and matching the Fe_3O_4 standard card (JCPDS card no. 19-0629). These patterns indicate that the samples are all Fd-3m counter spin Fe_3O_4 single crystals [28]. The SFBC in the Fe_3O_4@SFBC magnetic composite structure cannot alter the structure of Fe_3O_4. The main peak of SFBC 450 is located at 29.57°main peak (Figure 1a) matches the amorphous structure diffraction peak, linked to the crystalline structure of cellulose in the biomass structure. Additionally, metal salts affect biomass char yield. Additionally, the main peak, K, Ca, and Si diffraction peaks, are also present in SFBC 450 [29]. However, this sharp peak is found only in the nanoparticles of Fe_3O_4@SFBC 1:2 with the highest SFBC content, and the peak at 29.57° is not noted for the other samples, attributed to the high intensity of Fe_3O_4 crystals in other samples with a high strong crystallinity that masks the presence of amorphous structures [30].

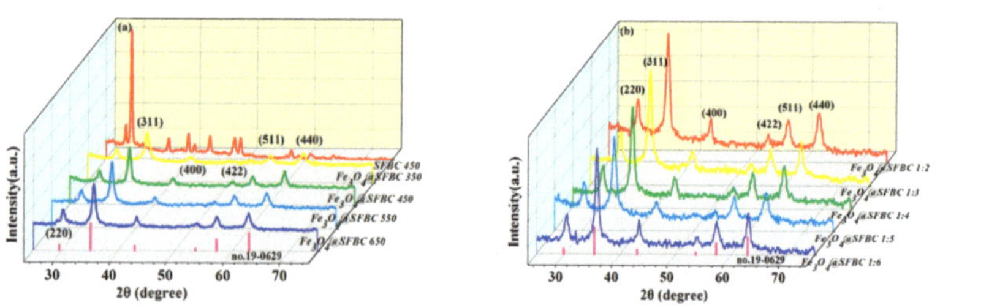

Figure 1. XRD spectra of Fe_3O_4 nanomaterials prepared by biochar at different calcination temperatures (**a**) and XRD spectra of Fe_3O_4 nanomaterials prepared by different SFBC/Fe mass ratios (**b**).

Figure 2a shows the FTIR patterns of Fe_3O_4@SFBC nanoparticles prepared by biochar synthesized at different pyrolysis temperatures (SFBC/Fe mass ratio of 1:5), and Figure 2b indicates the FTIR patterns of Fe_3O_4@SFBC nanoparticles manufactured according to diverse SFBC/Fe mass ratios (biochar calcination temperature of 450 °C). In both graphs, the broad peak observed around 3400 cm^{-1} denotes the hydroxyl group present in the sample. The FTIR spectra of Fe_3O_4@SFBC exhibit two peaks at 1634 cm^{-1} and 1200 cm^{-1} for bending of water and C-O stretching, respectively [31]. The peak at around 1400 cm^{-1} is associated with C-C and -COOH bond vibrations [32], and that bond vibration peak approves the formation of the carbonaceous structure of the composite. The peaks at about 583 cm^{-1} and 1140 cm^{-1} in the Fe_3O_4@SFBC spectra are the Fe-O stretching vibration peak and Fe-OH absorption peak [33]. The relative intensities of C-C, C-O, and Fe-O groups are, to some extent, altered as compared with the original SFBC and Fe_3O_4, demonstrating the successful synthesis of Fe_3O_4@SFBC nanoparticles. The peak intensities of the eight materials in both graphs are attributed to the different SFBC calcination temperatures and the mass ratios.

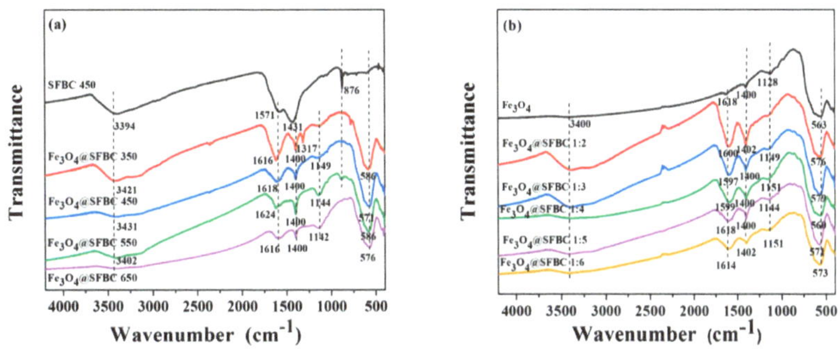

Figure 2. FTIR spectra of Fe_3O_4 nanomaterials prepared using biochar at different calcination temperatures (**a**) and FTIR spectra of Fe_3O_4 nanomaterials prepared by different SFBC/Fe mass ratios (**b**).

3.2. Arsenic Removal Experiments

3.2.1. Effect of pH of the Solution

According to previous studies, pH can affect the adsorption capacity and efficiency by influencing surface loading and ionization extent of different functional groups [34]. Hence, the adsorption efficiency of As(III/V) was determined by Fe_3O_4@SFBC (when the SFBC biochar is calcined at 450 °C with an SFBC/Fe mass ratio of 1:5) from the pH range of 3 to 11.

The initial pH effect on As(III/V) adsorption by Fe_3O_4@SFBC is shown in Figure 3. In brief, a change in the optimal pH of As(III/V) by Fe_3O_4@SFBC was observed, which is 6 and 4, respectively. The pKa1, pKa2, and pKa3 of H_3AsO_3 are 9.1, 12.1, and 13.4, respectively. The pKa1, pKa2, and pKa3 of H_3AsO_4 are 2.1, 6.7, and 11.2. Under most pH conditions, As(V) is present in a negative ionic form ($H_2AsO_4^-$ and $HAsO_4^{2-}$), whereas As(III) is in a nonionic form (H_3AsO_3).

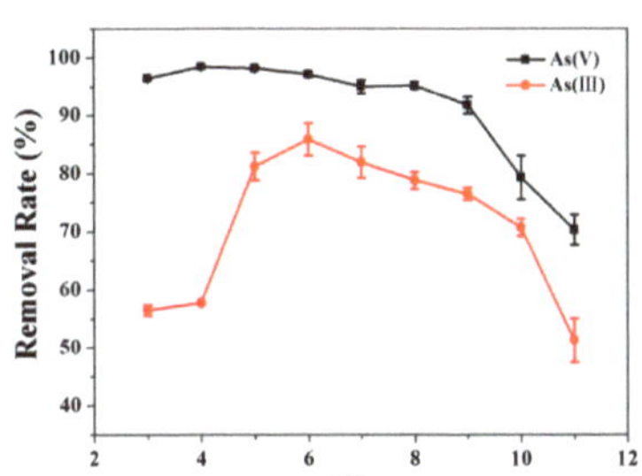

Figure 3. Effect of pH on the removal of As(III/V) (experimental conditions: pH = 3–11, dose = 1 g/L, initial concentration = 5 mg/L, T = 25 °C), *Y* error bars indicate the standard deviation of each data point (*n* = 3).

The rate of removal of As(III) increased directly with pH until pH 6 and decreased inversely with pH at greater than pH 6. This is attributed to the availability of As(III) in neutral form (H_3AsO_3) from pH 3 to 9. For As(III), the number of negatively charged arsenic species increases with increasing pH. However, the adsorption of As(III) on Fe_3O_4@SFBC composites did not decrease unidirectionally. The increase in the adsorption of As(III) in neutral solutions suggests that its adsorption process on Fe_3O_4@SFBC is followed by surface complexation rather than electrostatic interactions [35].

Moreover, greater As(V) removal efficiency is observed at low pH, while it decreases significantly at higher pH values. At pH 4, the maximum removal rate is 95.94%; at pH 11, it changes sharply to 67.67%. In general, the adsorption of As(V) on Fe_3O_4@SFBC composites is mainly governed by electrostatic attraction [36]. At smaller pH values, positively charged H^+ in the water leads to the strong electrostatic attraction between arsenate anions ($HAsO_4{}^{2-}$ and $AsO_4{}^{3-}$) and the positively charged adsorbent surface. However, at greater pH values, maximum -OH concentration makes the adsorbent negative charge, generating machine repulsion [37].

3.2.2. Adsorption Kinetics

The influence of contact time on As(III/V) adsorption was evaluated by atomic fluorescence spectrometry. A total of 0.5 g Fe_3O_4@SFBC was added to a 500 mL solution of 5 mg/L As(III) or As(V) solution, and pH 6 or 4 were maintained for As(III) or As(V) with contact time from 0 to 150 min. Samples were collected after adsorption, followed by filtration through a 0.45 μm membrane. The solids were dried for subsequent characterization. The adsorption process of Fe_3O_4@SFBC on arsenic were revealed by kinetic experiments. Models of pseudo-first-order (Equation (3)) and pseudo-second-order (Equation (4)) used are given below:

$$\ln(q_e - q_t) = \ln q_e - k_1 t \tag{3}$$

$$\frac{t}{q_t} = \frac{1}{k_2 {q_e}^2} + \frac{t}{q_e} \tag{4}$$

where t is the equilibration period (min), q_t and q_e are the adsorption amounts at time t and equilibrium (mg/g), respectively, k_1 (min^{-1}) and k_2 (g/(mg·min)) are the rate constants for pseudo-first-order and pseudo-second-order kinetics, respectively.

Figures 4 and 5 show the effect of contact time on As(III/V) adsorption of Fe_3O_4@SFBC prepared from biochar with different calcination temperatures and SFBC/Fe with different mass ratios. Regarding the adsorption of As by the above materials, compared to the pseudo-first-order, the experimental findings agree well with the pseudo-second-order. Obtained kinetics calculations are presented in Table 1. The higher value R^2 of the pseudo-second-order kinetics for the Fe_3O_4@SFBC nanomaterials indicates the experimental data are better and more accurate [38]. This also suggests that the process of adsorption follows chemisorption [39]. Furthermore, Figures 4 and 5 show that the adsorption of arsenic by Fe_3O_4@SFBC nanomaterials is a time-dependent process.

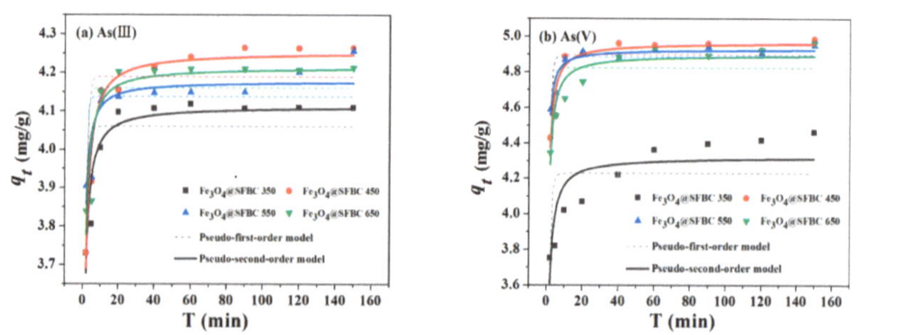

Figure 4. Adsorption kinetics of As(III) (**a**) and As (V) (**b**) on Fe_3O_4@SFBC (prepared from biochar at different calcination temperatures) (experimental conditions: pH = 6 and 4, dose = 1 g/L, initial concentration = 5 mg/L, T = 25 °C).

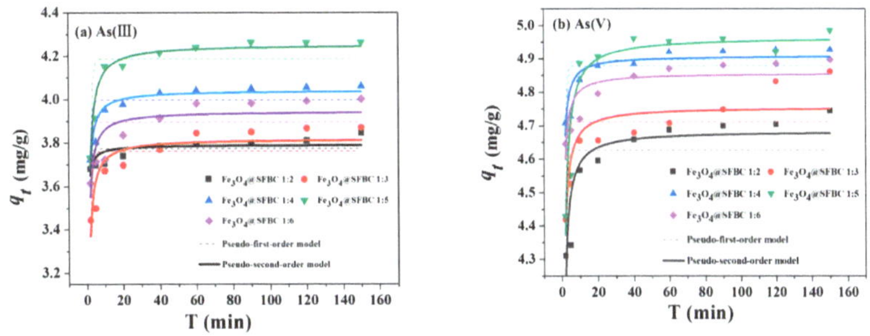

Figure 5. Adsorption kinetics of As(III) (**a**) and As(V) (**b**) on Fe_3O_4@SFBC (different SFBC/Fe mass ratios) (experimental conditions: pH = 6 and 4, dosage = 1 g/L, initial concentration = 5 mg/L, T = 25 °C).

Table 1. Fitting parameters of adsorption kinetics.

	Object	First-Order Equation			Second-Order Equation			Conditions
		K_1/min^{-1}	q_e/(mg/g)	R^2	K_2/(g/(mg·min))	q_e/(mg/g)	R^2	
Fe_3O_4@SFBC 350	As(III)	1.233	4.060	0.559	1.027	4.113	0.889	SFBC/Fe mass ratio 1:5
Fe_3O_4@SFBC 450	As(III)	1.080	4.189	0.684	0.768	4.254	0.951	
Fe_3O_4@SFBC 550	As(III)	1.422	4.138	0.435	1.432	4.178	0.777	
Fe_3O_4@SFBC 650	As(III)	1.254	4.161	0.496	1.025	4.215	0.828	
Fe_3O_4@SFBC 350	As(V)	1.051	4.228	0.371	0.584	4.323	0.727	
Fe_3O_4@SFBC 450	As(V)	1.148	4.896	0.609	0.741	4.966	0.898	
Fe_3O_4@SFBC 550	As(V)	1.391	4.884	0.613	1.245	4.927	0.901	
Fe_3O_4@SFBC 650	As(V)	1.136	4.822	0.578	0.706	4.898	0.887	
Fe_3O_4@SFBC 1:2	As(III)	1.851	3.772	0.280	3.427	3.791	0.603	calcination temperature of the SFBC is 450 °C
Fe_3O_4@SFBC 1:3	As(III)	1.323	3.763	0.447	0.979	3.821	0.808	
Fe_3O_4@SFBC 1:4	As(III)	1.323	3.998	0.575	1.265	4.042	0.906	
Fe_3O_4@SFBC 1:5	As(III)	1.080	4.189	0.684	0.768	4.254	0.951	
Fe_3O_4@SFBC 1:6	As(III)	1.301	3.896	0.414	1.141	3.946	0.750	
Fe_3O_4@SFBC 1:2	As(V)	1.319	4.627	0.450	1.006	4.684	0.819	
Fe_3O_4@SFBC 1:3	As(V)	1.380	4.710	0.493	1.201	4.756	0.779	
Fe_3O_4@SFBC 1:4	As(V)	1.670	4.879	0.447	1.986	4.909	0.820	
Fe_3O_4@SFBC 1:5	As(V)	1.148	4.896	0.609	0.741	4.966	0.898	
Fe_3O_4@SFBC 1:6	As(V)	1.644	4.824	0.386	1.844	4.857	0.746	

3.2.3. Adsorption Isotherm

Maximum adsorption of As(III/V) by Fe_3O_4@SFBC is evaluated by an adsorption isotherm. Initially, As(V) or As(III) solution concentrations are maintained at 0.1, 0.5, 1, 3, 5, 7, and 10 mg/L with same pH value as in kinetic experiment. The 0.1 g of Fe_3O_4@SFBC was mixed with 100 mL arsenic solution followed by 1 h shaking at constant temperature (25 °C). It was then placed in an external magnet for the separation of magnetic nanoparticles. A membrane filter was used for supernatant filtration and analyzed by atomic fluorescence spectrophotometer to assess the residual arsenic concentration. In order to study the adsorption behavior of As(III/V) after Fe_3O_4@SFBC, the Langmuir model (Equation (5)) and Freundlich model (Equation (6)) were used for fitting the experimental data.

$$q_e = \frac{q_m K_L C_e}{1 + K_L C_e} \tag{5}$$

$$q_e = K_f C_e^{1/n} \tag{6}$$

where C_e indicates As concentration (mg/L) at equilibrium, q_e is the adsorption capacity (mg/g) at adsorption equilibrium, and q_m represents the maximum adsorption capacity at single-molecule layer adsorption (mg/g). K_L is the Langmuir constant linked with the thermodynamics of the adsorption process, and n and K_f are the Freundlich constants.

The fitted curves of Langmuir and Freundlich are given in Figures 6–9. The isotherm parameters are revealed in Table 2. For As(III/V), the Langmuir isotherm and Freundlich isotherm fit the experimental data well, and the coefficient of determination R^2 does not differ significantly for both As(III/V).

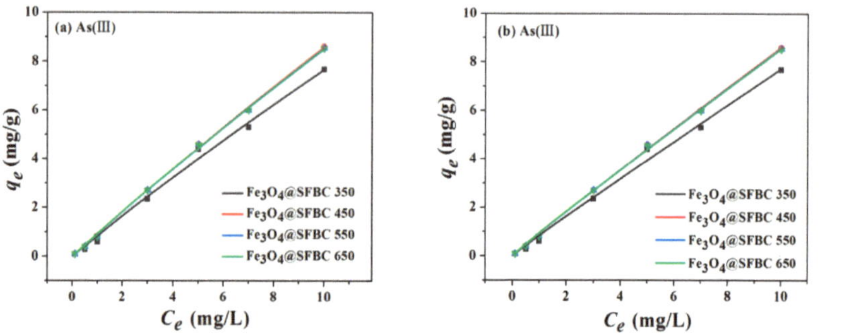

Figure 6. Adsorption isotherms of Langmuir (**a**) and Freundlich (**b**) on As(III) by Fe_3O_4@SFBC (prepared from biochar at different calcination temperatures) (experimental conditions: pH = 6, dose = 1 g/L, initial concentration = 0.1–10 mg/L, T = 25 °C).

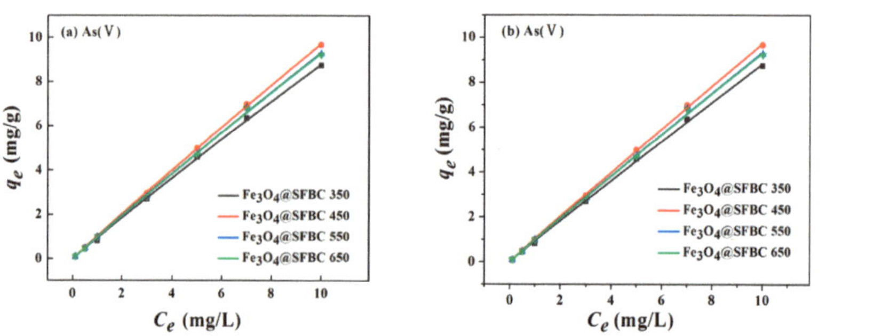

Figure 7. Adsorption isotherms of Langmuir (**a**) and Freundlich (**b**) on As(V) by Fe_3O_4@SFBC (prepared from biochar at different calcination temperatures) (experimental conditions: pH = 4, dose = 1 g/L, initial concentration = 0.1–10 mg/L, T = 25 °C).

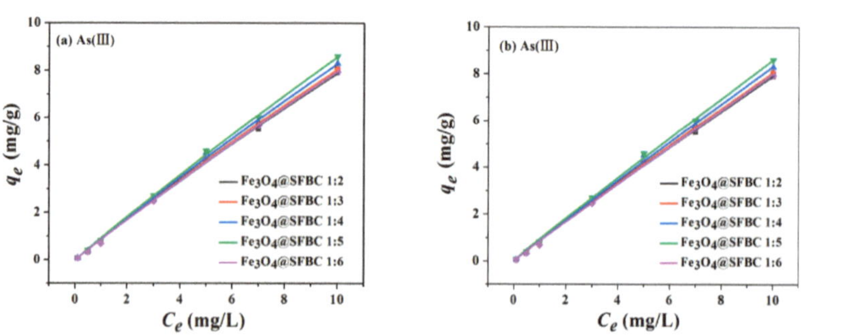

Figure 8. Adsorption isotherms of Langmuir (**a**) and Freundlich (**b**) on As(III) by Fe_3O_4@SFBC (different SFBC/Fe mass ratios) (experimental conditions: pH = 6, dose = 1 g/L, initial concentration = 0.1–10 mg/L, T = 25 °C).

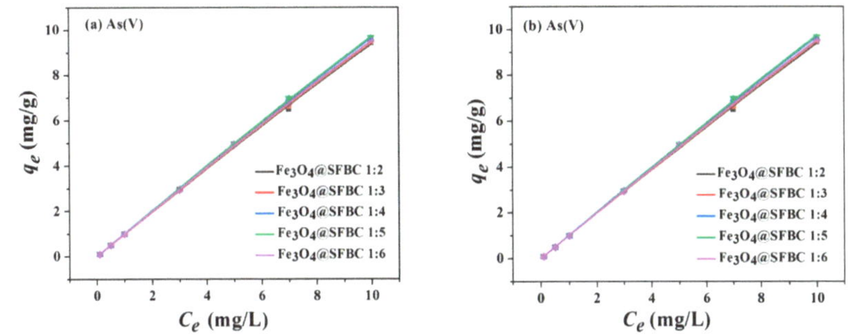

Figure 9. Adsorption isotherms of Langmuir (**a**) and Freundlich (**b**) on As(V) by Fe_3O_4@SFBC (different SFBC/Fe mass ratios) (experimental conditions: pH = 4, dose = 1 g/L, initial concentration = 0.1–10 mg/L, T = 25 °C).

Table 2. Optimized isotherm parameters for As(III/V) adsorption by Fe_3O_4@SFBC magnetic nanoparticles.

	Object	Langmuir			Freundlich			Conditions
		q_m/(mg/g)	K_L/(L/mg)	R^2	$1/n$	K_F/(m/mg)(L/mg)$^{1/n}$	R^2	
Fe_3O_4@SFBC 350	As(III)	86.565	0.00970	0.994	0.966	0.834	0.993	SFBC/Fe mass ratio 1:5
Fe_3O_4@SFBC 450	As(III)	121.374	0.00759	0.999	0.960	0.944	0.999	
Fe_3O_4@SFBC 550	As(III)	109.061	0.00847	0.999	0.961	0.937	0.998	
Fe_3O_4@SFBC 650	As(III)	107.977	0.00855	0.999	0.960	0.935	0.998	
Fe_3O_4@SFBC 350	As(V)	138.059	0.00680	0.999	0.974	0.938	0.999	
Fe_3O_4@SFBC 450	As(V)	188.753	0.00542	0.999	0.971	1.041	0.999	
Fe_3O_4@SFBC 550	As(V)	172.638	0.00571	0.999	0.974	0.994	0.999	
Fe_3O_4@SFBC 650	As(V)	166.117	0.00592	0.999	0.973	0.992	0.999	
Fe_3O_4@SFBC 1:2	As(III)	101.761	0.00840	0.999	0.956	0.878	0.999	calcination temperature of the SFBC is 450 °C
Fe_3O_4@SFBC 1:3	As(III)	105.290	0.00826	0.999	0.957	0.892	0.999	
Fe_3O_4@SFBC 1:4	As(III)	109.382	0.00819	0.998	0.957	0.919	0.999	
Fe_3O_4@SFBC 1:5	As(III)	121.374	0.00759	0.999	0.960	0.944	0.999	
Fe_3O_4@SFBC 1:6	As(III)	116.837	0.00731	0.999	0.961	0.874	0.999	
Fe_3O_4@SFBC 1:2	As(V)	170.653	0.00582	0.999	0.969	1.012	0.999	
Fe_3O_4@SFBC 1:3	As(V)	171.149	0.00584	0.999	0.969	1.019	0.999	
Fe_3O_4@SFBC 1:4	As(V)	181.723	0.00561	0.999	0.970	1.039	0.999	
Fe_3O_4@SFBC 1:5	As(V)	188.753	0.00542	0.999	0.971	1.041	0.999	
Fe_3O_4@SFBC 1:6	As(V)	176.037	0.00571	0.999	0.969	1.024	0.999	

Consequently, the removal of As(III/V) by Fe_3O_4@SFBC comprises both monolayer adsorption on a uniform surface and complex multilayer adsorption [40]. The optimal adsorption follows the Langmuir isotherm model based on the SFBC-modified Fe_3O_4@SFBC (SFBC/Fe of 1:5). The best adsorption system of Fe_3O_4@SFBC material prepared from SFBC material calcined at 450 °C according to the SFBC/Fe mass ratio of 1:5 was studied, and the absorption capacity of As(III) and As(V) is around 121.374 and 188.753 mg/g, respectively. Moreover, the adsorption of As(III/V) by Fe_3O_4@SFBC nanomaterials is greater than that of other adsorbents in previous articles (Table 3).

Biochar mostly contains elements such as C, H, O, and N, and its surface contains carboxyl, phenolic hydroxyl, carbonyl, anhydride, and other groups [41]. The physicochemical characteristics of biochar are closely linked to the pyrolysis temperature, and the pyrolysis process can be roughly divided into three temperature stages, 220–310 °C for hemicellulose pyrolysis and 315–400 °C for cellulose pyrolysis, and when the temperature exceeds 500 °C, the structure of biochar usually contains only an aromatic ring structure [42,43]. Moreover, the rise in the specific surface area of biochar with the increasing pyrolysis temperature and adsorption sites are useful for the adsorption of heavy metals; therefore, the temperature conditions should be controlled in the practical application to achieve the best results. These results suggest that the material prepared using biochar fired at 450 °C exhibits the best adsorption performance for the studied metal ions since it possesses more C=C and C-H functional groups at this temperature [44,45].

Table 3. Various adsorbent materials' Langmuir adsorption capacities.

Fe-Biochar-Based Sorbents	Object	Maximum Adsorption Capacity (mg/g)	Reference
Fe_3O_4@SFBC	As(III) As(V)	121.374 188.753	This study
Fe_3O_4@CSAC	As(III) As(V)	80.99 107.96	[46]
Iron-modified activated carbon	As(V)	43.6	[47]
Iron(III) loaded orange waste	As(V)	68.6	[48]
Bioinspired 2D-carbon flakes-Fe_3O_4	As(III)	57.47	[49]
Fe-Zr-BC	As(III) As(V)	107.57 40.79	[50]
magneticgelatin-modified biochar	As(V)	42.7	[51]

The use of optimal doses of biochar is essential to maximize arsenic removal from water. Biochar adsorption efficiency decreases with a high concentration above the optimum limit [22]. Therefore, changing the ratio of SFBC to iron ions in Fe_3O_4@SFBC material can improve optimal adsorption. This can be explained by the functional site saturation on the surface of biochar and partially due to available binding sites activation or pores on the biochar surface by the limited amount of Fe ions, which saturates the surface functional sites of the Fe_3O_4@SFBC material during the adsorption process.

3.3. *Discussion of Morphology and Adsorption Mechanism*

Fe_3O_4@SFBC (SFBC calcination temperature is 450 °C, SFBC/Fe is 1:5), which has the largest adsorption capacity, was taken as an example.

3.3.1. Morphology

The morphology of Fe_3O_4@SFBC was further examined by SEM and TEM. The surface of SFBC is fragmented, with uneven particle size and irregular surface shape, filled with porous structures and unevenness, as demonstrated in Figure 10a. SEM and TEM images of Fe_3O_4@SFBC show that Fe_3O_4 combines with SFBC biomass charcoal to form spherical nanoparticles of uniform size (Figure 10b,c). The histogram of the particle size achieved from the corresponding transmission electron microscopy image (upper right corner of Figure 10c) displays that in Fe_3O_4@SFBC, particle size is primarily homogeneously distributed in the range of 12–14 nm. The lattice stripe spacing diagram of Fe_3O_4@SFBC was obtained from Figure 10d with 0.24 nm a lattice gap, corresponding to the (222) crystal lattice plane in XRD, again confirming that the composition of the prepared sample is mainly Fe_3O_4 single crystal.

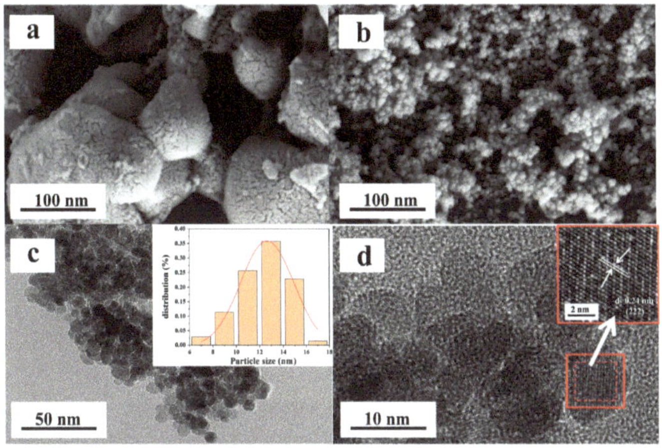

Figure 10. SEM images of SFBC (**a**) and Fe_3O_4@SFBC (**b**). TEM particle size distribution (**c**) and TEM lattice stripe (**d**) patterns of Fe_3O_4@SFBC.

3.3.2. Magnetic Properties

Magnetic strength is a key characteristic of magnetic adsorbents as it helps in the separation of the adsorbent from the aqueous solution and decreases the method cost. Fe_3O_4 nanoparticles and Fe_3O_4@SFBC magnetic nanocomposites were analyzed in the range of −2000 Oe to 2000 Oe, as shown in Figure 11. The saturation magnetizations (Ms) and magnetic remanence (Mr) values of Fe_3O_4 and Fe_3O_4@SFBC are 74.02 and 1.85, 54.52 and 1.22 emu/g, respectively, while the corresponding coercivity (Hc) values are 15.01 and 11.30 Oe. A non-magnetic matrix (SFBC) may be responsible for the reduction in magnetic saturation in the prepared magnetic nanocomposites.

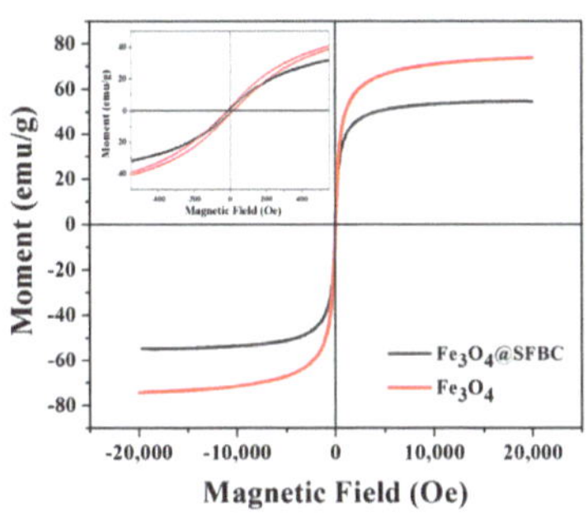

Figure 11. Magnetic hysteresis loop of Fe_3O_4 and Fe_3O_4@SFBC nanoparticles.

3.3.3. Adsorption Mechanism

The adsorption mechanism of As(III/V) using Fe_3O_4@SFBC was further evaluated by XPS. Changes in the chemical state of O, As, Fe, and C in Fe_3O_4@SFBC were determined before and after adsorption.

Figure 12a shows three peaks appearing in the O 1s spectra at binding energies of 530, 531, and 532–533 eV, corresponding to lattice oxygen (O_2), hydroxyl group, and adsorbed water in metal oxides, and the peaks at binding energies of 530.4 and 531.4 eV were assigned to Fe-O-H and Fe-O-C, respectively. These peaks are associated with the formation of the matching iron oxides during the adsorption of iron and oxygen. After adsorption of As(III/V), the main peak height of O 1s increased, attributed to the adsorption of arsenic by Fe_3O_4@SFBC (Figure 12a) [52]. The obtained findings show that the surface complexation between functional groups and heavy metals is complemented by the electrostatic attraction between O atoms and heavy metal ions.

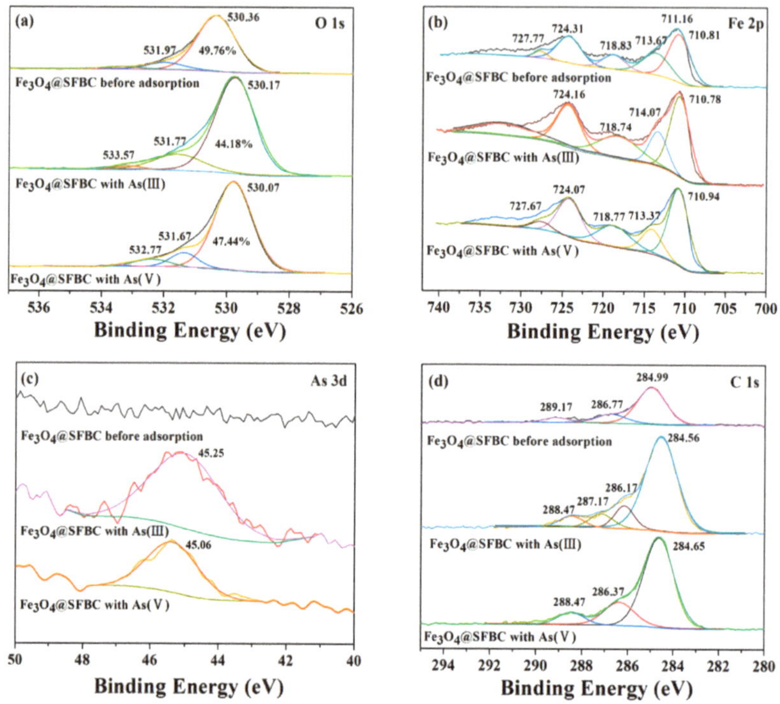

Figure 12. XPS spectra of Fe_3O_4@SFBC before and after arsenic adsorption: O 1s (**a**), Fe 2p (**b**), As 3d (**c**), C 1s (**d**).

Figure 12b depicts the spectrum of Fe 2p indicating the peaks at 711.05 eV and 724.85 eV representing Fe $2p_{3/2}$ and Fe $2p_{1/2}$, respectively, confirming the coexistence of Fe(II) and Fe(III) in the synthesized Fe_3O_4@SFBC [53]. The presence of Fe(II) and Fe(III) is attributed to the formation of Fe_2O_3 at 712.66 eV and FeOOH at 725.85 eV. Therefore, hydrogen bonding may have occurred during the adsorption process.

The peak of As 3d (45.25, 45.06 eV) after As(III/V) adsorption indicates that Fe_3O_4@SFBC has some adsorption capacity for As (Figure 12c) [54].

Figure 12d indicates the C 1s spectrum, with three main peaks, namely C-C (284 eV), C-O-C (286 eV), and O-C=O (287 eV) and C=O (288 eV) [55]. After As(III/V) adsorption, binding energies of C-C, C-O-C, and C=O reduced by 0.43/0.60/0.70 eV and 0.34/0.40/0.70 eV, respectively. The decrease in material binding energy after adsorption may be due to the formation of coordination complexes between Fe_3O_4@SFBC and As [56]. The results for C1s and O1s indicate the complex formation. The O atoms in C=O and C-O share an electron pair with As(III/V), thus affecting the density of the electronic cloud between the carbon and adjacent O atoms, thereby changing the binding energy [57]. Briefly, the adsorption of Fe_3O_4@SFBC on arsenic mainly includes electrostatic adsorption and ligand complexation.

To better understand the mechanism of arsenic adsorption, FTIR analysis was performed on the samples before and after the adsorption of As(III/V). Figure 13 shows that Fe_3O_4@SFBC has a stretching vibration peak of water at 3402 cm^{-1}. However, after adsorption of As(III/V), the bend at 3421 cm^{-1} shifted to 3421 and 3429 cm^{-1}, respectively, and these changes suggest that As(III/V) reacts with -OH in the protonation reaction. Moreover, the peaks in the range of 1400 cm^{-1} are significantly shifted, indicating that the vibrational correlation of C-C and -COOH is also closely related to the adsorption of As(III/V). The disappearance of the Fe-OH peak at 1144 cm^{-1} after adsorption of As(III/V) indicates that the arsenic compounds may have redox reactions with the active sites of Fe ion-related compounds during the adsorption process [58]. The main peaks after adsorption are shifted to changed degrees, representing that arsenic can be adsorbed on the surface of Fe_3O_4@SFBC by ion exchange, while the changes in -OH, -COOH, C-C, and Fe-OH indicate that the adsorbent experiences hydroxyl complexation on the surface [59].

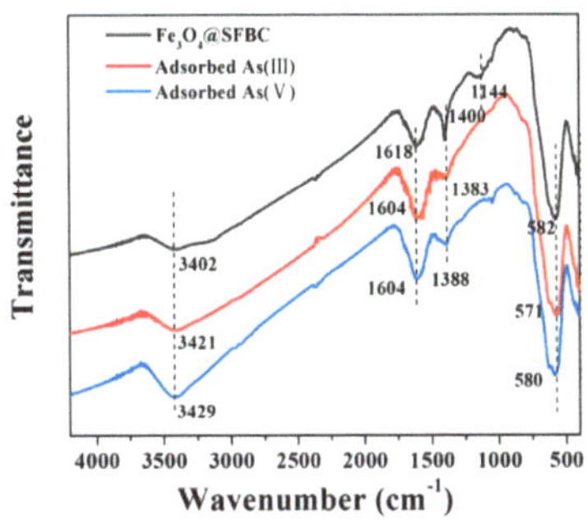

Figure 13. FTIR spectra of Fe_3O_4@SFBC before and after the adsorption of As(III/V).

3.3.4. Regeneration and Reusability

In order to ensure economic practicality, adsorption–desorption reproducibility experiments are performed on the Fe_3O_4@SFBC adsorbent. As(III/V) on the adsorbed Fe_3O_4@SFBC are desorbed with different concentrations of sodium hydroxide solution. According to the results reported by Jin et al. [60], the best desorption of As is achieved by 1 mol/L NaOH. Thus, 1 mol/L NaOH was used for desorption. The arsenic removal efficiency of the regenerated adsorbent is subjected to five consecutive cycles, as shown in Figure 14. The adsorbent exhibits good removal rates in five consecutive cycles, with As(III) and As(V) decreasing from 89.20% to 77.89% and 98.35% to 76.27%, respectively. Therefore, the adsorbent has an efficient renewal potential, and it may be effectively and continuously utilized in arsenic-contaminated wastewater treatment.

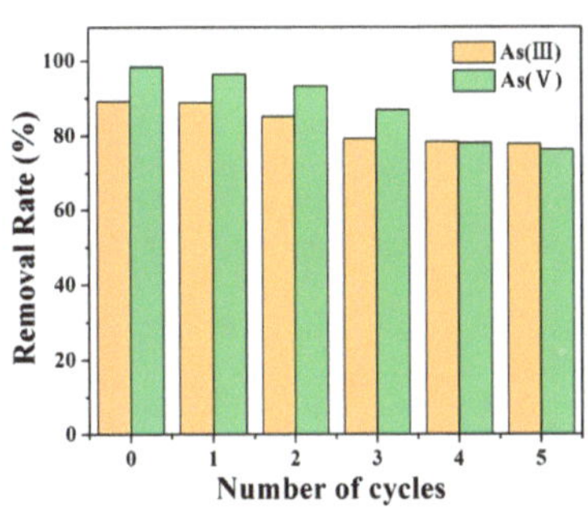

Figure 14. Recycling performance of different Fe_3O_4@SFBC materials (pH = 6 and 4, dose = 1 g/L, initial concentration = 5 mg/L, T = 25°C).

4. Conclusions

In this study, sunflower straw biochar was combined with Fe_3O_4 to prepare an efficient adsorbent for arsenic treatment. The adsorption batch experiments were carried out using Fe_3O_4@SFBC materials prepared calcinated at different temperatures and different SFBC/Fe mass ratios, respectively, and the results revealed that all materials conformed to the quasi-secondary kinetic equation and were consistent with the chemisorption process. As stated by the Langmuir model, the 450 °C fired SFBC showed the highest adsorption effect at pH 6 and 4 when SFBC/Fe ratio was 1:5, with maximum adsorption capacities of 121.374 and 188.753 mg/g for As(III) and As(V), respectively. Fe_3O_4@SFBC, compared to activated carbon materials and metal oxide composites, exhibited a stronger adsorption capacity for arsenic. The morphology and adsorption mechanism of Fe_3O_4@SFBC were studied, and it was found that most of the particles were 12–14 nm in size with a small size effect and good magnetization performance. Based on XPS and FTIR results, the proposed mechanism of interaction was based on chemisorption, complexation, and electrostatic interactions. Functional groups containing O, Fe, and C intrinsic to Fe_3O_4@SFBC played an important role in the removal of As(III/V).

Furthermore, adsorption–desorption treatment of Fe_3O_4@SFBC after five consecutive cycles was around 77.89% and 76.27% for As(III) and As(V) removal, respectively. Therefore, Fe_3O_4@SFBC can be used well as an adsorbent for arsenic removal in wastewater treatment.

Author Contributions: F.Y. and Y.Z. conceived and designed the experiment; Y.Z. and H.S. performed the experiment; F.Y., Y.Z., X.T., J.Z., D.K. and H.S. analyzed the data and prepared the manuscript. All authors have read and agreed to the published version of the manuscript.

Funding: This work was supported by the Natural Science Foundation of Xinjiang Uygur Autonomous Region, China [grant number 2022D01B20], the 67th batch of general project funding from China Postdoctoral Science Foundation [grant number 2020M673643XB], Xinjiang Key Laboratory of Water Resources Engineering Safety and Water Disaster Prevention and Control 2022 Open Subjects [grant number ZDSYS-JS-2022-13], the National Natural Science Foundation of China (NNSFC) [grant number 42067035, 42007161] and 2022 Graduate Research Innovation Program of Xinjiang Agricultural University [grant number XJAUGRI2022044].

Institutional Review Board Statement: Not applicable.

Informed Consent Statement: Not applicable.

Data Availability Statement: The data presented in this article are available on request from the corresponding authors.

Conflicts of Interest: The authors declare no conflict of interest.

References

1. Singh, S.; Parihar, P.; Singh, V.P.; Prasad, S.M. Arsenic contamination, consequences and remediation techniques: A review. *Ecotoxicol. Environ. Saf.* **2015**, *112*, 247–270. [CrossRef] [PubMed]
2. Agrawal, A.; Petersen, M.R. Detecting Arsenic Contamination Using Satellite Imagery and Machine Learning. *Toxics* **2021**, *9*, 333. [CrossRef] [PubMed]
3. Nordstrom, D.K. Worldwide occurrences of arsenic in ground water. *Science* **2002**, *296*, 2143–2145. [CrossRef]
4. Bissen, M.; Frimmel, F.H. Arsenic—A review. Part I: Occurrence, toxicity, speciation, mobility. *Acta Hydroch. Hydrob.* **2003**, *31*, 9–18. [CrossRef]
5. World Health Organization. *Guidelines for Drinking-Water Quality*, 4th ed.; WHO: Geneva, Switzerland, 2011; ISBN 978-92-4-154815-1.
6. He, Y.; Liu, J.; Han, G.; Chung, T.S. Novel thin-film composite nanofiltration membranes consisting of a zwitterionic co-polymer for selenium and arsenic removal. *J. Membr. Sci.* **2018**, *555*, 299–306. [CrossRef]
7. Yadanaparthi, S.K.R.; Graybill, D.; von Wandruszka, R. Adsorbents for the removal of arsenic, cadmium, and lead from contaminated waters. *J. Hazard. Mater.* **2009**, *171*, 1–15. [CrossRef]
8. Alvarado, S.; Guédez, M.; Lué-Merú, M.P.; Nelson, G.; Alvaro, A.; Jesús, A.C.; Gyula, Z. Arsenic removal from waters by bioremediation with the aquatic plants Water Hyacinth and Lesser Duckweed. *Bioresour. Technol.* **2008**, *99*, 8436–8440. [CrossRef]
9. Gecol, H.; Ergican, E.; Fuchs, A. Molecular level separation of arsenic (V) from water using cationic surfactant micelles and ultrafiltration membrane. *J. Membr. Sci.* **2004**, *241*, 105–119. [CrossRef]
10. Arcibar-Orozco, J.A.; Josue, D.B.; Rios-Hurtado, J.C.; Rangel-Mendez, J.R. Influence of iron content, surface area and charge distribution in the arsenic removal by activated carbons. *Chem. Eng. J.* **2014**, *249*, 201–209. [CrossRef]
11. Matusik, J. Arsenate, orthophosphate, sulfate, and nitrate sorption equilibria and kinetics for halloysite and kaolinites with an induced positive charge. *Chem. Eng. J.* **2014**, *246*, 244–253. [CrossRef]
12. Faria, M.C.; Rosemberg, R.S.; Bomfeti, C.A.; Monteiro, D.S.; Barbosa, F.; Oliveira, L.C.; Rodriguez, M.; Pereira, M.C.; Rodrigues, J.L. Arsenic removal from contaminated water by ultrafine δ-FeOOH adsorbents. *Chem. Eng. J.* **2014**, *237*, 47–54. [CrossRef]

13. Fox, D.I.; Pichler, T.; Yeh, D.H.; Alcantar, N.A. Removing heavy metals in water: The interaction of cactus mucilage and arsenate (As (V)). *Environ. Sci. Technol.* **2012**, *46*, 4553–4559. [CrossRef] [PubMed]
14. Arora, R. Nano adsorbents for removing the arsenic from waste/ground water for energy and environment management—A review. *Mater. Today Proc.* **2021**, *45*, 4437–4440. [CrossRef]
15. Lata, S.; Samadder, S.R. Removal of arsenic from water using nano adsorbents and challenges: A review. *J. Environ. Manag.* **2016**, *166*, 387–406. [CrossRef] [PubMed]
16. Awual, M.R.; Hasan, M.M.; Asiri, A.M.; Rahman, M.M. Cleaning the arsenic(V) contaminated water for safe-guarding the public health using novel composite material. *Compos. Part B* **2019**, *171*, 294–301. [CrossRef]
17. Shahat, A.; Hassan, H.M.; Azzazy, H.M.; Hosni, M.; Awual, M.R. Novel nano-conjugate materials for effective arsenic(V) and phosphate capturing in aqueous media. *Chem. Eng. J.* **2018**, *331*, 54–63. [CrossRef]
18. Wang, T.; Zhang, L.; Wang, H.; Yang, W.; Fu, Y.; Zhou, W.; Yu, W.; Xiang, K.; Su, Z.; Dai, S.; et al. Controllable synthesis of hierarchical porous Fe_3O_4 particles mediated by poly and their application in arsenic removal. *ACS Appl. Mater. Interfaces* **2013**, *5*, 12449–12459. [CrossRef]
19. Zhan, H.; Bian, Y.; Yuan, Q.; Ren, B.; Hursthouse, A.; Zhu, G. Preparation and potential applications of super paramagnetic nano-Fe_3O_4. *Processes* **2018**, *6*, 33. [CrossRef]
20. Maity, D.; Agrawal, D.C. Synthesis of iron oxide nanoparticles under oxidizing environment and their stabilization in aqueous and non-aqueous media. *J. Magn. Magn. Mater.* **2007**, *308*, 46–55. [CrossRef]
21. Wang, C.; Chen, S.; Ye, C.; Wang, X. Research progress on monodisperse Fe_3O_4 magnetic nanoparticles. *Chem. Ind. Eng. Pro.* **2016**, *35*, 242–247.
22. Chen, X.; Chen, G.; Chen, L.; Chen, Y.; Lehmann, J.; McBride, M.B.; Hay, A.G. Adsorption of copper and zinc by biochars produced from pyrolysis of hardwood and corn straw in aqueous solution. *Bioresour. Technol.* **2011**, *102*, 8877–8884. [CrossRef] [PubMed]
23. Li, R.; Wang, J.J.; Gaston, L.A.; Zhou, B.; Li, M.; Xiao, R.; Wang, Q.; Zhang, Z.; Huang, H.; Liang, W.; et al. An overview of carbothermal synthesis of metal–biochar composites for the removal of oxyanion contaminants from aqueous solution. *Carbon* **2018**, *129*, 674–687. [CrossRef]
24. Wang, Y.; Zheng, K.; Jiao, Z.; Zhan, W.; Ge, S.; Ning, S.; Ning, S.; Fang, S.; Ruan, X. Simultaneous Removal of Cu(II), Cd(II) and Pb(II) by Modified Wheat Straw Biochar from Aqueous Solution: Preparation, Characterization and Adsorption Mechanism. *Toxics* **2022**, *10*, 316. [CrossRef] [PubMed]
25. Wen, T.; Wang, J.; Yu, S.; Chen, Z.; Hayat, T.; Wang, X. Magnetic porous carbonaceous material produced from tea waste for efficient removal of As (V), Cr (VI), humic acid, and dyes. *ACS Sustain. Chem. Eng.* **2017**, *5*, 4371–4380. [CrossRef]
26. Wang, S.; Gao, B.; Li, Y.; Creamer, A.E.; He, F. Adsorptive removal of arsenate from aqueous solutions by biochar supported zero-valent iron nanocomposite: Batch and continuous flow tests. *J. Hazard. Mater.* **2017**, *322*, 172–181. [CrossRef]
27. Koomson, B.; Asiam, E.K. Arsenic adsorption by some iron oxide minerals: Influence of interfacial chemistry. *J. Gha. Min.* **2020**, *20*, 43–48. [CrossRef]
28. Lima, M.J.; Sampaio, M.J.; Silva, C.G.; Silva, A.M.; Faria, J.L. Magnetically recoverable Fe_3O_4/g-C_3N_4 composite for photocatalytic production of benzaldehyde under UV-led radiation. *Catal. Today* **2019**, *328*, 293–299. [CrossRef]
29. He, J.; He, C.; Guo, H.; Ming, L. Adsorption of methylene blue by five straw biochars and its performance comparison. *J. Nanjing Agric. Univ.* **2019**, *42*, 382–388. [CrossRef]
30. Mahmoud, M.E.; Saad, S.R.; El-Ghanam, A.M.; Mohamed, R.H.A. Developed magnetic Fe_3O_4–MoO_3-AC nanocomposite for effective removal of ciprofloxacin from water. *Mater. Chem. Phys.* **2021**, *257*, 123454. [CrossRef]
31. Malkoc, E.; Nuhoglu, Y. Fixed bed studies for the sorption of chromium(VI) onto tea factory waste. *Chem. Eng. Sci.* **2006**, *61*, 4363–4372. [CrossRef]
32. Marriam, I.; Xu, F.; Tebyetekerwa, M.; Gao, Y.; Liu, W.; Liu, X.; Qiu, Y. Synergistic effect of CNT films impregnated with CNT modified epoxy solution towards boosted interfacial bonding and functional properties of the composites. *Compos. Part A Appl. S.* **2018**, *110*, 1–10. [CrossRef]
33. Lin-Vien, D.; Colthup, N.B.; Fateley, W.G.; Grasselli, J.G. *The Handbook of Infrared and Raman Characteristic Frequencies of Organic Molecules*; Elsevier: Amsterdam, The Netherlands, 1991.
34. Baidya, K.S.; Kumar, U. Adsorption of brilliant green dye from aqueous solution onto chemically modified areca nut husk. *S. Afr. J. Chem. Eng.* **2021**, *35*, 33–43. [CrossRef]
35. Chandra, V.; Park, J.; Chun, Y.; Lee, J.W.; Hwang, I.C.; Kim, K.S. Water-dispersible magnetite-reduced graphene oxide composites for arsenic removal. *ACS Nano* **2010**, *4*, 3979–3986. [CrossRef] [PubMed]
36. Guo, L.; Ye, P.; Wang, J.; Fu, F.; Wu, Z. Three-dimensional Fe_3O_4-graphene macroscopic composites for arsenic and arsenate removal. *J. Hazard. Mater.* **2015**, *298*, 28–35. [CrossRef]
37. Zhu, H.; Jia, Y.; Wu, X.; Wang, H. Removal of arsenic from water by supported nano zero-valent iron on activated carbon. *J. Hazard. Mater.* **2009**, *172*, 1591–1596. [CrossRef]
38. Cui, H.; Li, Q.; Gao, S.; Shang, J.K. Strong adsorption of arsenic species by amorphous zirconium oxide nanoparticles. *J. Ind. Eng. Chem.* **2012**, *18*, 1418–1427. [CrossRef]
39. Qiu, H.; Lv, L.; Pan, B.C.; Zhang, Q.J.; Zhang, W.M.; Zhang, Q.X. Critical review in adsorption kinetic models. *J. Zhejing Univ. Sci. A* **2009**, *10*, 716–724. [CrossRef]

40. Li, Y.; Kang, X.; Qiao, Y.; Chen, M. Preparation of clay biochar from spent bleaching earth and durian shell and its adsorption for Cr(VI). *Fine. Chem.* **2022**, *39*, 178–186. [CrossRef]
41. Keiluweit, M.; Nico, P.S.; Johnson, M.G.; Kleber, M. Dynamic molecular structure of plant biomass-derived black carbon. *Environ. Sci. Technol.* **2010**, *44*, 1247–1253. [CrossRef]
42. Xiao, X.; Chen, B.; Zhu, L. Transformation, morphology, and dissolution of silicon and carbon in rice straw-derived biochars under different pyrolytic temperatures. *Environ. Sci. Technol.* **2014**, *48*, 3411–3419. [CrossRef]
43. Yang, H.; Yan, R.; Chen, H.; Lee, D.H.; Zheng, C. Characteristics of hemicellulose, cellulose and lignin pyrolysis. *Fuel* **2007**, *86*, 1781–1788. [CrossRef]
44. Li, J.H.; Lv, G.H.; Bai, W.B.; Liu, Q.; Zhang, Y.C.; Song, J.Q. Modification and use of biochar from wheat straw for nitrate and phosphate removal from water. *Desalin. Water Treat.* **2016**, *57*, 4681–4693. [CrossRef]
45. Qian, K.; Kumar, A.; Zhang, H.; Bellmer, D.; Huhnke, R. Recent advances in utilization of biochar. *Renew. Sustain. Energy Rev.* **2015**, *42*, 1055–1064. [CrossRef]
46. Sahu, U.K.; Sahu, S.; Mahapatra, S.S.; Patel, R.K. Cigarette soot activated carbon modified with Fe_3O_4 nanoparticles as an effective adsorbent for As(III) and As(V): Material preparation, characterization and adsorption mechanism study. *J. Mol. Liq.* **2017**, *243*, 395–405. [CrossRef]
47. Chen, W.; Parette, R.; Zou, J.; Cannon, F.S.; Dempsey, B.A. Arsenic removal by iron-modified activated carbon. *Water Res.* **2007**, *41*, 1851–1858. [CrossRef] [PubMed]
48. Ghimire, K.N.; Inoue, K.; Yamaguchi, H.; Makino, K.; Miyajima, T. Adsorptive separation of arsenate and arsenite anions from aqueous medium by using orange waste. *Water Res.* **2003**, *37*, 4945–4953. [CrossRef]
49. Venkateswarlu, S.; Lee, D.; Yoon, M. Bioinspired 2D-carbon flakes and Fe_3O_4 nanoparticles composite for arsenite removal. *ACS Appl. Mater. Interfaces* **2016**, *8*, 23876–23885. [CrossRef]
50. Peng, Y.; Azeem, M.; Li, R.; Xing, L.; Li, Y.; Zhang, Y.; Guo, Z.; Wang, Q.; Ngo, H.H.; Qu, G.; et al. Zirconium hydroxide nanoparticle encapsulated magnetic biochar composite derived from rice residue: Application for As (III) and As (V) polluted water purification. *J. Hazard. Mater.* **2022**, *423*, 127081. [CrossRef]
51. Zhou, Z.; Liu, Y.G.; Liu, S.B.; Liu, H.Y.; Zeng, G.M.; Tan, X.F.; Yang, C.P.; Ding, Y.; Yan, Z.L.; Cai, X.X. Sorption performance and mechanisms of arsenic (V) removal by magnetic gelatin-modified biochar. *Chem. Eng. J.* **2017**, *314*, 223–231. [CrossRef]
52. Hu, X.; Ding, Z.; Zimmerman, A.R.; Wang, S.; Gao, B. Batch and column sorption of arsenic onto iron-impregnated biochar synthesized through hydrolysis. *Water Res.* **2015**, *68*, 206–216. [CrossRef]
53. Cai, Z.; Liu, Q.; Li, H.; Wang, J.; Tai, G.; Wang, F.; Han, J.; Zhu, Y.; Wu, G. Waste-to-Resource Strategy to Fabricate Functionalized MOFs Composite Material Based on Durian Shell Biomass Carbon Fiber and Fe_3O_4 for Highly Efficient and Recyclable Dye Adsorption. *Int. J. Mol. Sci.* **2022**, *23*, 5900. [CrossRef] [PubMed]
54. Alchouron, J.; Navarathna, C.; Chludil, H.D.; Dewage, N.B.; Perez, F.; Hassand, E.B.; Pittman Jr, C.U.; Vegaae, A.S.; Mlsna, T.E. Assessing South American Guadua chacoensis bamboo biochar and Fe_3O_4 nanoparticle dispersed analogues for aqueous arsenic (V) remediation. *Sci. Total Environ.* **2020**, *706*, 135943. [CrossRef]
55. Yi, Y.; Tu, G.; Zhao, D.; Tsang, P.E.; Fang, Z. Key role of FeO in the reduction of Cr(VI) by magnetic biochar synthesised using steel pickling waste liquor and sugarcane bagasse. *J. Clean. Prod.* **2020**, *245*, 118886. [CrossRef]
56. Baig, S.A.; Zhu, J.; Muhammad, N.; Sheng, T.; Xu, X. Effect of synthesis methods on magnetic Kans grass biochar for enhanced As(III, V) adsorption from aqueous solutions. *Biomass Bioenergy* **2014**, *71*, 299–310. [CrossRef]
57. Zhou, J.; Liu, Y.; Zhou, X.; Ren, J.; Zhong, C. Removal of mercury ions from aqueous solution by thiourea-functionalized magnetic biosorbent: Preparation and mechanism study. *J. Colloid Interface Sci.* **2017**, *507*, 107–118. [CrossRef] [PubMed]
58. Balint, R.; Bartoli, M.; Jagdale, P.; Tagliaferro, A.; Memon, A.S.; Rovere, M.; Martin, M. Defective bismuth oxide as effective adsorbent for arsenic removal from water and wastewater. *Toxics* **2021**, *9*, 158. [CrossRef] [PubMed]
59. Wu, J.; Huang, D.; Liu, X.; Meng, J.; Tang, C.; Xu, J. Remediation of As(III) and Cd(II) co-contamination and its mechanism in aqueous systems by a novel calcium-based magnetic biochar. *J. Hazard. Mater.* **2018**, *348*, 10–19. [CrossRef]
60. Jin, Y.; Liu, F.; Tong, M.; Hou, Y. Removal of arsenate by cetyltrimethylammonium bromide modified magnetic nanoparticles. *J. Hazard. Mater.* **2012**, *227*, 461–468. [CrossRef]

toxics

Article

Study on Tritium and Iodine Species Transport through Porous Granite: A Non-Sorption Effect by Anion Exclusion

Yunfeng Shi [1], Song Yang [1], Wenjie Chen [1], Weijia Xiong [2], Aiming Zhang [1], Zhixiang Yu [1], Bing Lian [1,*] and Chuan-Pin Lee [3,*]

1 Department of Nuclear Environmental Science, China Institute for Radiation Protection (CIRP), Taiyuan 030006, China
2 CNNC Environmental Protection Corporation (ECPC), Beijing 100045, China
3 School of Nuclear Science and Engineering, East China University of Technology, Nanchang 330013, China
* Correspondence: bennis6723@ecut.edu.cn (B.L.); bennis6723@gmail.com (C.-P.L.); Tel.: +86-151-03418769 (B.L.)

Abstract: The safety of deep geological repositories is important in the disposal of high-level radioactive waste (HLW). In this study, advection–dispersion experiments were designed to build a transport model through a calibration/validation process, and the transport behavior of tritiated water (HTO) and various iodine species (iodide: I^- and iodate: IO_3^-) was studied on a dynamic compacted granite column. Breakthrough curves (BTCs) were plotted under various flow rates (1–5 mL/min). BTCs showed that the non-sorption effect by anion exclusion was observed only in I^- transport because the retardation factor (R) of I^- was lower than that of HTO (R = 1). Moreover, equilibrium and nonequilibrium transport models were used and compared to identify the mobile/immobile zones in the compacted granite column. The anion exclusion effect was influenced by the immobile zones in the column. The non-sorption effect by anion exclusion (R < 1) was only observed for I^- at 5.0 ± 0.2 mL/min flow rate, and a relatively higher Coulomb's repulsive force may be caused by the smaller hydration radius of I^-(3.31 Å) than that of IO_3^-(3.74 Å).

Keywords: non-sorption; anion exclusion; iodine; two-region non-equilibrium; granite

Citation: Shi, Y.; Yang, S.; Chen, W.; Xiong, W.; Zhang, A.; Yu, Z.; Lian, B.; Lee, C.-P. Study on Tritium and Iodine Species Transport through Porous Granite: A Non-Sorption Effect by Anion Exclusion. *Toxics* **2022**, *10*, 540. https://doi.org/10.3390/toxics10090540

Academic Editors: Yuezhou Wei, Yan Wu and Xinpeng Wang

Received: 24 August 2022
Accepted: 13 September 2022
Published: 16 September 2022

Publisher's Note: MDPI stays neutral with regard to jurisdictional claims in published maps and institutional affiliations.

Copyright: © 2022 by the authors. Licensee MDPI, Basel, Switzerland. This article is an open access article distributed under the terms and conditions of the Creative Commons Attribution (CC BY) license (https://creativecommons.org/licenses/by/4.0/).

1. Introduction

It is necessary to carry out safety evaluations for deep geological disposal site of high-level radioactive waste (HLW) to evaluate the migration behavior of nuclides in surrounding rock [1]. In general, a common cause of radionuclide migration is groundwater intrusion into HLW repositories due to the failure of engineering barriers. Granite is a potential crystalline host rock in HLW geological disposal [2–5].

Considerable research efforts worldwide have increased our knowledge and understanding of how underground disposal systems function over long periods of time. Recently, several studies have investigated the advection–dispersion behavior of cationic radionuclides, such as ^{238}Pu, ^{237}Np, ^{137}Cs, and ^{90}Sr [6–10], to evaluate the safety of the migration of various radionuclides in deep geological environments. At present, cationic radionuclides in liquid show high adsorption ability on mineral surfaces because of the permanently negative charges by isomorphous substitution in the crystal lattice of a mineral. Migration of cationic radionuclides in host rocks would be retarded, but anionic radionuclides (i.e., ^{36}Cl, ^{99}Tc, ^{129}I) show exclusive or repulsive behavior by Coulomb's force [11–13]. Few studies on safety assessments (SA) for HLW disposal focused on the anion exclusion effect of anionic nuclide transport in granite and other host rocks [14–16].

In addition to transuranic radionuclides (TRUs) in HLW, the long-lived radionuclides that exhibit a relatively high mobility under deep geochemical conditions must be considered when building an HLW repository. As a primary fission product, ^{129}I is characterized by a long half-life ($t_{1/2}$ = 1.57 × 10^7 years), high fission yield, easy volatilization, easy migration, high radioactive toxicity, high bioavailability, and weakly adsorbed radionuclide anions [17–20]. Actually, iodine species that have various valence states (iodide (I^-), iodate (IO_3^-), I_2, and organic iodine) depend on pH and redox conditions in aqueous solution. Thus, ^{129}I is the main source of the potential risk posed not only by HLW repositories but also by nuclear accidents, such as the 2011 Fukushima Daiichi nuclear disaster. This threat was recently made obvious when ^{129}I was observed in the Pacific Ocean around Japan, where fast dispersion by seawater movement and deposition caused it to migrate into the marine sediments [21].

The in-house dynamic column technique and numerical analyses have been simultaneously applied to simulate the transport of different contaminated solutes under natural groundwater conditions to provide a safety assessment for environmental impact on industrial or radioactive waste storage and disposal. According to different experimental conditions, the tests are designed to the filling mode of geotechnical medium (undisturbed or broken), solute injection mode (pulse or continuous injection), and water content maintenance mode (unsaturated or aquifer zone) [22–24]. To understand the anion exclusion effect, a previous study injected tritiated water (HTO) and anions into the same column and expressed the "accelerated" migration of anions by comparing the experimental breakthrough curves (BTCs) with the effluent concentration [25]. By comparing the BTCs of HDO and Cl^- in the dual-porosity structure of peat, McCarter obtained the anion exclusion of Cl^- and found that anion exclusion weakens with increasing Cl^- diffusion in pores [16]. Rao and Jain [25] determined the "accelerated" transport of $[CO(CN)_6]^-$ under the anion exclusion effect by analyzing the migration concentration distribution curves of HTO and $[CO(CN)_6]_3^-$ in clay and established a numerical model. In addition, some studies proposed "accelerated" migration by comparing the peak position of BTCs with pore volume (PV) (if PV < 1, then accelerated transport occurs). Shukla studied the transport behavior of Cl^- in soil columns of different particle sizes and indicated the anion exclusion by comparing the peak position. However, the homogeneity of the filling soils must be guaranteed to prevent the preferential flow (PV < 1) in the column.

In 1972, a two-region numerical model including the immobile/mobile zone in saturated porous soil based on diffusion double layer (DDL) theory was first developed by Krupp for the solute's pore velocity distribution formed by anion exclusion [26,27]. In addition, a physical nonequilibrium ion exclusion model was proposed and modified from DDL, and the porewater in porous soil was separated and expressed as flowing and nonflowing areas. Moreover, field experimental results by Gvirtzman [28] explained a dispersion model without ion exchange and an exchange model with ion exchange and obtained a good fitting.

In a previous work [29,30], individual and coexisting I^- and IO_3^- were successfully analyzed by applying ion chromatography (IC) coupled with a convenient and rapid detection tool for inductively coupled plasma optical emission spectrometry (ICP-OES), and it has become a quick and effective method in analyzing anion speciation in environmental samples. In the present study, the advection-dispersion behavior of I^- and IO_3^- in granite was studied and used to simulate their transport behavior in a micropore system. The BTCs of HTO, I^-, and IO_3^- at different flow rates were plotted and compared, and the key transport parameters were estimated from STudio of ANalytical MODels (STANMOD) and HYDRUS-1D under various conditions. The R, water content in the immobile regions (θ_{im}), first-order mass transfer coefficient (α), and other parameters were obtained to quantify the anion exclusion effect, and the influence of flow rate and iodine speciation on anion exclusion was explored. The results will support safety assessments of the impact of "accelerated migration" of anionic radionuclides for HLW repositories.

2. Theory of Advection and Dispersion

2.1. Equilibrium Transport

The majority of studies focused on radionuclide transport in saturated porous media, such as compacted crushed granite, and the one-dimensional governing solute transport equation is:

$$\frac{\partial C}{\partial t} = D\frac{\partial^2 C}{\partial x^2} - v\frac{\partial C}{\partial x}, \tag{1}$$

where C is radionuclide concentration in solution [Bq/mL^3], v is the average pore-water velocity [cm/min], t is time [min], and x is the distance [cm]. D is the dispersion coefficient of the granite media [cm ^{2}/min], and it can be further defined and written as

$$D = D_L v + D^*, \tag{2}$$

where D_L[cm] is the dispersivity and D^* the diffusion coefficient in free water [cm^2/min]. For the case where the adsorption equilibrium can be reached instantaneously or it takes a long time to reach the adsorption equilibrium, the distribution coefficient K_d is expressed as

$$S = K_d C, \tag{3}$$

where K_d is the distribution coefficient [mL^3/g]. In this study, we only take account of instant adsorption of radionuclides on the pores of crushed granite, and Equation (2) can be re-written as

$$\frac{\partial C}{\partial t} + \frac{\rho_b}{\theta}\frac{\partial S}{\partial t} = D\frac{\partial^2 C}{\partial x^2} - v\frac{\partial C}{\partial x}, \tag{4}$$

where θ is the porosity or volumetric water content [-], and ρ_b is the bulk density of porous medium [mL^3/g]. It can be combined by a retardation factor given by $R = 1 + \frac{\rho_b K_d}{\theta}$, and Equation (3) can be rewritten as

$$R\frac{\partial C}{\partial t} = D\frac{\partial^2 C}{\partial x^2} - v\frac{\partial C}{\partial x} \tag{5}$$

In addition, the boundary conditions and initial conditions are

$$C(x,0) = 0,\ 0 < x < L \tag{6}$$

$$C(0,t) = C_0 \tag{7}$$

$$\frac{\partial C(\infty,t)}{\partial t} = 0 \tag{8}$$

where C_0 is inlet radionuclide concentration [Bq/mL^3], and L is the length of the column [cm].

2.2. Two-Region Nonequilibrium Transport

A two-region transport model was proposed and assumed that the liquid phase can be partitioned into mobile (flowing) and immobile (stagnant) regions (which are produced by anion exclusion). The solute exchange between the two liquid regions is simulated as a first-order kinetic process. The two-region solute transport model is given by

$$(\theta_m + f\rho_b K_d)\frac{\partial c_m}{\partial t} = \theta_m D_m\frac{\partial^2 c_m}{\partial x^2} - J_w\frac{\partial c_m}{\partial x} - a(c_m - c_{im}) - (\theta_m\mu_{l,m} + f\rho_b K_d\mu_{s,m})c_m + \theta_m r_{l,m}(x) + f\rho_b r_{s,m}(x) \tag{9}$$

$$(\theta_{im} + (1-f)\rho_b K_d)\frac{\partial c_{im}}{\partial t} = a(c_m - c_{im}) - (\theta_{im}\mu_{l,im} + (1-f)\rho_b K_d\mu_{s,im})c_{im} + \theta_{im} r_{l,im}(x) + (1-f)\rho_b r_{s,im}(x), \tag{10}$$

where the subscripts m and im refer to the mobile and immobile liquid regions, respectively, $J_w = v \times \theta = v_m \times \theta_m$ is the volumetric water flux density, f represents the fraction of adsorption sites that equilibrates with the mobile liquid phase, and a is the first-order mass transfer coefficient governing the rate of solute exchange between the mobile and immobile liquid regions. θ is equal to $\theta_m + \theta_{im}$. $\mu_{l,m}$ and $\mu_{l,im}$ are first-order decay coefficients for the mobile and immobile liquid phases, respectively; $\mu_{s,m}$ and $\mu_{s,im}$ are first-order decay coefficients for the mobile and immobile adsorbed phases, respectively; $r_{l,m}$ and $r_{l,im}$ are zero-order production for the mobile and immobile liquid phases, respectively; and $r_{s,m}$ and $r_{s,im}$ are zero-order production terms for the mobile and immobile adsorbed phases, respectively.

3. Experiments

3.1. Experimental Device

Figure 1 shows the dynamic column experimental device used in this experiment. The experimental devices consisted of a peristaltic pump (MASTERFLEX L/S, Cole-Parmer Instrument Co., Barrington, IL, USA), a high-performance silica glass cylinder (modified type, Shanghai SuKe Industrial Co., Ltd. Shanghai, China), an automatic liquid sample collector (MODEL 2110, Bio-Rad Laboratories, Inc., Hercules, CA, USA), and four reservoirs: namely, No. 1, synthetic groundwater (SGW); No. 2, radiotracer (HTO); No. 3, I^-; and No. 4, IO_3^- and glass microfiber filters (0.4 mm thickness) with a pore size of 0.7 um (GF/F, Associated Design & Manufacturing Co., Alexandria, VA, USA). Table 1 lists the parameters of the filled granite, the composition of the synthetic groundwater, and the size of the column. The inner diameter of the experimental column was 1.6 cm, and the length was 30 cm. A total of 85.40 $\pm$ 0.50 g crushed granite was filled, with a density of 1.45 $\pm$ 0.05 g/cm^3 and a porosity of 0.47 $\pm$ 0.03. In order to provide a stable experimental flow rate, the device was also equipped with a pressure sensor and switch drain valves.

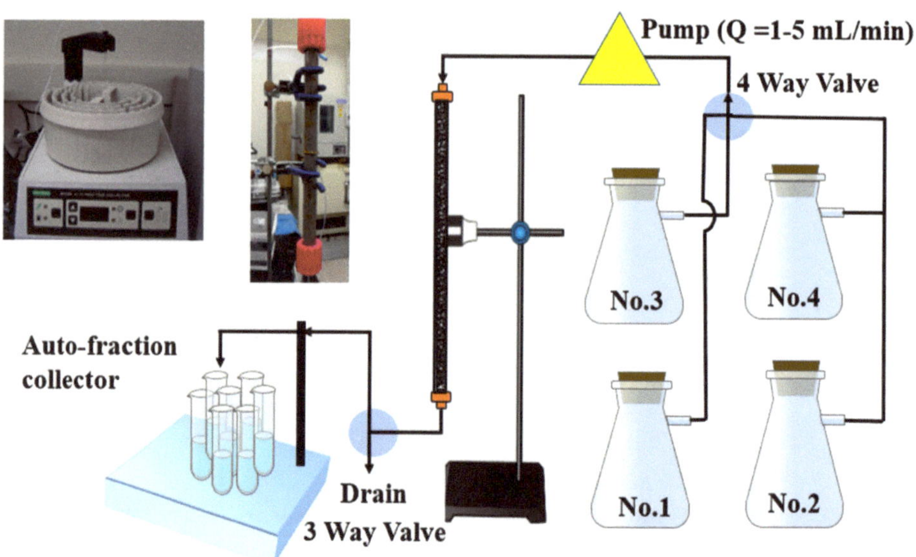

Figure 1. ADE experimental device in this work (schematic diagram).

3.2. Rocks and Liquids

A discrete granitic island surrounding Fujian Province and local synthetic GW were sampled and investigated in previous studies [6,10]. All granite samples were crushed with a grinder to a particle size of 1 mm. Before filling the soil column to crush granite samples, all granite samples were washed three times with deionized water, dried at 100 $\pm$ 10 °C for 12–48 h, and stored in a vacuum dryer for future use.

Table 1. The experimental crushed granite and synthetic groundwater (SGW) in this study.

Item	Granite	SGW (M)			
Location	Fujian Province, China	Ca^{2+}	3.76×10^{-4}	F^-	1.85×10^{-4}
Particle Size (mm)	<1	Mg^{2+}	6.38×10^{-5}	HCO_3^-	1.21×10^{-3}
Length (cm)	30.0	Na^+	1.57×10^{-3}	pH	8.0 ± 0.2
Diameter (cm)	1.6	K^+	9.22×10^{-5}	Eh	220 ± 20
Weight (g)	85.40 ± 0.50	Cl^-	9.77×10^{-4}	T(°C)	20 ± 2
bulk density (g/cm^3)	1.45 ± 0.05	SO_4^{2-}	1.05×10^{-4}	I (M)	3.11×10^{-3}
Porosity	0.47 ± 0.03	I: Ionic Strength			

3.3. Mineral Composition and Elemental Analysis

A mineral analysis of granite in rock was conducted using X-ray diffraction (XRD) as previously described. XRD spectra and micro-polar microscopy images showed that the main minerals included quartz, plagioclase, K-feldspars, amphibole, and biotite [31,32]. In addition, the crushed granite samples (~2 g) were analyzed under scanning electron microscopy coupled with energy dispersive spectrometry (SEM–EDS) with an accelerating voltage of 20 kV and a current of 10 μA. EDS was used to analyze the major elements of the granite samples.

3.4. Pre-Equilibrium: Water Saturation in Column

In order to simulate the migration behavior of nuclides under aquifer conditions, it was necessary to saturate the crushed granite column. The valve was opened to make the groundwater flow into the column slowly at a flow rate of 3 mL/min. During the water filling, 10 mL of effluent from each column was extracted every 30 min. The concentrations of K, Ca, Mg and Na in the effluent were measured by ICP-OES. When the concentration changed within 5% of the corresponding liquid phase concentration, it indicated that water saturation had been reached.

3.5. Nonreactive Tests of HTO

In order to characterize the internal structure of the crushed granite column, the dispersion and effective porosity were obtained by convection dispersion experiments with nonreactive tracer (HTO). By regularly collecting the effluent at different flow rates, the HTO content in the effluent was analyzed and measured, and the data were used to draw the penetration curve BTC. The water circulation was divided into two stages. The first stage was the manual operation stage, and its basic process was as follows. As shown in Figure 1, set water circulation under different flow rate conditions, 3 PV (about 2100 Bq) each time, including up/down flooding. The liquid sample was collected by an automatic sampler, and a 5 mL aliquot of the effluent was sampled to measure HTO activities with a liquid scintillation counter (Packard 3170 AB/TR, Shelton, CT, USA). During the first upflooding process, both control valves were open to reservoir No. 2, and the flow rates through the column were controlled at approximately 5.0 ± 0.1, 3.0 ± 0.1, and 1.0 ± 0.1 mL/min. After 3 PV of the liquid phase passed through the column in the upflooding process, the HTO was flushed out from the column in a downflooding process where switch valves were open to reservoir No. 1 (GW). The effluent was collected using an auto-fraction collector every 60, 100, and 300 s. Table 2 shows that this up/downflooding process was set up and repeated two more times. A 10 mL aliquot of the sample was mixed with 10 mL of scintillation cocktail in a 20 mL polyethylene counting vial. All chemicals used in the experiments were of analytical purity, and de-ionized water (DIW) was used throughout the experiments.

Table 2. ADE experimental process under different flow rates.

Batch / Flow Rate (mL/min) \ RN	No. 1 HTO: 50 Bq/mL (A_0)	No. 2 I^- 7.78 (C_0: ppm)	No. 3 IO_3^- 5.41 (C_0: ppm)	Remark
5.0 ± 0.1	U	U	U	U: up-flooding
	D	D	D	
3.0 ± 0.1	U	U	U	D: down-flooding
	D	D	D	
1.0 ± 0.1	U	U	U	V_0 = 2000 mL
	D	D	D	

3.6. Reactive Tests of I^- and IO_3^- (Anion Exclusion)

The transport model was calibrated/validated by designing ADE experiments at different flow rates [33]. I^- and IO_3^-, as well as disodium potassium iodide (KI) and potassium iodate (KIO_3), were injected into the column as stable isotope tracers after the HTO in the column was completely washed out. The corresponding concentrations (C(t)) of I^- and IO_3^- were determined via ion chromatography (IS-600, Thermo Fisher Scientific Inc., Waltham, MA, USA) for I speciation (I^- or IO_3^-), and the iodine concentrations in the solution were measured with an induced coupled plasma optic emission spectrometer (iCAP 7000, Thermo Fisher Scientific Inc., Waltham, MA, USA). Table 2 lists the results of a multi-stage ADE column test, which included a series of upflooding/downflooding processes, for I^- and IO_3^-. During the advection–dispersion experiments (ADEx) period for I^- and IO_3^-, the frequency of collection depended on the flow rate, that is, every 60 s for approximately 3 mL/min. The corresponding concentrations (C(t)) of I^- and IO_3^- in the effluent were then determined, and the respective breakthroughs were plotted for analysis. To ensure that the ADEx was finished, the upflooding process was stopped when the relative concentration (C/C_0) reached approximately 1.0, and the downflooding process was stopped when C/C_0 decreased to 0.05–0.1.

3.7. Mathematical Model and Parameter Estimations

STANMOD is a Windows-based computer software package for evaluating solute transport in porous media using analytical solutions of the convection-dispersion solute transport equation. The software package includes a modified and updated version of the CXTFIT code for estimating solute transport parameters using a nonlinear least-squares parameter optimization method. Residuals are a measure of the degree to which experimental results (*Ce*) deviate from the predicted values (*Cp*), *N* is the number of experimental data, and the root mean square error (RMSE) is defined as

$$RMSE = \sqrt{\frac{\Sigma_i^N (C_p - C_e)^2}{N}}. \quad (11)$$

The HYDRUS-1D program numerically solves the Richards equation for variably saturated water flow and advection-dispersion type equations for heat and solute transport. It uses a numerical solution for numerical fitting. The transport equations include provisions for nonlinear nonequilibrium reactions between the solid and liquid phases. In addition, physical nonequilibrium solute transport can be accounted for by assuming a two-region, dual-porosity type formulation that partitions the liquid phase into mobile and immobile regions. As part of the inverse solution, HYDRUS produces a correlation matrix that specifies the degree of correlation between the fitted coefficients. An important measure of the goodness of fit is the r^2 value for regression of the observed $\hat{y}_i$ versus fitted y_i values:

$$r^2 = \frac{\left[\sum w_i \widehat{y_i}\, y_i - \frac{\sum \widehat{y_i} \sum y_i}{\sum w_i}\right]}{\left[\sum w_i \widehat{y_i}^2 - \frac{(\sum \widehat{y_i})^2}{\sum w_i}\right]\left[\sum y_i^2 - \frac{\sum y_i^2}{\sum w_i}\right]}. \tag{12}$$

The r^2 value is a measure of the relative magnitude of the total sum of squares associated with the fitted equation; a value of 1 indicates a perfect correlation between the fitted and observed values. where w_i is weighting factor for the two overlapping regions of the model, $\hat{y}_i$ is fitting value, y_i is observed value.

4. Results and Discussion

4.1. Elemental Analysis by SEM–EDS

XRD, micro-X-ray computed tomography, polar microscopy, and analysis in previous works [31,32,34] showed that granite contains crystalline particles because it is igneous rock. In addition, we obtained and compared the corresponding images of SEM-EDS and elemental mapping analysis (Figure 2). Comparison of the mapping area and mineral components in granite showed that O, Si, Al, Mg Fe, K, Na, Ti, Mn, and Ca are the major mineral components, which agrees with previous studies that used different analysis methods, such as Rutherford backscattering spectrometry [34].

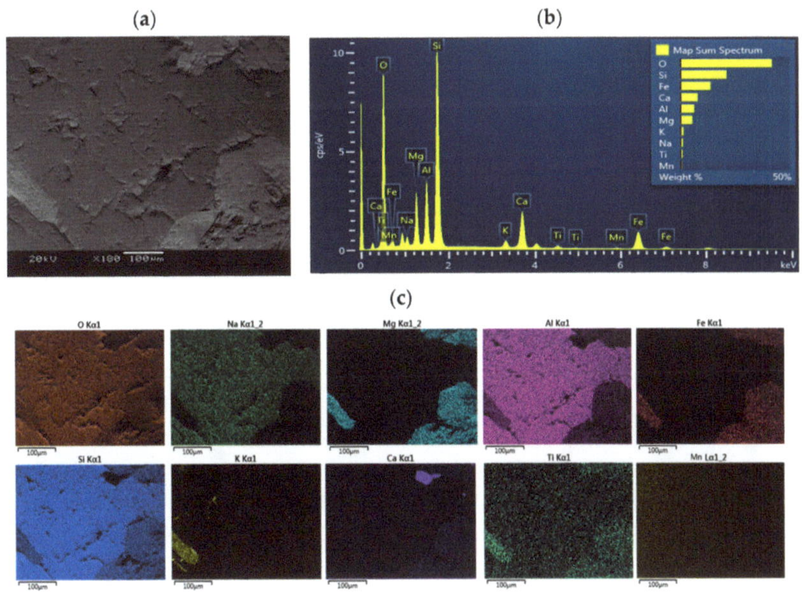

Figure 2. The SEM photos in granite and the EDS mapping of surface of granite obtained at 10 µm spatial resolution. (**a**) SEM (**b**) EDS spectrum (**c**) elemental mapping.

4.2. Experimental BTCs of HTO, I^-, and IO_3^-

Before using I^- and IO_3^-, a nonreactive radiotracer (HTO) was applied to characterize the major physical transport processes in the proposed dynamic column system with a PV (1 PV = 28 mL). Figure 3 shows that the HTO, I^-, and IO_3^- BTCs at different flow rates reached 1 (C/C_0 = 1) from a series of sorption/desorption (up/downflooding) processes. Our experimental results indicated the absence of dead-end or dead pores to block the HTO, I^-, and IO_3^- pathways, and water was saturated in each pore space in the compacted granite powder.

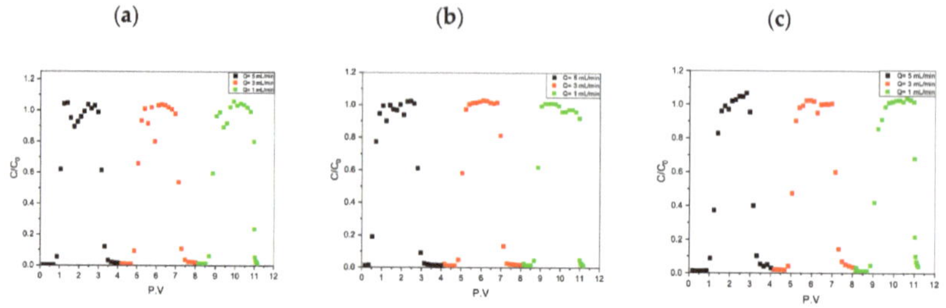

Figure 3. The ADE experimental breakthrough curves (BTCs). (**a**) HTO; (**b**) I^-; (**c**) IO_3^-.

It can be seen from the figure that since the injection method of solute is to continue to inject for a period of time and then stop the injection, all penetration curves are "s" type (divided into "s" type in the injection adsorption stage of the first half and "s" type in the desorption stage of the second half). Specifically, when the injection amount of solute reaches about one column pore volume, the ratio of solute activity concentration in the effluent quickly reaches the peak value, and then the concentration rapidly decreases to zero after continuous injection of clean water.

4.3. Fitting BTCs of Nonreactive Tracer (HTO)

The nonreactive tracer (HTO) was designed and applied in multi-stage ADEx to identify the reliability of our ADE device to build adequate reaction models through a calibration/validation process. Figure 4 (Fit-1 to -3) and Table 3 show that the equilibrium model in STANMOD- Fit (S) and HYDRUS-1D- Fit (H) were applied to fit the experimental data to obtain the parameters for HTO transport in the granite. The parameters (Fit-1 to -3) obtained by the two fitting methods showed similar results, indicating that the filling granite in the column showed good homogeneity. In addition, the highly reliable and accurate experimental data for HTO were identified according to the dispersivity (D_L = 0.270–0.289 cm), retardation factor R (=1), and distribution coefficient K_d (=0) in our ADE system by fitting with STANMOD and HYDRUS-1D.

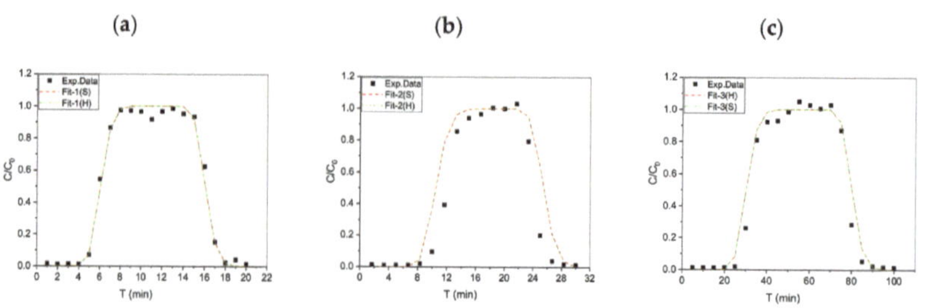

Figure 4. The HTO breakthrough curves (BTCs) at various flow rates. (**a**) 5.0 ± 0.1 mL/min; (**b**) 3.0 ± 0.1 mL/min; (**c**) 1.0 ± 0.1 mL/min.

Table 3. The parameters for fitting HTO BTCs.

HTO	STANMOD (S) Fit 1	STANMOD (S) Fit 2	STANMOD (S) Fit 3	HYDRUS-1D (H) Fit 1	HYDRUS-1D (H) Fit 2	HYDRUS-1D (H) Fit 3
Q (mL/min)	5.0 ± 0.1	3.0 ± 0.1	1.0 ± 0.1	5.0 ± 0.1	3.0 ± 0.1	1.0 ± 0.1
D_L (cm)	0.268	0.331	0.268	0.270	0.270	0.269
R	1.11	1.07	1.29	1.17	1.21	1.16
K_d (mL^3/g)	=0	=0	=0	=0	=0	=0
RMSE/r^2	4.05×10^{-2}	1.62×10^{-1}	8.58×10^{-2}	0.99	0.90	0.97

$R = 1 + \frac{\rho_b K_d}{\theta}$; Bulk Density = 1.45 ± 0.05 g/cm^3; porosity = 0.47 ± 0.03

4.4. Fitting BTCs of Anionic I^- and IO_3^-

Figure 5 shows the BTCs of anionic I^- and IO_3^- in the granite column at different flow rates, and only the BTCs of IO_3^- (Fit-4 to -6) were similarly displayed and compared with the BTC of HTO. When the elution concentration ratio (C/C_0) of IO_3^- reached approximately 0.5 at 1 PV, R was almost equal to 1 and similar to HTO. Table 4 shows the fitting parameters of IO_3^-. The R (STANMOD) of IO_3^- and distribution coefficient K_d (HYDRUS-1D) were almost equal 0 (=0), showing the non-reactive or adsorptive behavior of IO_3^- in granite.

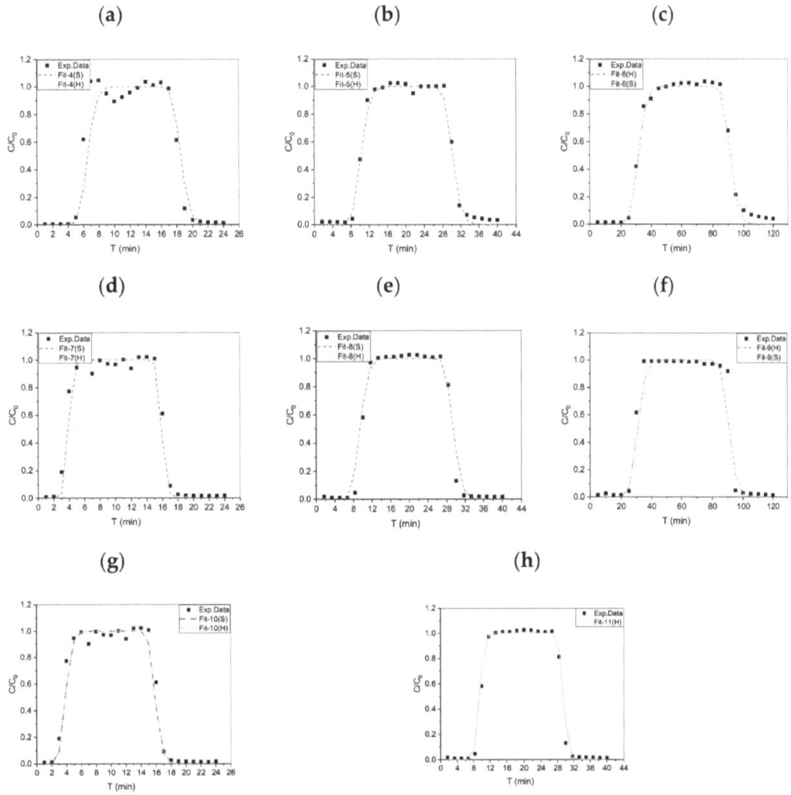

Figure 5. The experimental breakthroughs of I^- and IO_3^- fitting curves. (**a**) IO_3^-: 5.0 ± 0.1 mL/min; (**b**) IO_3^-: 3.0 ± 0.1 mL/min; (**c**) IO_3^-:1.0 ± 0.1 mL/min; (**d**) I^-: 5.0 ± 0.1 mL/min; (**e**) I^-: 3.0 ± 0.1 mL/min; (**f**) I^-: 1.0 ± 0.1 mL/min; (**g**) I^-: 5.0 ± 0.1 mL/min; (**h**) I^-: 3.0 ± 0.1 mL/min.

As shown in Figure 5d and Table 5, the BTC C/C_0 was less than 1 PV (Fit-7 and -8) of I^-, and the R of I^- was lower than 1. An obvious anion repulsive effect occurred, which accelerated the I^- transport compared with HTO and IO_3^- transport. In fact, for the BTC of I^-, when the equilibrium model was used for fitting, when Q = 5.0 ± 0.1 mL/min, the analytical solution could be used for fitting to obtain a good effect. When the numerical solution was used for fitting, the fitting effect was poor. This phenomenon did not occur when fitting the penetration curve with Q = 1.0 ± 0.1 and 3.0 ± 0.1 mL/min.

Table 4. The parameters for fitting IO_3^- BTCs.

	No	Flow Rate (mL/min)	v (cm/min)	D (cm^2/min)	R	D_L (cm)	RMSE
STANMOD (S)	Fit 4	5.0 ± 0.1	5.41	1.45	1.19	0.268	0.14×10^{-1}
	Fit 5	3.0 ± 0.1	3.02	1.00	1.04	0.331	0.15×10^{-2}
	Fit 6	1.0 ± 0.1	1.27	0.34	1.34	0.268	0.15×10^{-2}
	No	Flow Rate (mL/min)	Ks (cm/min)	K_d (mL^3/g)	D_L (cm)		r^2
HYDRUS-1D (H)	Fit 4	5.0 ± 0.1	2.488	0.050	0.270		0.98
	Fit 5	3.0 ± 0.1	1.492	0.044	0.270		0.99
	Fit 6	1.0 ± 0.1	0.498	0.069	0.269		0.99

Table 5. The parameters for fitting I^- BTCs (equilibrium model).

	No	Flow Rate (mL/min)	v (cm/min)	D (cm^2/min)	R	D_L (cm)	RMSE
STANMOD (S)	Fit 7	5.0 ± 0.1	5.41	1.45	0.71	0.268	0.55×10^{-2}
	Fit 8	3.0 ± 0.1	3.02	1.00	0.96	0.331	0.37×10^{-2}
	Fit 9	1.0 ± 0.1	1.27	0.34	1.31	0.268	0.87×10^{-2}
	No	Flow Rate (mL/min)	Ks (cm/min)	K_d (mL^3/g)	D_L (cm)		r^2
HYDRUS-1D (H)	Fit 7	5.0 ± 0.1	2.488	=0	0.270		0.77
	Fit 8	3.0 ± 0.1	1.492	=0	0.270		0.99
	Fit 9	1.0 ± 0.1	0.498	=0	0.269		0.96

In terms of I^- fitting parameters, the R of I^- obtained by fitting 5.0 ± 0.1 mL/min (Fit-7) and Q = 3.0 ± 0.1 (Fit-8) curves with analytical solution was less than 1, possibly indicating obvious anion exclusion. After fitting with the numerical solution in Table 5, the K_d of the curve with Q = 5.0 ± 0.1 mL/min was greater than 1 by default (Fit-7). Thus, the fitting correlation was not good enough (r^2 = 0.77) compared with Fit-8 and -9 ($r^2 > 0.95$). The K_d of I^- Q = 3.0 ± 0.1 mL/min was smaller than that of HTO, and I^- transport in granite could not be explained by fitting the equilibrium models (STANMOD and HYDRUS−1D).

Table 6, Fit-10 and 11 ($r^2 > 0.95$) show that the two-region nonequilibrium model was developed [33] and applied in this work to fit the BTC curves of Q = 5.0 ± 0.1 and 3.0 ± 0.1 mL/min (STANMOD) and Q = 5.0 ± 0.1 mL/min (HYDRUS-1D) and understand the I^- transport in granite. The fitting parameters showed that the water contents of immobile liquid phases (θ_{im}) increased with increasing flow velocity (Fit-10), indicating that the anion exclusion effect was more obvious. In addition, the θ_{im} was larger than the analytical solution possibly because of the numerical algorithm.

The migration behavior of HTO and I^- in the same medium was compared microscopically to explain the influence of anion exclusion of I^-. Figure 6a describes the migration of HTO in the groundwater environment. As shown in the figure, HTO can pass through all water channels under the action of convection dispersion and finally fill the whole medium. Figure 6b shows the migration of I^- in the groundwater environment. According to the "electric DDL theory", the anion concentration in a single pore increases exponentially with the distance from the pore wall. When I^- migrates in a wide water passage, it will converge to the center due to anion exclusion and will not pass through a narrow passage, forming "immobile regions". Finally, I^- will select the channel with a wide water surface to migrate, resulting in acceleration.

Table 6. The parameters for fitting I^- BTCs (non-equilibrium model).

	No	Flow Rate (mL/min)	v (cm/min)	D (cm^2/min)	θ_m	θ_{im}	D_L (cm)	α (min^{-1})	RMSE
STANMOD (S)	Fit 10	5.0 ± 0.1	7.66	3.95	0.368	0.067	0.516	2.70	0.54×10^{-2}
	Fit 11	3.0 ± 0.1	3.17	0.52	0.471	0.019	0.164	3.85	0.27×10^{-2}
	No	**Flow rate (mL/min)**	**K_s (cm/min)**		**θ_m**	**θ_{im}**	**D_L (cm)**	**A (min^{-1})**	**r^2**
HYDRUS-1D (H)	Fit 10	5.0 ± 0.1	3.279		0.358	0.102	0.530	3.04	0.98

(a)

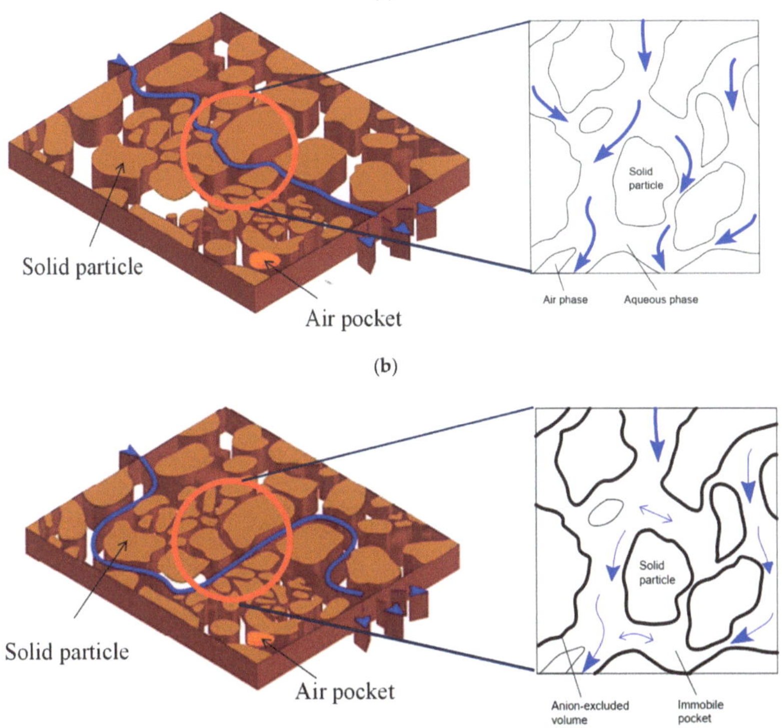

Figure 6. Schematic model of HTO and I^- transport. (**a**) HTO; (**b**) I^-.

Figure 7 shows a comparison of the migration behaviors of I^- and IO_3^- under different flow rates. According to the theory of the "electric double layer model", a negative ion layer forms when the dielectric surface is negatively charged, and a positive ion layer forms outside due to electrical attraction. The closer it is to the dielectric surface, the more negative its potential becomes (the more obvious the repulsion). Comparison of the ion radii of I^- and IO_3^- showed that the ion radius of I^- (3.31 Å) was smaller than that of IO_3^- (3.74 Å) [35]. Therefore, I^- can pass through the diffusion layer and become closer to the medium surface during migration. At the same time, the greater the migration speed of ions, the greater the kinetic energy generated, the closer the energy to the medium surface, the stronger the anion exclusion effect, and the more obvious the acceleration during migration. However, it is difficult for IO_3^- to become close to the medium surface because of its large ion radius and weak anion exclusion. Thus, the acceleration during migration was not obvious.

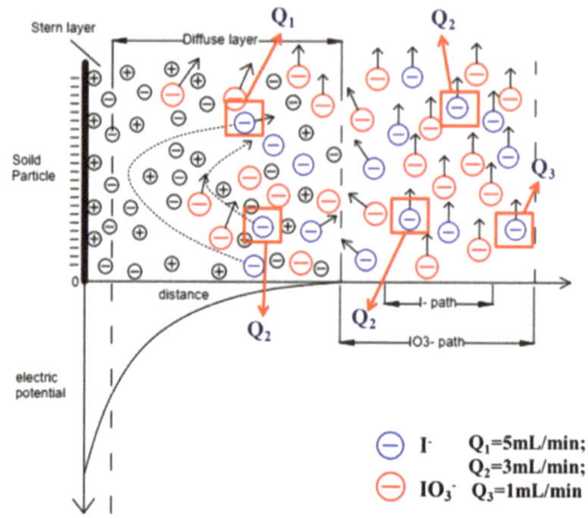

Figure 7. Schematic model of I^- and IO_3^- transport at various flow velocities.

5. Conclusions

The transport behavior of HTO, I^-, and IO_3^- in granite was studied by ADE column experiments. Comparison of the BTCs of HTO, I^-, and IO_3^- under different flow rates showed that the anion exclusion effect occurred in I^- at Q = 5.0 ± 0.1 mL/min during fitting STANMOD and HYDRUS-1D when R < 1. The key parameters of HTO, I^-, and IO_3^- were obtained by fitting the equilibrium and nonequilibrium models of the analytical method (STANMOD) and numerical inversion (HYDRUS-1D). The conclusions are as follows:

1. The BTCs for HTO were symmetrical at various flow rates (1.0 ± 0.1–5.0 ± 0.1 mL/min), and no significant differences were found in the accessible porosity and dispersivity values of HTO. Therefore, the experimental apparatus was highly reliable, and the granite samples filled and compacted in the column were almost homogeneous.
2. Comparison with the BTCs of HTO, I^-, and IO_3^- at different flow rates showed that obvious anion exclusion only occurred in I^- transport by increasing the flow rate from 1 to 5. In fact, the anion exclusion (R < 1) only occurred for I^- at a flow rate of 5 mL/min, and a relative Coulomb's repulsive force may be caused by the smaller hydration radius of I^-(3.31 Å) than that of IO_3^-(3.74 Å) according to electric DDL theory.
3. The equilibrium and nonequilibrium transport models were used and compared to identify the mobile/immobile zones in the compacted granite column. The anion exclusion effect depended on the immobile zones in the column. In sum, the nonequilibrium model can well characterize the immobile regions of anion exclusion, which is obviously proportional to the value of immobile regions.
4. Since the experiment was conducted in laboratories, the environmental conditions were quite different from the actual environment. Whether the research results can be applied to the actual plant site needs to be demonstrated. It is recommended to carry out field experiments as soon as possible.

Author Contributions: Conceptualization, Y.S. and B.L.; methodology, Y.S. and S.Y.; software, Y.S., W.C. and W.X.; validation, B.L. and A.Z.; formal analysis, Y.S. and S.Y.; investigation, W.C. and Z.Y.; resources, B.L. and A.Z.; data curation, Y.S., C.-P.L., W.C. and W.X.; writing—original draft preparation, Y.S. and S.Y.; writing—review and editing, Y.S., C.-P.L. and S.Y.; visualization, Y.S. and S.Y.; supervision, Y.S. and B.L.; project administration, Y.S. and B.L.; funding acquisition, A.Z. and B.L. All authors have read and agreed to the published version of the manuscript.

Funding: This project was mainly supported by the The Safety Case and Safety Assessment for Feifeng mountain low and intermediate level solid waste disposal site (No.TY19040302), Department of Nuclear Environmental Science of the China Institute for Radiation Protection (CIRP), and Shanxi Province Basic research Project (No.SN22010301), Start up funds for doctoral research of East China University of Technology (No.1410000434), CIRP Open Fund of Radiation Protection Laboratories.

Institutional Review Board Statement: Not applicable for studies not involving humans or animals.

Informed Consent Statement: Not applicable.

Data Availability Statement: Not applicable.

Acknowledgments: The experimentation conducted in this study was assisted by Neng-Chuan Tien, members of Nuclear Science and Technology Development Center, National Tsing Hua University, Hsinchu 30013, Taiwan (ROC). Assistance in conducting the experimental and instrumental analyses of this study was provided by the East China University of Technology, Nanchang, Jiangxi, China.

Conflicts of Interest: The authors declare no conflict of interest.

References

1. IAEA. *Joint Convention on the Safety of Spend Fuel Management and on the Safety of Radioactive Waste Management*; INFCIRC/546; IAEA: Vienna, Austria, 1997.
2. U.S. Nuclear Waste Technical Review Board. *Survey of National Programs for Managing High−Level Radioactive Waste and Spent Nuclear Fuel: Update, A Report to Congress and the Secretary of Energy*; U.S. Nuclear Waste Technical Review Board: Arlington, VA, USA, 2016.
3. Hedin, A. *Data Report for the Safety Assessment SR-Site TR 10-52*; Svensk Kärnbränslehantering AB: Stockholm, Sweden, 2010.
4. Japan Nuclear Cycle Development Institute. *Japan, H12: Project to Establish the Scientific and Technical Basis for HLW Disposal in Japan*; Japan Nuclear Cycle Development Institute: Naka, Japan, 2000.
5. Wang, J. Progress of Geological Disposal of High−level Radioactive Waste in China in the 21st Century. *J. At. Energy Sci. Technol.* **2019**, *53*, 2072.
6. Tsai, S.C.; Wang, T.H.; Jan, Y.L. Comparison of different methods to determine the retardation factor of ^{137}Cs transport through granite in column experiments. *J. Radioanal. Nucl. Chem.* **2008**, *275*, 351–354. [CrossRef]
7. Li, Y.; Zhang, A.; Chen, H. Migration Experiment of ^{238}Pu in Sandy Soil in Vadose Zone of Chinese Arid Area. *Environ. Sci. Technol.* **2017**, *41*, 97–99.
8. Yu, F.; Zuo, U.; Gou, Z. Migration Parameter Optimization Red Clay from Zhejiang of ^{237}Np in Saturated Province. *Earth Environ.* **2014**, *42*, 502–504.
9. Zhu, J.; Denu, A.; Zhang, A. Experiment on Nuclide Migration of Aquifer by Three−Dimensional Model in Laboratory Scale. *J. Nucl. Radiochem.* **2019**, *42*, 290–296.
10. Lee, C.P.; Chen, D.; Hu, Y.; Jan, Y.L.; Shi, Y.; Wang, Z.; Wu, E.; Tien, N.C.; Sun, Y.; Tsai, S.C. An application of safety assessment for radioactive waste repository: Non-equilibrium transport of Tritium, Selenium and Cesium in crushed granite with different path lengths. *Appl. Sci.* **2021**, *11*, 9750. [CrossRef]
11. Allred, B.J. Effects of nitrate concentration and ionic strength on nitrate anion exclusion under unsaturated flow conditions. *Soil Sci.* **2007**, *172*, 842–860. [CrossRef]
12. Allred, B.J. Cation effects on nitrate mobility in an unsaturated soil. *Trans. ASABE* **2008**, *51*, 1997–2012. [CrossRef]
13. Allred, B.J.; Brown, G.O.; Martinez, L.R. Laboratory investigation of boundary condition impacts on nitrate anion exclusion in an unsaturated soil. *Soil Sci. Soc. Am. J.* **2015**, *79*, 742–756. [CrossRef]
14. Phillips, F.M.; Mattick, J.L.; Duval, T.A.; Elmore, D.; Kubik, P.W. Chlorine 36 and tritium from nuclear weapons fallout as tracers for long−term liquid and vapor movement in desert soils. *Water Resour. Res.* **1988**, *24*, 1877–1891. [CrossRef]
15. Mäder, U. Advective displacement method for the characterization of pore water chemistry and transport properties in Claystone. *Geofluids* **2018**, *2018*, 819876. [CrossRef]
16. McCarter, C.; Rezanezhad, F.; Gharedaghloo, B.; Price, J.S.; Van Cappellen, P. Transport of chloride and deuterated water in peat: The role of anion exclusion, diffusion, and anion adsorption in a dual porosity organic media. *J. Contam. Hydrol.* **2019**, *225*, 103497. [CrossRef]
17. Kimmig, S.R.; Thompson, C.; Baum, S.; Brown, C.F. Evaluation of iodine speciation and $^{129}I/^{127}I$ ratios at low concentrations in environmental samples using IC−ICP−MS J. *Radioanal. Nucl. Chem.* **2021**, *327*, 929–937. [CrossRef]
18. Van Loon, L.R.; Soler, J.M.; Jakob, A.; Bradbury, M.H. Effect of confining pressure on the diffusion of HTO, $^{36}Cl^-$ and $^{125}I^-$ in a layered argillaceous rock (Opalinus Clay): Diffusion perpendicular to bedding. *Appl. Geochem.* **2003**, *18*, 1653–1662. [CrossRef]
19. Zhang, S.; Xu, C.; Creely, D.; Ho, Y.F.; Li, H.P.; Grandbois, R.; Schwehr, K.A.; Kaplan, D.I.; Yeager, C.M.; Wellman, D. Iodine-129 and Iodine-127 speciation in groundwater at the Hanford Site, U.S: Iodate Incorporation into Calcite. *Environ. Sci. Technol.* **2013**, *47*, 9635–9642. [CrossRef]

20. Takeda, A.; Tsukada, H.; Takaku, Y.; Satta, N.; Baba, M.; Shinata, T.; Hasegawa, H.; Unno, Y.; Hisamatsu, S. Determination of iodide, iodate and total iodine in natural water samples by HPLC with amperometric and spectrophotometric detection, and off-line UV irradiation. *Anal. Sci.* **2016**, *32*, 839–845. [CrossRef]
21. Fan, Y.; Hou, X.; Fukuda, M.; Zheng, J.; Aono, T.; Chen, N.; Zhang, L.; Zhou, W. ^{129}I in a sediment core offshore Fukushima: Distribution, source and its implication. *Chemosphere* **2020**, *252*, 126524. [CrossRef]
22. Cheng, H.P.; Li, M.H.; Li, S. A sensibility analysis of model selection in modeling the reactive transport of cesium in crushed granite. *J. Contam. Hydrol.* **2003**, *61*, 371–385. [CrossRef]
23. Van Genuchten, M.T.; Wagenet, R.J. Two site/Two region models for pesticides transport and degradation: Theoretical development and analytical solutions. *Soil Sci. Soc. Am. J.* **1989**, *53*, 1303–1310. [CrossRef]
24. Palagyi, Š.; Stamberg, K.; Vodickova, H.; Hercik, M. Transport parameters of I^- and $IO3^-$ determined in crushed rock column and groundwater system under dynamic flow conditions. *J. Radioanal. Nucl. Chem.* **2014**, *302*, 647–653. [CrossRef]
25. Rao, S.M.; Jain, S.K. Transport Modelling for Unsaturated Zone-Effect of Anion Exclusion and Immobile Water. *Isot. Isot. Environ. Health Stud.* **1985**, *21*, 433–438. [CrossRef]
26. Babcock, K.L. Theory of the chemical properties of soil colloidal systems at equilibrium. *Hilgardia* **1963**, *34*, 417–542. [CrossRef]
27. Krupp, H.K.; Biggar, J.W.; Nielsen, D.R. Relative flow rates of salt and water in soil. *Soil Sci. Soc. Am. Proc.* **1972**, *36*, 412–417. [CrossRef]
28. Gvirtzman, H.; Gorelick, S.M. Dispersion and advection in unsaturated porous media enhanced by anion exclusion. *Nature* **1991**, *352*, 793. [CrossRef]
29. Glaus, M.A.; Müller, W.; Van Loon, L.R. Diffusion of iodide and iodate through Opalinus Clay: Monitoring of the redox state using an anion chromatographic technique. *Appl. Geochem.* **2008**, *23*, 3612–3619. [CrossRef]
30. Lee, C.P.; Hu, Y.; Chen, D.; Wu, E.; Wang, Z.; Wen, Z.; Tien, N.C.; Yang, F.; Tsai, S.C.; Shi, Y.; et al. An Improved Speciation Method Combining IC with ICPOES and its Application to Iodide and Iodate Diffusion Behavior in Compacted Bentonite Clay. *Materials* **2021**, *14*, 7056. [CrossRef]
31. Lee, C.P.; Tsai, S.C.; Jan, Y.L.; Wei, Y.Y.; Teng, S.P.; Hsu, C.N. Sorption and diffusion of HTO and cesium in crushed granite compacted to different lengths. *J. Radioanal. Nucl. Chem.* **2008**, *275*, 371–378. [CrossRef]
32. Lee, C.P.; Tsai, S.C.; Wu, M.C.; Tsai, T.L.; Tu, Y.L.; Kang, L.J. A comparative study on sorption and diffusion of Cs in crushed argillite and granite investigated in batch and through-diffusion experiment. *J. Radioanal. Nucl. Chem.* **2017**, *311*, 1155–1162. [CrossRef]
33. Toride, N.; Leij, F.J.; van Genuchten, M.T. *The CXTFIT Code for Estimating Transport Parameters from Laboratory on Field Tracer Experiments Version 2.1*; Research Report No. 137; U.S. Salinity Laboratory, USDA, ARS: Riverside, CA, USA, 1999.
34. Tsai, S.C.; Lee, C.P.; Tsai, T.L.; Yu, Y.C. Characterization of cesium diffusion behavior into granite matrix using Rutherford backscattering spectrometry. *Nucl. Instrum. Methods Phys. Res. B* **2017**, *409*, 305–308. [CrossRef]
35. Nightinggale, E.R. Phenomenological theory of ion solvation: Effective radii of hydrated ions. *J. Phys. Chem.* **1959**, *63*, 1381–1387. [CrossRef]

toxics MDPI

Article

Highly Selective Adsorption of $^{99}TcO_4^-/ReO_4^-$ by a Novel Polyamide-Functionalized Polyacrylamide Polymer Material

Ben Qin [1,2], Yanqin Hu [2], Meiying Xie [2,3], Liyan Xue [2,3,4], Chunfa Liao [1,*] and Fan Yang [3,4,5,*]

1 Faculty of Materials, Metallurgy and Chemistry, Jiangxi University of Science and Technology, Ganzhou 341000, China
2 Xiamen Institute of Rare Earth Materials, Haixi Institute, Chinese Academy of Sciences, Xiamen 361021, China
3 Fujian Province Joint Innovation Key Laboratory of Fuel and Materials in Clean Nuclear Energy System, Fujian Institute of Research on the Structure of Matter, Chinese Academy of Sciences, Fuzhou 350002, China
4 Fujian Science & Technology Innovation Laboratory for Optoelectronic Information of China, Fuzhou 350108, China
5 Sichuan Jcc Rare Earth Matals New Material Co., Ltd., Chengdu 610213, China
* Correspondence: liaochfa@163.com (C.L.); fanyang2013@fjirsm.ac.cn (F.Y.)

Abstract: The treatment of radioactive wastewater is one of the major problems in the current research. With the development of nuclear energy, the efficient removal of $^{99}TcO_4^-$ in radioactive wastewater has attracted the attention of countries all over the world. In this study, a novel functional polyamide polymer p-(Amide)-PAM was synthesized by the two-step method. The experimental results show that p-(Amide)-PAM has good adsorptive properties for $^{99}TcO_4^-/ReO_4^-$ and has good selectivity in the nitric acid system. The kinetics of the reaction of p-(Amide)-PAM with $^{99}TcO_4^-/ReO_4^-$ was studied. The results show that p-(Amide)-PAM has a fast adsorption rate for $^{99}TcO_4^-/ReO_4^-$, the saturated adsorption capacity reaches 346.02 mg/g, and the material has good reusability. This new polyamide-functionalized polyacrylamide polymer material has good application prospects in the removal of $^{99}TcO_4^-$ from radioactive wastewater.

Keywords: polyacrylamide; polyamidation; $^{99}TcO_4^-/ReO_4^-$ removal; polymer adsorption material

Citation: Qin, B.; Hu, Y.; Xie, M.; Xue, L.; Liao, C.; Yang, F. Highly Selective Adsorption of $^{99}TcO_4^-/ReO_4^-$ by a Novel Polyamide-Functionalized Polyacrylamide Polymer Material. *Toxics* **2022**, *10*, 630. https://doi.org/10.3390/toxics10100630

Academic Editors: Yuezhou Wei, Yan Wu and Xinpeng Wang

Received: 27 September 2022
Accepted: 19 October 2022
Published: 21 October 2022

Publisher's Note: MDPI stays neutral with regard to jurisdictional claims in published maps and institutional affiliations.

Copyright: © 2022 by the authors. Licensee MDPI, Basel, Switzerland. This article is an open access article distributed under the terms and conditions of the Creative Commons Attribution (CC BY) license (https://creativecommons.org/licenses/by/4.0/).

1. Introduction

With the advent of international "carbon peaking and carbon neutrality", advanced nuclear energy has attracted the attention of countries all over the world, but with the development of advanced nuclear energy, the treatment of technetium 99 (^{99}Tc) in nuclear waste has become a major problem [1,2]. A large amount of radioactive wastewater is produced in the process of nuclear waste material storage [3]. ^{99}Tc is the most potentially problematic radionuclide in radioactive wastewater because it decays by emitting β particles, and its half-life is as long as 2.13×10^5 years [4,5]. ^{99}Tc is a super hydrophilic radionuclide that mainly exists in radioactive wastewater in the form of $^{99}TcO_4^-$. $^{99}TcO_4^-$ has high water solubility (11.3 mol/L, 293.15 K) and almost no complexation properties, resulting in rapid migration in the environment and thus making it a dangerous radioactive pollutant [6,7]. Therefore, it is of great significance to remove $^{99}TcO_4^-$ from radioactive wastewater.

It has always been a difficult point for researchers to remove and capture $^{99}TcO_4^-$ in radioactive wastewater under the condition of strong acid, strong alkali, and a large number of anions [8]. The current research methods for removing and capturing $^{99}TcO_4^-$ in radioactive wastewater can be divided into solvent extraction and solid-phase extraction [9–12]. Compared with solvent extraction, solid phase extraction has a faster extraction rate, better reusability, simpler operation, and easier separation [13]. Therefore, solid phase extraction is a method that is widely used and studied in practice, and solid-phase extraction is a commonly used adsorbent. The most studied adsorbents are metal-organic frameworks (MOFs) and covalent organic frameworks (COFs). They have a high specific surface area and many pores, so they have high a saturated adsorption capacity for $^{99}TcO_4^-$, but they are insufficient for the treatment of $^{99}TcO_4^-$ in environments with high acidity and acidity [14–19]. Ion exchange resins are other polymer materials that have been widely studied. Such resins can effectively remove $^{99}TcO_4^-$ from radioactive wastewater by anion substitution and can maintain a good adsorption performance under highly acidic and alkaline conditions. Ion exchange resins have simple operation and a high recovery rate, but the removal efficiency and selectivity become very poor in the presence of a large number of competitive anions (SO_4^{2-}, NO_3^-, Cl^-, etc.) [20,21]. Some people have also studied the removal of $^{99}TcO_4^-$ by natural zeolites. They modified natural zeolites with cationic surfactants to convert the negative charges of the surface framework of zeolites into positive charges, thus enhancing the affinity for $^{99}TcO_4^-$. However, the saturated adsorption content of $^{99}TcO_4^-$ on zeolites is generally low [22,23]. By aminating the functional groups of chitosan, researchers removed $^{99}TcO_4^-$ under high-acidity conditions, but chitosan has the same disadvantages as zeolite [24,25]. High-molecular-weight polymers are also adsorbents that can effectively remove $^{99}TcO_4^-$ from radioactive wastewater. Many polymers have been developed and used to remove $^{99}TcO_4^-$ in radioactive wastewater [26–30]. After polymers are modified by polyamidation, the polymer can selectively remove $^{99}TcO_4^-$ by electrostatic interaction or hydrogen bonding [28,31,32]. Polyacrylamide (PAM) is a widely used polymer that is often used in biomedicine, sewage treatment, and other fields. As an adsorbent, PAM can effectively remove anions in wastewater. Therefore, the removal of $^{99}TcO_4^-$ from radioactive wastewater using PAM is also an important research topic [33–35].

$^{99}TcO_4^-$ has strong radioactivity, and ReO_4^-, which has similar physical and chemical properties, is used to replace $^{99}TcO_4^-$ in experiments [36]. In this study, a new type of polyamide-functionalized polyacrylamide polymer material was prepared by a two-step method, and its adsorption effect on $^{99}TcO_4^-/ReO_4^-$ was studied. In a certain range of pH values, p-(Amide)-PAM has a good effect on the removal of $^{99}TcO_4^-/ReO_4^-$. Under the influence of different competitive anions, p-(Amide)-PAM has excellent selectivity for $^{99}TcO_4^-/ReO_4^-$. The adsorption kinetics of p-(Amide)-PAM for $^{99}TcO_4^-/ReO_4^-$ was studied, and the material has a fast adsorption rate and high saturated adsorption capacity (346.02 mg/g). This study provides a new material design direction for the treatment of $^{99}TcO_4^-$ in radioactive wastewater.

2. Materials and Methods

2.1. Materials and Reagents

Polyacrylamide ($(C_3H_5NO)_n$, PAM, cationic; molecular weight 1800) was purchased from Beijing Huawei Ruike Chemical Co., Ltd, Beijing, China. Ammonium perrhenate (NH_4ReO_4, $\geq$ 99.99%) was purchased from Shanghai Dibo Biotechnology Co., Ltd, Shanghai, China. Ethylenediamine ($C_2H_8N_2$, analytical purity) was provided by Shanghai Aladdin Limited Chemical Reagent Co., Ltd, Shanghai, China. Ammonia ($NH_3 \cdot H_2O$), nitric acid (HNO_3), and dimethylformamide (C_3H_7NO, DMF, 0.945 g/mL) were all analytically pure and were purchased from Sinopharm Chemical Reagent Co., Ltd, Beijing, China. The ultra-pure water (18.2 m Ω cm) used in this experiment was obtained from a Direct-Q3UV purification system (Research Water Purification Technology Co., Ltd, Xiamen City, Fujian Province, China).

2.2. Synthesis of p-(Amide)-PAM

The synthesis process of p-(Amide)-PAM is shown in Figure 1. Step 1: According to published articles, the p-(Amide)-PAM preparation process was as follows [37,38]: 4.2 g PAM was added to 30 mL of ethylenediamine at 373 K stirred and heated for 72 h, and the resulting product was washed 3 times with ethanol. Then, the resulting solid product was placed in a 353 K vacuum oven for 48 h, and the solid is named N-PAM.

Step 2: Amide acid was synthesized according to the literature [39]. The amide acid was mixed with N-PAM in 30 mL DMF. The mixture was heated and stirred at 373 K for

12 h. Then the product was washed 3 times with ethanol before dried at 353 K for 48h, resulting in a yellow solid powder named p-(Amide)-PAM.

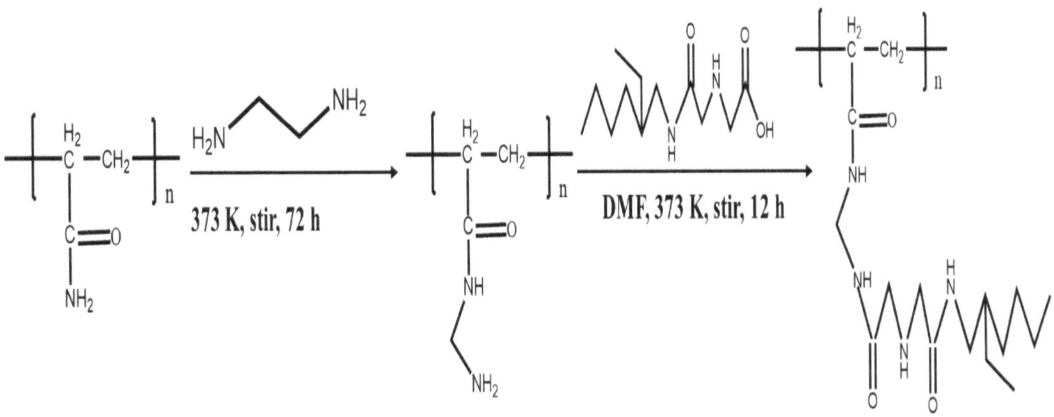

Figure 1. Preparation flow chart of p-(Amide)-PAM.

2.3. Characterization of PAM, N-PAM, and p-(Amide)-PAM

The elemental (C, H, O, and N) contents of PAM, N-PAM, and p-(Amide)-PAM were determined by elemental analysis (Vario El Cube, Germany) and X-ray photoelectron spectroscopy (XPS, K-alpha+, U.K.). The surface morphologies of PAM, N-PAM, and p-(Amide)-PAM were observed by field emission scanning electron microscopy (Apreo S LoVac, Czech Republic). The surface functional groups in the range of 4000-500 cm^{-1} were obtained by Fourier transform infrared spectroscopy (FT-IR, Nicolet iS 50, USA). The specific surface area and porosity of N-PAM and p-(Amide)-PAM were measured by an automatic specific surface area and porosity analyzer (Quantachrome Autosorb IQ, USA). FT-IR and XPS were used to analyze p-(Amide)-PAM before and after adsorption of ReO_4^- to explore its potential adsorption mechanism.

2.4. Batch Adsorption Experiments

The original solution of ReO_4^- with a concentration of 1000 mg/L was prepared with ammonium perrhenate, and other desired concentrations of ReO_4^- were prepared by further dilution of this solution. All adsorption experiments were carried out in 15 mL centrifuge tubes on a constant temperature shaker with a rotational speed of 250 rpm. The pH value of the solution was adjusted with 0.1 M HNO_3 and $NH_3 \cdot H_2O$. The initial concentration of ReO_4^- was 0–1000 mg/L, the contact time was 0–15 h, the pH value was 1–14, and the initial pH was 5.5. After adsorption, the liquid was filtered through a 0.22 μm nylon filter. The initial metal concentration and residual metal concentrations of the samples were determined by inductively coupled plasma optical emission spectrometry (ICP-OES, Ultima2, France).

The adsorption capacity of the adsorbent was calculated by Q_e (mg/g) and removal rate (R) was calculated by the following formulas (Formulas (1) and (2)):

$$Q_e = (C_0 - C_e) \times \frac{V}{m} \tag{1}$$

$$R\% = \frac{C_0 - C_e}{C_0} \times 100\% \tag{2}$$

where C_0 and C_e are the initial concentration and equilibrium concentration (mg/L) of ReO_4^-, Q_e is the adsorption capacity at equilibrium (mg/g), V is the volume of the aqueous phase (L), and m is the mass of the adsorbent (g).

The Langmuir model and Freundlich model were used to fit the isothermal adsorption data and are expressed by Formulas (3) and (4) [40]:

$$\frac{C_e}{Q_e} = \frac{1}{K_L Q_{\max}} + \frac{C_e}{Q_{\max}} \quad (3)$$

$$\ln Q_e = \frac{1}{n} \ln C_e + \ln K_F \quad (4)$$

where C_e refers to the concentration of ReO_4^- at equilibrium (mg/L), Q_{max} is the theoretical maximum adsorption capacity (mg/g), K_L is the Langmuir constant (L/mg), and K_F (mg/g (L/mg)1/n) and $1/n$ are Freundlich constants.

The pseudo-first-order kinetic model and pseudo-second-order kinetic model of adsorption kinetics are expressed by Formula (5) and Formula (6), respectively [41]:

$$\ln(Q_e - Q_t) = \ln(Q_e) - k_1 t \quad (5)$$

$$\frac{t}{Q_t} = \frac{t}{Q_e} + \frac{1}{k_2 Q_e^2} \quad (6)$$

where C_e refers to the concentration of ReO_4^- at equilibrium (mg/L); Q_e and Q_t are the adsorption capacity (mg/g) of ReO_4^- at equilibrium and at time t (min), respectively; and k_1 and k_2 are the pseudo-first-order and pseudo-second-order kinetic model constants.

3. Results and Discussion

3.1. Characterizations of PAM, N-PAM, and p-(Amide)-PAM

The scanning electron microscopy pictures of PAM (Figure 2a), N-PAM (Figure 2b), and p-(Amide)-PAM (Figure 2c) are shown in the figure and their elemental content is shown in Table 1. It can be seen from the diagram that the morphology of PAM is large particles that are relatively regular; N-PAM and p-(Amide)-PAM show irregular small particles after the reaction. By comparing the N_2 adsorption-desorption isotherms of N-PAM (Figure S1a) and p-(Amide)-PAM (Figure S1b), it can be seen that the specific surface area of p-(Amide)-PAM is 90 times higher than the specific surface area of N-PAM, and the pores of p(Amide)-PAM (Figure S1c) are microporous. Figure 2d shows the dispersion of PAM, N-PAM, and p-(Amide)-PAM in water. PAM dissolved in water and formed a hydrogel while N-PAM and p-(Amide)-PAM were insoluble in water. In the FT-IR spectrum (Figure 2e), the characteristic peaks of $-NH_2$ at 3345 and 3183 cm^{-1} are greatly weakened after the reaction, indicating that most of the $-NH_2$ is involved in the reaction process. The peak at 1650 cm^{-1} is characteristic of C=O, and the intensity of the peak of synthesized p-(Amide)-PAM increases substantially, which indicates that the functional modification of N-PAM by amide acid was successful. In the XPS wide scan spectrum of N-PAM (Figure S2a) and p-(Amide)-PAM (Figure S2b), there are N1s, C1s, and O1s spectrum. In the XPS N 1s spectrum (Figure 2f), the -NH (400.8 eV) peak of p-(Amide)-PAM is significantly increased, indicating a significant increase in the number of -NH groups on p-(Amide)-PAM. The binding energies of the modified -NH and C-N increase from 400.5 and 399.0 to 400.8 and 399.2 eV, respectively, indicating that the modification reaction took place on the amino group [42–44]. These results further indicate that the functional modification of N-PAM by amide acid was successful.

Table 1. The elemental content of PAM, N-PAM, and p-(Amide)-PAM.

Elemental	PAM	N-PAM	p-(Amide)-PAM
C [%]	45.140	42.667	39.926
H [%]	7.580	8.670	7.559
O [%]	30.577	29.696	36.562
N [%]	16.517	18.994	15.508

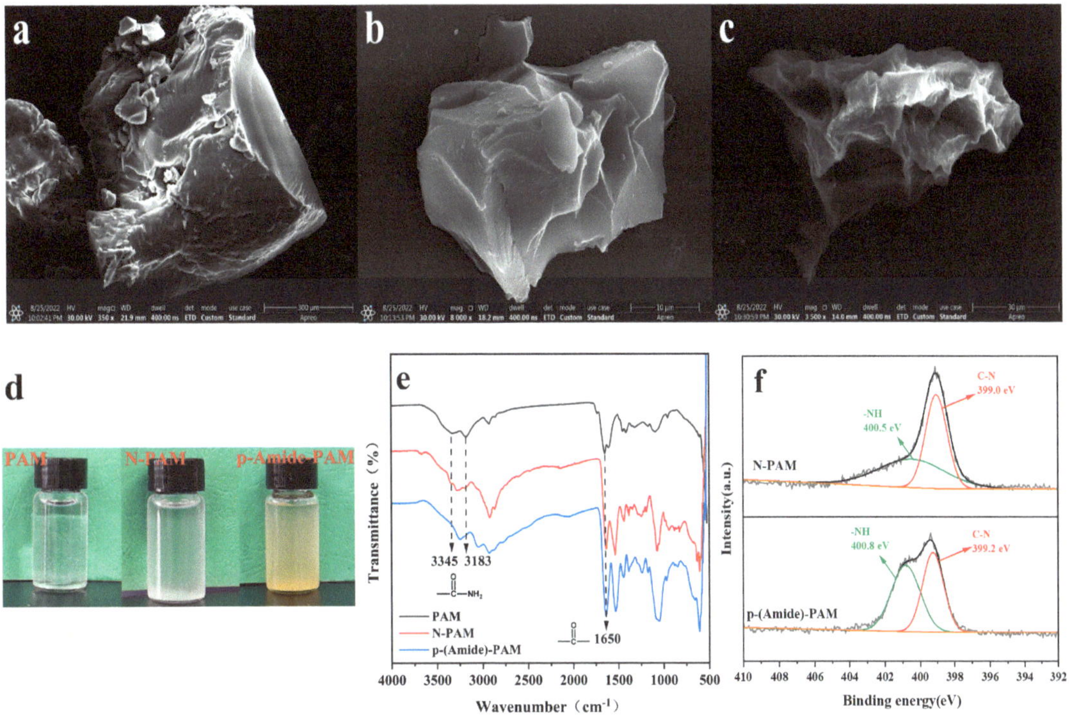

Figure 2. SEM images of PAM (**a**), N-PAM (**b**), and p-(Amide)-PAM (**c**); dispersion of PAM, N-PAM, and p-(Amide)-PAM in water (**d**); FT-IR spectrum of PAM, N-PAM, and p-(Amide)-PAM (**e**); and XPS N 1s spectrum of N-PAM and p-(Amide)-PAM (**f**).

3.2. Adsorption Experiment of ReO_4^- by p-(Amide)-PAM

3.2.1. Influence of Different Molar Reaction Ratios

The effect of the amount of amide acid on the removal of ReO_4^- in the synthesis of p-(Amide)-PAM was investigated. Five groups of samples were prepared with the molar ratios of amide acid to N-PAM of 0.5:1, 0.75:1, 1:1, 1.25:1, and 1.5:1. In Figure 3, the removal rate of ReO_4^- by the intermediate N-PAM is only 29%. The removal rate of p-(Amide)-PAM significantly improves after the functionalization of polyamides. When the molar ratio of the reaction is 0.5:1, the removal rate of ReO_4^- is the highest (up to 90%). With the increase in the amide acid molar ratio, the removal rate of ReO_4^- by p-(Amide)-PAM decreases. This may be due to the N-H functional groups of the adsorbents occupying the adsorption sites of ReO_4^-. These results show that the addition of amide acid has an effect on the removal of ReO_4^- by p-(Amide)-PAM, especially when the molar amount of amide acid is greater than that of N-PAM.

3.2.2. Effect of Initial pH

The effect of p-(Amide)-PAM on ReO_4^- removal under acid-base conditions was explored. In this study, adsorption experiments under different pH conditions were performed, as shown in Figure 4. As the pH increases from 1.3 to 4.0, the removal rate of ReO_4^- by p-(Amide)-PAM gradually increases (from 4% to 80%). When the pH is in the range of 4.0 to 8.0, the removal rate of ReO_4^- by p-(Amide)-PAM is more than 80%. When the pH is 5.5, the maximum removal rate is 88%. From pH 8.0 to 11.0, the removal rate of ReO_4^- by p-(Amide)-PAM decreases gradually, and the removal rate is only 3% when the pH is 11.0. In the case of low pH, the removal rate of ReO_4^- by p-(Amide)-PAM is low, which may be due to the high concentration of NO_3^- and the lack of protonation of -NH on p-(Amide)-PAM. As the pH increases, -NH can protonate to produce a positive charge, resulting in electrostatic interactions with ReO_4^-. In an alkaline environment, OH^- in the aqueous phase will be attracted by protonated -NH, which occupies the adsorption sites, resulting in a decline in the adsorption effect of ReO_4^-. Compared to amino triazole-modified microcrystalline cellulose microsphere and ionic liquid-MIMDIDOA, p-(Amide)-PAM can efficiently remove ReO_4^- over a wide range of pH values [42,45].

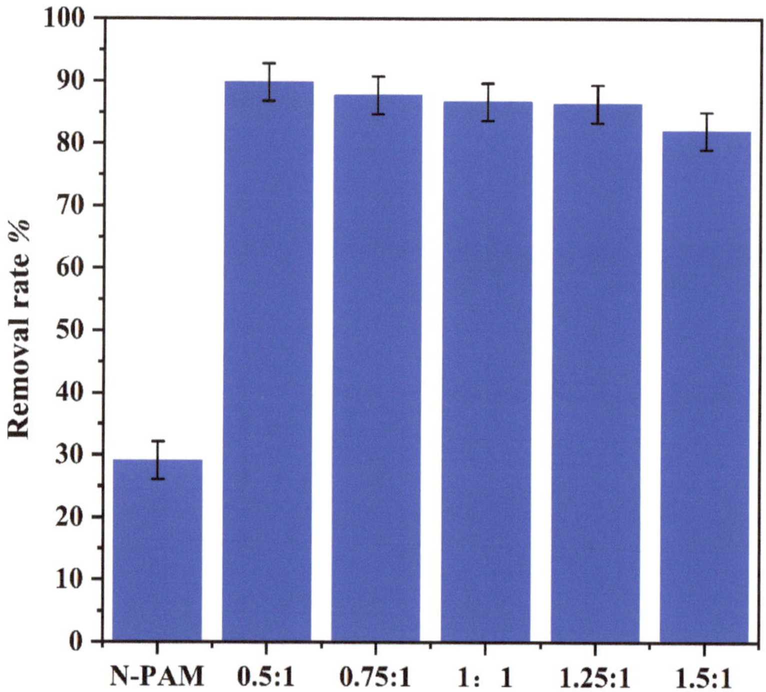

Figure 3. Removal rate of ReO_4^- by p-(Amide)-PAM generated by different reaction molar ratios. The dosage of the absorbent was 1 g/L, pH was 5.5, time was 240 min, initial concentration of ReO_4^- was 100 mg/L, and the temperature was 298.15 K.

3.2.3. Influence of Competitive Anions

There are a large number of competitive anions (NO_3^-, Cl^-, SO_4^{2-}, etc.) in radioactive wastewater, which will adversely affect the adsorption of ReO_4^-. As shown in Figure 5, when the molar ratio of ReO_4^- to competing anions is 1:1, the removal rate of ReO_4^- by p-(Amide)-PAM is 95.7–96.7%. When the molar ratio of ReO_4^- to competitive anions is 1:100, the removal rate of ReO_4^- by p-(Amide)-PAM still reaches 60%. The selectivity of p-(Amide)-PAM may be attributable to its hydrophobic surface and ReO_4^- has a relatively low hydration energy (−170 kJ/mol). Compared with other anions such as NO_3^- (−275 kJ/mol) and Cl^- (−340 kJ/mol), the hydrophobic surface of p-(Amide)-PAM more easily adsorbs ReO_4^-. In addition, the negative charge of SO_4^{2-} (−1080 kJ/mol) is higher than that of ReO_4^-, and SO_4^{2-} is a more favorable compound for adsorption via electrostatic interactions [28]. According to the research report of existing adsorbents, in the anionic system of SO_4^{2-} or Cl^-, the adsorbent is more effective in removing $^{99}TcO_4^-/ReO_4^-$ from radioactive wastewater, but it is interesting that in this work, the effect is better under the system of NO_3^-.

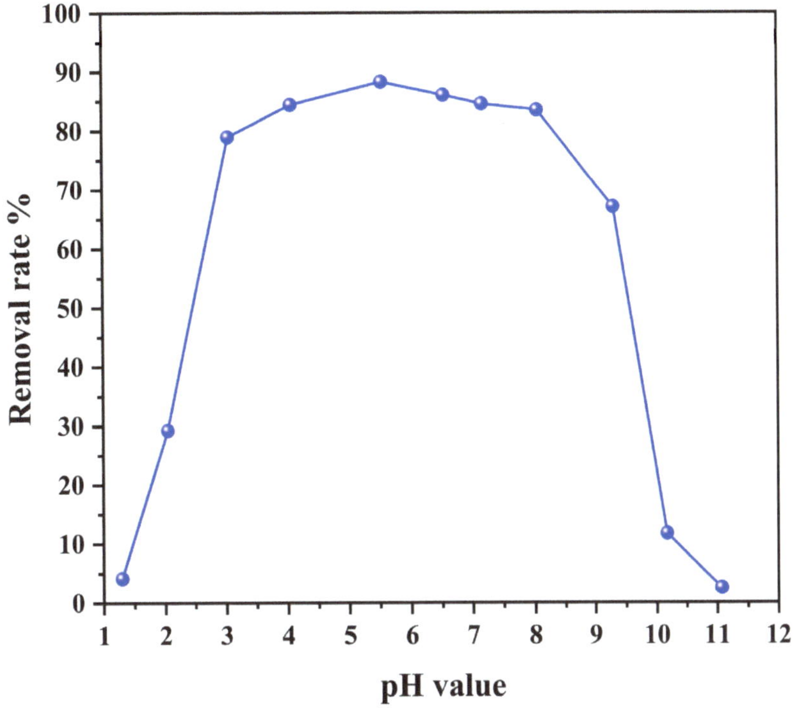

Figure 4. The pH effect on the removal efficiency for ReO_4^- by p-(Amide)-PAM. The dosage of the absorbent was 1 g/L, time was 240 min, initial concentration of ReO_4^- was 100 mg/L, and the temperature was 298.15 K.

3.2.4. Adsorption Isotherm

In order to explore the adsorption performance, the adsorption isotherm of ReO_4^- by p-(Amide)-PAM (Figure 6a) was tested, and the adsorption isotherm data were fitted by the Langmuir model (Figure 6b, Table 2) and Freundlich model (Figure 6c, Table 2). Through the comparison of the two models, the adsorption of ReO_4^- by p-(Amide)-PAM is more consistent with the Langmuir model (R^2 = 0.99452), indicating that the adsorption is monolayer chemisorption on a homogeneous surface. The results show that the saturated adsorption capacity of p-(Amide)-PAM for ReO_4^- is as high as 346.02 mg/g. Compared with the reported adsorbent materials (Table 3), the saturated adsorption capacity of p-(Amide)-PAM exceeds that of most adsorbent materials.

Table 2. Fitting parameters of the Langmuir model and Freundlich model.

Models	Parameters	p-(Amide)-PAM
Langmuir	K_L	0.0247
	Q_{max} (mg/g)	346.02
	R^2	0.99452
Freundlich	K_F	25.319
	n	2.28
	R^2	0.95520

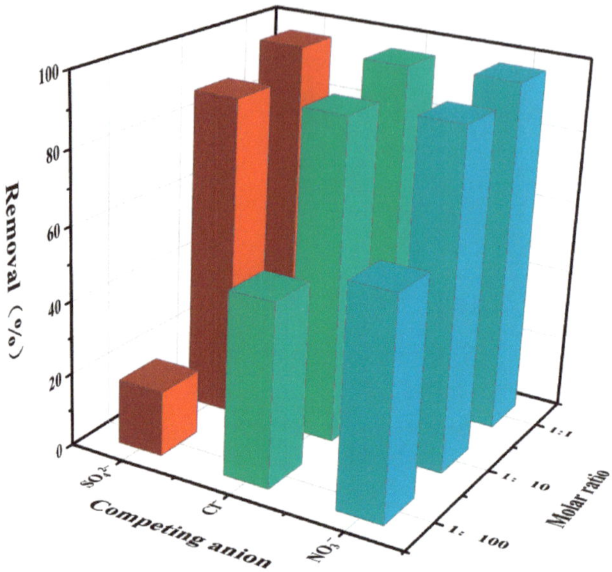

Figure 5. Effect of competing anions on the removal percentage of ReO_4^- by p-(Amide)-PAM. The dosage of the absorbent was 4 g/L, pH was 5.5, time was 240 min, initial concentration of ReO_4^- was 10 mg/L, and the temperature was 298.15 K.

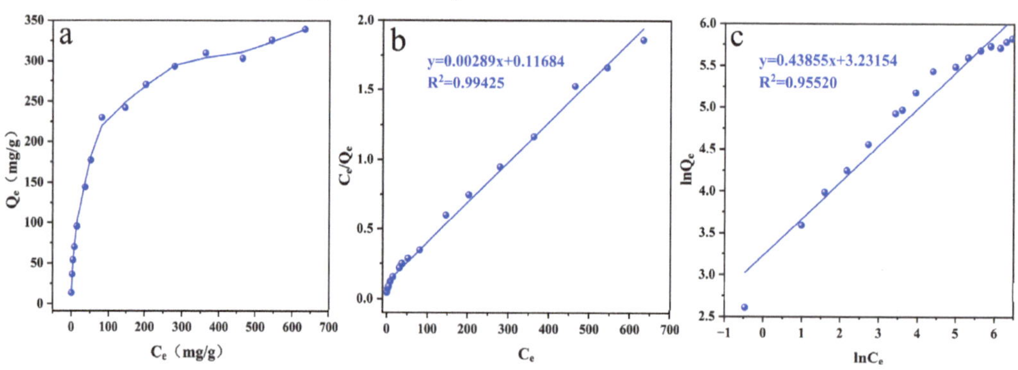

Figure 6. Adsorption isotherm of ReO_4^- by p-(Amide)-PAM (**a**): fitting the Langmuir model (**b**) and fitting the Freundlich model (**c**). The dosage of the absorbent was 1 g/L, pH was 5.5, time was 12 h, initial concentration of ReO_4^- was 0–1000 mg/L, and the temperature was 298.15 K.

Table 3. Adsorption capacity of $^{99}TcO_4^-/ReO_4^-$ by different adsorption materials.

Adsorbent	Adsorption Capacity (mg/g)	References
p-(Amide)-PAM	346.02	This work
GO-DEADIBA	140.82	[32]
SCU-100	541	[15]
SCU-101	217	[46]
SCU-102	291	[17]
SCU-103	318	[47]
Ag-TPPE	251	[19]
ZJU-X6	507	[18]
3-ATAR	146.4	[42]
DNOA-GO-CS	90.33	[25]
CSN	222	[48]
SCU-CPN-4	437	[30]

3.2.5. Adsorption Kinetics

The adsorption properties were further explored, and the adsorption kinetics of ReO_4^- by p-(Amide)-PAM were determined (Figure 7a); the data were fitted by a pseudo-first-order kinetic model (Figure 7b, Table 4) and pseudo-second-order kinetic model (Figure 7c, Table 4). Figure 7a shows that the adsorption rate is fast during the initial stage of adsorption, more than 80% of ReO_4^- is adsorbed in approximately 60 s, and adsorption equilibrium is gradually reached after 120 s. In the first stage, the rapid adsorption process is mainly controlled by physical diffusion, and ReO_4^- quickly occupies the effective adsorption sites. The slow adsorption in the second stage mainly depends on chemical adsorption, which continues until the adsorption equilibrium is reached. Comparing the pseudo-first-order kinetic model and the pseudo-second-order kinetic model, the adsorption of ReO_4^- by p-(Amide)-PAM is more consistent with the pseudo-second-order kinetic model, which indicates that the adsorption process is controlled by chemical adsorption such as surface complexation and metal coprecipitation [49]. The adsorption rate and equilibrium time of adsorbents are important factors for evaluating the performance of adsorbents, where a fast rate and short equilibrium time correspond to a good performance. Therefore, p-(Amide)-PAM has good prospects for the removal of $^{99}TcO_4^-$ in radioactive wastewater.

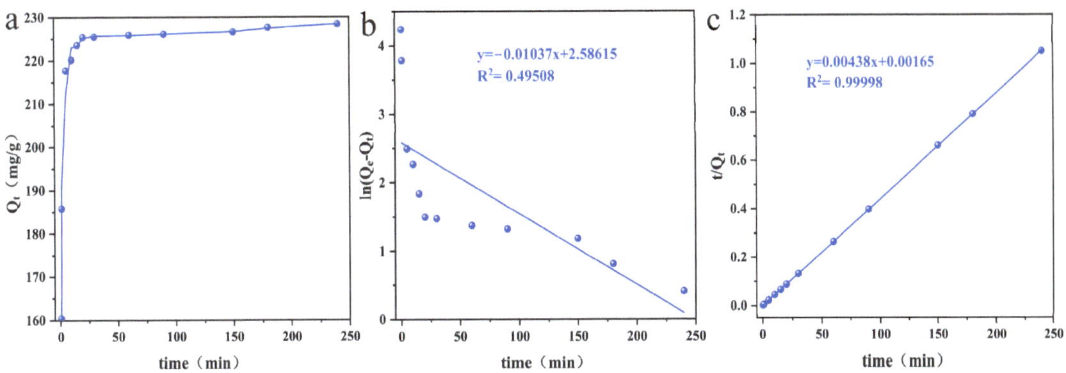

Figure 7. Adsorption kinetics of p-(Amide)-PAM for ReO_4^- (**a**): fitting pseudo-first-order kinetic model (**b**) and fitting pseudo-second-order kinetic model (**c**). The dosage of the absorbent was 1 g/L, pH was 5.5, time was 0–240 min, initial concentration of ReO_4^- was 300 mg/L, and the temperature was 298.15 K.

Table 4. Fitting parameters of the pseudo-first-order model and pseudo-second-order model.

Models	Parameters	p-(Amide)-PAM
Pseudo-first-order	k_1	0.01037
	Q_{e1}(mg/g)	13.279
	R^2	0.49508
Pseudo-second-order	k_2	0.0116
	Q_{e2}(mg/g)	228.311
	R^2	0.99998

3.3. Sorption Mechanism

The sorption mechanism of ReO_4^- by p-(Amide)-PAM was studied by XPS and FT-IR. The comparison of the FT-IR spectra (Figure 8a) before and after the adsorption of ReO_4^- by p-(Amide)-PAM shows a new peak at 903 cm^{-1} after adsorption, which corresponds to the stretching vibration of Re-O formed by electrostatic interactions between ReO_4^- and protonated -NH groups [25,27]. Comparing the XPS (Figure 8b) patterns before and after the adsorption of ReO_4^- by p-(Amide)-PAM, the peaks of Re $4f_{5/2}$ (47.8 eV) and Re $4f_{7/2}$

(45.4 eV) in Re 4f of p-(Amide)-PAM@Re (Figure 8c) are visible, which indicates that the removed ReO_4^- still exists in the form of ions. In the N 1s pattern of p-(Amide)-PAM@Re (Figure 8d), the binding energies of -NH (400.2 eV) and C-N (399.3 eV) have changed, indicating electrostatic interactions between amino groups and ReO_4^- after protonation.

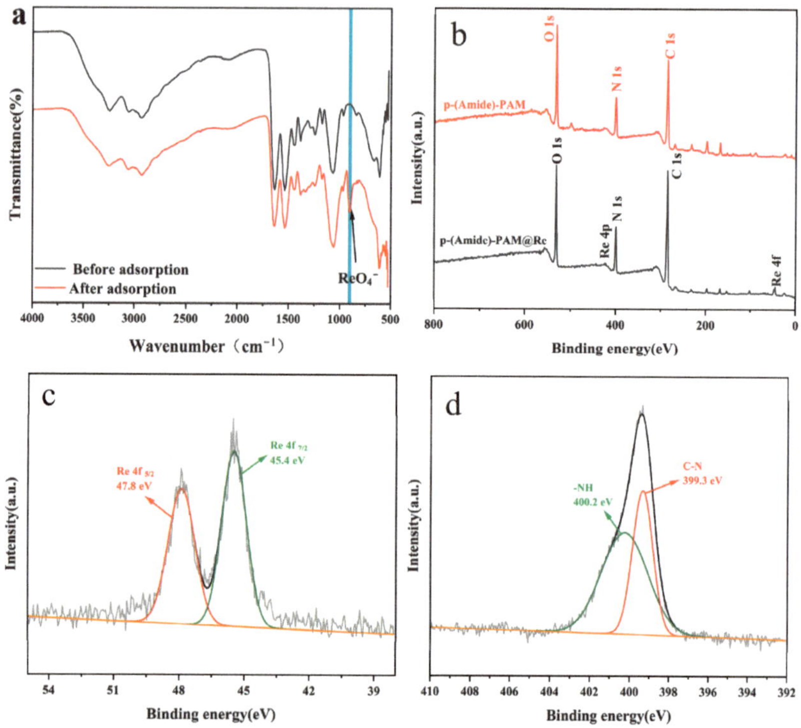

Figure 8. FT-IR spectra (**a**) and XPS wide scan spectrum (**b**) of p-(Amide)-PAM and p-(Amide)-PAM@Re; XPS Re 4f spectrum (**c**) and XPS N 1s spectrum (**d**) of p-(Amide)-PAM@Re.

4. Conclusions

In this study, the polymer material p-(Amide)-PAM with polyamide functionalization was successfully synthesized. The results show that p-(Amide)-PAM has good selective adsorption properties for $^{99}TcO_4^-$/ ReO_4^-. The removal of $^{99}TcO_4^-/ReO_4^-$ by p-(Amide)-PAM has a relatively wide pH window (3.0-8.0) and maintains an excellent adsorption performance. P-(Amide)-PAM maintains good selectivity in environments with a large number of competitive anions (NO_3^-, Cl^-, SO_4^{2-}), and was best under the NO_3^- system. P-(Amide)-PAM has a fast adsorption rate (adsorption equilibrium after 120 s) and high saturated adsorption capacity (346.02 mg/g) for $^{99}TcO_4^-/ReO_4^-$. The predominant $^{99}TcO_4^-/ReO_4^-$ sorption mechanism by p-(Amide)-PAM was the electrostatic interaction of amino groups with $^{99}TcO_4^-/ReO_4$. These results all indicate that p-(Amide)-PAM has good application prospects in the rapid and deep removal of $^{99}TcO_4^-/ReO_4^-$ from radioactive wastewater.

Supplementary Materials: The following supporting information can be downloaded at: https://www.mdpi.com/article/10.3390/toxics10100630/s1, Figure S1a: N_2 adsorption-desorption isotherms of the N-PAM, Figure S1b: N_2 adsorption-desorption isotherms of the p-(Amide)-PAM, Figure S1c: pore size distribution of N-PAM and p-(Amide)-PAM; Figure S2a: XPS wide scan spectrum of N-PAM, Figure S2b: XPS wide scan spectrum of p-(Amide)-PAM.

Author Contributions: Conceptualization: B.Q. and F.Y.; methodology and writing—review and editing, F.Y. and C.L.; writing—original draft preparation, validation and formal analysis, B.Q. and Y.H.; data curation, B.Q. and M.X.; visualization, M.X. and L.X.; supervision, F.Y. and C.L.; All authors have read and agreed to the published version of the manuscript.

Funding: This research was funded by The National Key Research and Development Program of China (No. 2019YFC0605000); the independent deployment project of Ganjiang Innovation Research Institute of Chinese Academy of Sciences (No. E055A002); The key deployment project of the Chinese Academy of Sciences (No. ZDRW-CN-2021-3); the independent deployment project of Fujian Science & Technology Innovation Laboratory for Optoelectronic Information of China (No. 2021ZZ109); Supported by the Fund of Science and Technology on Reactor Fuel and Materials Laboratory (No. STRFML-2021-07).

Institutional Review Board Statement: Not applicable.

Informed Consent Statement: Not applicable.

Data Availability Statement: Not applicable.

Conflicts of Interest: The authors declare no conflict of interest.

References

1. Zheng, K.; Zhang, F.; Xu, Z.; Feng, J.; Chen, X. Research on the Prospect of Nuclear Power Development in Mid-and-Long Term. *IOP Conf. Ser. Earth Environ. Sci.* **2020**, *546*, 022028. [CrossRef]
2. Zhan, L.; Bo, Y.; Lin, T.; Fan, Z. Development and outlook of advanced nuclear energy technology. *Energy Strategy Rev.* **2021**, *34*, 100630. [CrossRef]
3. Zhang, X.; Ping, G.; Liu, Y. Decontamination of radioactive wastewater: State of the art and challenges forward. *Chemosphere* **2018**, *215*, 543–553. [CrossRef] [PubMed]
4. Gephart, R.E. A short history of waste management at the Hanford Site. *Phys. Chem. Earth Parts A/B/C* **2010**, *35*, 298–306. [CrossRef]
5. Smirnov, I.V.; Karavan, M.D.; Logunov, M.V.; Tananaev, I.G.; Myasoedov, B.F. Extraction of Radionuclides from Alkaline and Carbonate Media. *Radiochemistry* **2018**, *60*, 470–487. [CrossRef]
6. Zhou, X.; Ye, G.-A.; Zhang, H.; Li, L.; Luo, F.; Meng, Z. Chemical behavior of neptunium in the presence of technetium in nitric acid media. *Radiochim. Acta* **2014**, *102*, 111–116. [CrossRef]
7. Lee, M.S.; Um, W.; Wang, G.; Kruger, A.A.; Lukens, W.W.; Rousseau, R.; Glezakou, V.A. Impeding ^{99}Tc(IV) mobility in novel waste forms. *Nat. Commun.* **2016**, *7*, 12067. [CrossRef]
8. Li, J.; Dai, X.; Zhu, L.; Xu, C.; Zhang, D.; Silver, M.A.; Li, P.; Chen, L.; Li, Y.; Zuo, D.; et al. $^{99}TcO_4^-$ remediation by a cationic polymeric network. *Nat. Commun.* **2018**, *9*, 3007. [CrossRef]
9. Shu, X.; Xu, Y.; Liu, L.; Fan, Y.; Zhuang, X.; Huang, C.; Chen, S.; Zheng, C.; Jin, Y.; Xia, C. Synthesis of new carbazole derivative extractants and study on extraction of perrhenate/pertechnetate. *Polyhedron* **2022**, *214*, 115641. [CrossRef]
10. Patra, K.; Sengupta, A.; Boda, A.; Ali, M.; Mittal, V.K.; Valsala, T.P.; Kaushik, C.P. Mechanism unravelling for highly efficient and selective $^{99}TcO_4^-$ sequestration utilising crown ether based solvent system from nuclear liquid waste: Experimental and computational investigations. *RSC Adv.* **2022**, *12*, 3216–3226. [CrossRef]
11. Wang, X.; Ding, S.; Wang, Z.; Song, L.; Yang, X.; Xiao, Q.; Xu, H.; Wang, J.; Shen, Z.; Wang, H. A H-Bonding and Electrostatic Interaction Combined Strategy for TcO_4^- Separation by a Nitrotriacetate-Derived Amine–Amide Extractant. *Inorg. Chem.* **2021**, *60*, 10899–10908. [CrossRef] [PubMed]
12. Fiskum, S.K.; Thompson, C.J.; Riley, R.G. Preconcentration and analysis of strontium-90 and technetium-99 from Hanford groundwater using solid phase extraction. *J. Radioanal. Nucl. Chem.* **2000**, *245*, 261–272. [CrossRef]
13. Kameo, Y.; Katayama, A.; Hoshi, A.; Haraga, T.; Nakashima, M. Simple determination of ^{99}Tc in radioactive waste using Tc extraction disk and imaging plates. *Appl. Radiat. Isot.* **2010**, *68*, 139–143. [CrossRef] [PubMed]
14. Wang, Y.; Lan, J.; Yang, X.; Zhong, S.; Yuan, L.; Li, J.; Peng, J.; Chai, Z.; Gibson, J.K.; Zhai, M.; et al. Superhydrophobic Phosphonium Modified Robust 3D Covalent Organic Framework for Preferential Trapping of Charge Dispersed Oxoanionic Pollutants. *Adv. Funct. Mater.* **2022**, *32*, 2205222. [CrossRef]

15. Sheng, D.; Zhu, L.; Xu, C.; Xiao, C.; Wang, Y.; Wang, Y.; Chen, L.; Diwu, J.; Chen, J.; Chai, Z.; et al. Efficient and Selective Uptake of TcO_4^- by a Cationic Metal-Organic Framework Material with Open Ag^+ Sites. *Environ. Sci. Technol.* **2017**, *51*, 3471–3479. [CrossRef]
16. Hao, M.; Chen, Z.; Yang, H.; Waterhouse, G.I.N.; Ma, S.; Wang, X. Pyridinium salt-based covalent organic framework with well-defined nanochannels for efficient and selective capture of aqueous $^{99}TcO_4^-$. *Sci. Bull.* **2022**, *67*, 924–932. [CrossRef]
17. Sheng, D.; Zhu, L.; Dai, X.; Xu, C.; Li, P.; Pearce, C.I.; Xiao, C.; Chen, J.; Zhou, R.; Duan, T.; et al. Successful Decontamination of $^{99}TcO_4^-$ in Groundwater at Legacy Nuclear Sites by a Cationic Metal-Organic Framework with Hydrophobic Pockets. *Angew. Chem. Int. Ed. Engl.* **2019**, *58*, 4968–4972. [CrossRef]
18. Kang, K.; Shen, N.; Wang, Y.; Li, L.; Zhang, M.; Zhang, X.; Lei, L.; Miao, X.; Wang, S.; Xiao, C. Efficient sequestration of radioactive $^{99}TcO_4^-$ by a rare 3-fold interlocking cationic metal-organic framework: A combined batch experiments, pair distribution function, and crystallographic investigation. *Chem. Eng. J.* **2022**, *427*, 130942. [CrossRef]
19. Kang, K.; Liu, S.; Zhang, M.; Li, L.; Liu, C.; Lei, L.; Dai, X.; Xu, C.; Xiao, C. Fast Room-Temperature Synthesis of an Extremely Alkaline-Resistant Cationic Metal-Organic Framework for Sequestering TcO_4^- with Exceptional Selectivity. *Adv. Funct. Mater.* **2022**, 1–12. [CrossRef]
20. Li, J.; Zhu, L.; Xiao, C.; Chen, L.; Chai, Z.; Wang, S. Efficient uptake of perrhenate/pertechnenate from aqueous solutions by the bifunctional anion-exchange resin. *Radiochim. Acta* **2018**, *106*, 581–591. [CrossRef]
21. Fu, L.; Zu, J.; He, L.; Gu, E.; Wang, H. An adsorption study of ^{99}Tc using nanoscale zero-valent iron supported on D001 resin. *Front. Energy* **2019**, *14*, 11–17. [CrossRef]
22. Milićević, S.; Matović, L.; Petrović, Đ.; Đukić, A.; Milošević, V.; Đokić, D.; Kumrić, K. Surfactant modification and adsorption properties of clinoptilolite for the removal of pertechnetate from aqueous solutions. *J. Radioanal. Nucl. Chem.* **2016**, *310*, 805–815. [CrossRef]
23. Dickson, J.; Conroy, N.A.; Xie, Y.; Powell, B.A.; Seaman, J.C.; Boyanov, M.I.; Kemner, K.M.; Kaplan, D.I. Surfactant-modified siliceous zeolite Y for pertechnetate remediation. *Chem. Eng. J.* **2020**, *402*, 126268. [CrossRef]
24. Xiong, Y.; Song, Y.; Tong, Q.; Zhang, P.; Wang, Y.; Lou, Z.; Zhang, F.; Shan, W. Adsorption-controlled preparation of anionic imprinted amino-functionalization chitosan for recognizing rhenium(VII). *Sep. Purif. Technol.* **2017**, *177*, 142–151. [CrossRef]
25. Zhang, H.; Yang, F.; Lu, C.; Du, C.; Bai, R.; Zeng, X.; Zhao, Z.; Cai, C.; Li, J. Effective decontamination of $^{99}TcO_4^-/ReO_4^-$ from Hanford low-activity waste by functionalized graphene oxide-chitosan sponges. *Environ. Chem. Lett.* **2020**, *18*, 1379–1388. [CrossRef]
26. Banerjee, D.; Kim, D.; Schweiger, M.J.; Kruger, A.A.; Thallapally, P.K. Removal of TcO_4^- ions from solution: Materials and future outlook. *Chem. Soc. Rev.* **2016**, *45*, 2724–2739. [CrossRef]
27. Hu, Q.-H.; Jiang, W.; Liang, R.-P.; Lin, S.; Qiu, J.-D. Synthesis of imidazolium-based cationic organic polymer for highly efficient and selective removal of ReO_4^-/TcO_4^-. *Chem. Eng. J.* **2021**, *419*, 129546. [CrossRef]
28. Huang, Y.; Ding, M.; Ding, J.; Kang, J.; Yan, Z.; Zhao, P.; Zhou, X.; Jin, Y.; Chen, S.; Xia, C. Targeted synthesis of a high-stability cationic porous aromatic framework for highly efficient remediation of $^{99}TcO_4^-$. *Chem. Eng. J.* **2022**, *435*, 134785. [CrossRef]
29. Han, D.; Li, X.; Cui, Y.; Yang, X.; Chen, X.; Xu, L.; Peng, J.; Li, J.; Zhai, M. Polymeric ionic liquid gels composed of hydrophilic and hydrophobic units for high adsorption selectivity of perrhenate. *RSC Adv.* **2018**, *8*, 9311–9319. [CrossRef]
30. Li, J.; Li, B.; Shen, N.; Chen, L.; Guo, Q.; Chen, L.; He, L.; Dai, X.; Chai, Z.; Wang, S. Task-Specific Tailored Cationic Polymeric Network with High Base-Resistance for Unprecedented $^{99}TcO_4^-$ Cleanup from Alkaline Nuclear Waste. *ACS Cent. Sci.* **2021**, *7*, 1441–1450. [CrossRef]
31. Sarri, S.; Misaelides, P.; Zamboulis, D.; Gaona, X.; Altmaier, M.; Geckeis, H. Rhenium(VII) and technetium(VII) separation from aqueous solutions using a polyethylenimine-epichlorohydrin resin. *J. Radioanal. Nucl. Chem.* **2015**, *307*, 681–689. [CrossRef]
32. Xiong, Y.; Cui, X.; Zhang, P.; Wang, Y.; Lou, Z.; Shan, W. Improving Re(VII) Adsorption on Diisobutylamine-Functionalized Graphene Oxide. *ACS Sustain. Chem. Eng.* **2016**, *5*, 1010–1018. [CrossRef]
33. Mo, L.; Zhang, S.; Qi, F.; Huang, A. Highly stable cellulose nanofiber/polyacrylamide aerogel via in-situ physical/chemical double crosslinking for highly efficient Cu(II) ions removal. *Int. J. Biol. Macromol.* **2022**, *209*, 1922–1932. [CrossRef]
34. Xiong, B.; Loss, R.D.; Shields, D.; Pawlik, T.; Hochreiter, R.; Zydney, A.L.; Kumar, M. Polyacrylamide degradation and its implications in environmental systems. *Npj Clean Water* **2018**, *1*, 17. [CrossRef]
35. Tuzikov, A.; Chinarev, A.; Shilova, N.; Gordeeva, E.; Galanina, O.; Ovchinnikova, T.; Schaefer, M.; Bovin, N. 40 years of glyco-polyacrylamide in glycobiology. *Glycoconj. J.* **2021**, *38*, 89–100. [CrossRef]
36. Farrell, D.; Gloe, K.; Gloe, K.; Goretzki, G.; Mckee, V.; Nelson, J.; Nieuwenhuyzen, M.; Pál, I.; Stephan, H.; Town, R.M.; et al. Towards promising oxoanion extractants: Azacages and open-chain counterparts. *Dalton Trans.* **2003**, *2003*, 1961–1968. [CrossRef]
37. Tamami, B.; Fadavi, A. Amino group immobilized on polyacrylamide: An efficient heterogeneous catalyst for the Knoevenagel reaction in solvent-free and aqueous media. *Catal. Commun.* **2005**, *6*, 747–751. [CrossRef]
38. Zare, M.A.; Husain, S.W.; Tehrani, M.S.; Azar, P.A. Pentaazatetraethylene supported polyacrylamide (PAA-N5) as a novel adsorbent for the efficient removal of industrial dyes from aqueous solutions: Adsorption isotherms and kinetics. *Mon. Fur Chem. Chem. Mon.* **2016**, *148*, 191–197. [CrossRef]
39. Baba, Y.; Kubota, F.; Kamiya, N.; Goto, M. Development of Novel Extractants with Amino Acid Structure for Efficient Separation of Nickel and Cobalt from Manganese Ions. *Ind. Eng. Chem. Res.* **2014**, *53*, 812–818. [CrossRef]

40. Wang, Y.; Zheng, K.; Jiao, Z.; Zhan, W.; Ge, S.; Ning, S.; Fang, S.; Ruan, X. Simultaneous Removal of Cu^{2+}, Cd^{2+} and Pb^{2+} by Modified Wheat Straw Biochar from Aqueous Solution: Preparation, Characterization and Adsorption Mechanism. *Toxics* **2022**, *10*, 316. [CrossRef]
41. Lin, W.; Zhao, Z.; Yang, F.; Liu, Z.; Tan, F.; Xie, M.; Ma, Y.; Meng, L. Promising priority separation of europium from lanthanide by novel DGA-functionalized metal organic frameworks. *Miner. Eng.* **2021**, *164*, 106831. [CrossRef]
42. Wen, D.; Hua, R.; Dong, Z.; Xie, K.; Qi, W.; Zhao, L. Efficient separation and recovery of Re(VII) from Re/U bearing acidic solutions using aminotriazole modified cellulose microsphere adsorbents. *Environ. Sci. Pollut. Res. Int.* **2021**, *28*, 52225–52235. [CrossRef]
43. Zhang, W.; Wang, S.; Ji, J.; Li, Y.; Zhang, G.; Zhang, F.; Fan, X. Primary and tertiary amines bifunctional graphene oxide for cooperative catalysis. *Nanoscale* **2013**, *5*, 6030–6033. [CrossRef]
44. Ding, M.; Chen, L.; Xu, Y.; Chen, B.; Ding, J.; Wu, R.; Huang, C.; He, Y.; Jin, Y.; Xia, C. Efficient capture of Tc/Re(VII, IV) by a viologen-based organic polymer containing tetraaza macrocycles. *Chem. Eng. J.* **2020**, *380*, 122581. [CrossRef]
45. Huang, Y.; Li, X.; Wu, F.; Yang, S.; Dong, F.; Zhi, X.; Chen, X.; Tian, G.; Shen, Y. A Novel Functionalized Ionic Liquid for Highly Selective Extraction of TcO_4^-. *Inorg. Chem.* **2022**, *61*, 10609–10617. [CrossRef]
46. Zhu, L.; Sheng, D.; Xu, C.; Dai, X.; Silver, M.A.; Li, J.; Li, P.; Wang, Y.; Wang, Y.; Chen, L.; et al. Identifying the Recognition Site for Selective Trapping of $^{99}TcO_4^-$ in a Hydrolytically Stable and Radiation Resistant Cationic Metal-Organic Framework. *J. Am. Chem. Soc.* **2017**, *139*, 14873–14876. [CrossRef]
47. Shen, N.; Yang, Z.; Liu, S.; Dai, X.; Xiao, C.; Taylor-Pashow, K.; Li, D.; Yang, C.; Li, J.; Zhang, Y.; et al. $^{99}TcO_4^-$ removal from legacy defense nuclear waste by an alkaline-stable 2D cationic metal organic framework. *Nat. Commun.* **2020**, *11*, 5571. [CrossRef]
48. Li, W.; Dong, X.; Zhu, L.; Tang, H. Highly selective separation of Re(VII) from Mo(VI) by using biomaterial-based ionic gel adsorbents: Extractive adsorption enrichment of Re and surface blocking of Mo. *Chem. Eng. J.* **2020**, *387*, 124078. [CrossRef]
49. Xia, Y.; Li, Y.; Xu, Y. Adsorption of Pb(II) and Cr(VI) from Aqueous Solution by Synthetic Allophane Suspension: Isotherm, Kinetics, and Mechanisms. *Toxics* **2022**, *10*, 291. [CrossRef]

toxics

MDPI

Article

Study on the Effect of Gamma-Ray Irradiation on the Adsorption of ^{99}Tc and Re by a Silica-Based Pyridine Resin

Hongji Sang, Cong Mao, Yan Wu * and Yuezhou Wei

School of Nuclear Science and Engineering, Shanghai Jiao Tong University, Shanghai 200240, China
* Correspondence: wu_yan@sjtu.edu.cn

Abstract: A silica-based anion exchange resin was synthesized and used to remove ^{99}Tc from real radioactive liquid waste. The adsorbent had a uniform particle size and exhibited good thermal stability up to 100 °C, which is promising for large-scale column experiments. In accordance with the chemical similarity with Tc, Re was used as a surrogate in this study. The N 1s high-resolution XPS spectra of the adsorbent before and after the adsorption of Re indicated that the ion exchange reaction was the controlling mechanism in the process. After γ-ray irradiation, the changing trend of the K_d was consistent, which showed that the competitive adsorption of NO_3^- led to a decrease in K_d. The adsorption capacity for the Re decreased slightly from 35.8 to 31.9 mg/g with the increase in the absorbed dose from 0 to 50 kGy. The separation and recovery of Re and the coexisting ions were achieved by chromatographic separation experiments, and the recovery percentage of Re was 86%. In real radioactive liquid waste, N3/SiO_2 exhibited good selectivity toward ^{99}Tc over the coexisting metals, namely, ^{90}Sr, ^{137}Cs, ^{241}Am, and U, and the decontamination efficiency of ^{99}Tc attained 65%.

Keywords: ^{99}Tc; adsorption; pyridine resin; γ-ray absorbed doses

Citation: Sang, H.; Mao, C.; Wu, Y.; Wei, Y. Study on the Effect of Gamma-Ray Irradiation on the Adsorption of ^{99}Tc and Re by a Silica-Based Pyridine Resin. *Toxics* **2022**, *10*, 638. https://doi.org/10.3390/toxics10110638

Academic Editor: Alexei Konoplev

Received: 30 September 2022
Accepted: 21 October 2022
Published: 25 October 2022

Publisher's Note: MDPI stays neutral with regard to jurisdictional claims in published maps and institutional affiliations.

Copyright: © 2022 by the authors. Licensee MDPI, Basel, Switzerland. This article is an open access article distributed under the terms and conditions of the Creative Commons Attribution (CC BY) license (https://creativecommons.org/licenses/by/4.0/).

1. Introduction

The removal of long-lived nuclides from high-level liquid waste (HLLW) could improve the safety of the spent fuel reprocessing and waste disposal process [1]. The radionuclide ^{99}Tc, with a high yield and a long half-life, is generated by the uranium fission reaction, and it is a hazardous nuclide commonly found in HLLW [2–4]. The behavior of Tc in the environment is influenced by biochemical processes and is therefore very complex. The pertechnetate anion (TcO_4^-) is the main existence form for ^{99}Tc in radioactive wastewater, which has a high solubility and mobility in ground water and soil, posing a potential threat to human health [5–8]. In addition, technetium metal can be used to prepare β radiation sources of low neutron and superconductors, while TcO_4^- can improve the corrosion resistance of steel [9]. Note that technetium is mostly considered to be a dangerous waste at present due to its environmental mobility and long half-life. To protect the environment and human health, the separation and recovery of ^{99}Tc from HLLW is highly valuable [9–11]. However, the effective removal of ^{99}Tc from wastewater has always been challenging. Furthermore, in the absence of an electron transfer reaction, Re is usually used as a non-radioactive analogue for ^{99}Tc because of their similar geochemical properties and ion size [2,12,13].

Under the plutonium and uranium redox extraction (PUREX) conditions, technetium can coextract with other species in solution, including zirconium and uranium [14]. The separation of technetium usually requires additional processes. Moreover, a large amount of secondary organic waste is generated, which increases the cost of the reprocessing process. The extraction behavior of technetium is complex due to the versatile chemical properties, so efficiency extractants with good selectivity need to be developed. With the characteristics of large exchange capacity, excellent selectivity, low cost, and less waste accumulation [15–17], ion exchange as an environmentally friendly technology has become an optimal choice. Until now, various ion exchangers including thorium borates [18], layered double hydroxides (LDHs) [19], metal sulfides [20], poly-electrolytes [15] and ion exchange resins [9,21–26] have been studied for the selective separation of ^{99}Tc. Among these, anion exchange resin with amine groups is one kind of promising material, which can capture negatively charged TcO_4^- in an aqueous solution by an anion exchange reaction. Our group prepared a pyridine type weakly basic anion exchange resin through copolymerization without chloromethylation and amination [27], simplifying the synthesis process of resins.

In this study, the pyridine type weakly basic anion exchange resin (N3), with high affinity to ^{99}Tc, was impregnated as the adsorbent into the pores of a rigid support material SiO_2 (N3/SiO_2) to enhance the hydraulic properties and radiation stability (Figure 1). The effect of the γ-ray, nitric acid concentration, initial metal concentration, and coexisting ions on the adsorption performance toward technetium and its analogue (rhenium) was investigated. The chromatographic separation was performed by dynamic column, and the selective separation of ^{99}Tc from real radioactive liquid waste was evaluated.

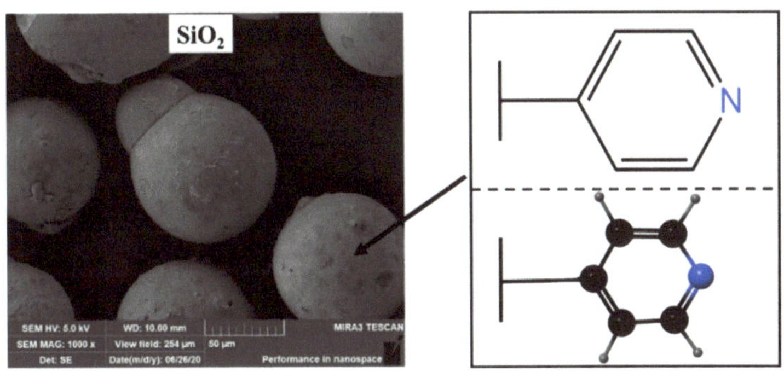

Figure 1. Schematic representation of the N3/SiO_2 structure.

2. Materials and Methods

2.1. Materials

The chemical reagents α,α'-azobisisobutyronitrile (AIBN) with chemical pure purity, 1,1′-azobiscyclohexane-1-carbonitrile (ACCN) with 98% purity, 4-vinylpyridine (4 VP) with 95% purity, tert-butyl-catechol (TBC) with 98% purity and divinylbenzene (DVB) with 80% purity were purchased from Beijing InnoChem Science & Technology Co., Ltd., Beijing, China. Ammonium rhenate (NH_4ReO_4), ammonium chloride (NH_4Cl), molybdenum pentachloride ($MoCl_5$), sodium hydroxide (NaOH), thiourea, methanol, acetophenone, diethyl-o-phthalate, hydrochloric acid, nitric acid, and the other nitrates were from Sinopharm Chemical Reagent Co., Ltd., Shanghai, China, and they were of analytical grade.

2.2. Preparation of the N3/SiO$_2$ Adsorbent

The N3/SiO_2 was prepared as follows. The SiO_2 was washed three times with concentrated hydrochloric acid and methanol, respectively, to remove the metal impurities and increase its activity. After filtration, the SiO_2 was dried at 90 °C for 24 h. Then, the SiO_2 was placed in a flask of a rotary evaporator, and a mixture of initiators (AIBN and ACCN), diluents (acetophenone and diethyl-o-phthalate), and monomers (4 VP and DVB) was sucked into the abovementioned flask under the action of a vacuum pump. The flask was gradually heated to 90 °C and rotated slowly. During the process, the monomers after successful polymerization were impregnated into the pores of SiO_2 under the protection of the nitrogen atmosphere [27]. Finally, the product was washed and dried.

2.3. Irradiation Experiments

$N3/SiO_2$ was irradiated at room temperature using a ^{60}Co source with an irradiation dose rate of 5 kGy/h. Then, batch adsorption experiments were carried out with the $N3/SiO_2$ after irradiation in different concentrations of nitric acid solution.

2.4. Characterization

A laser diffraction particle size analyzer (S3500, Microtrac Inc., Largo, FL, USA) was used to measure the particle-size distribution characteristics of the $N3/SiO_2$. The average pore size and pore size distribution were tested by the mercury intrusion method (AutoPore IV 9500 V1.09, Micromeritics Inc., Norcross, GA, USA). The thermal stability of the $N3/SiO_2$ was examined by thermogravimetry–differential thermal analysis (DTG-60, Shimadzu Corporation, Kyoto, Japan) with a heating rate of 10 °C/min in an air atmosphere. The composition and valence states of the main elements were tested through X-ray photoelectron spectroscopy (XPS, AXIS Ultra DLD, Shimadzu Corporation, Kyoto, Japan).

2.5. Batch Experiments

In the batch experiments, 50 mg of the adsorbent and 5 mL of HNO_3 solution were added to a glass vial with a plastic lid. The vial was shaken mechanically at 120 rpm and 25 °C for a certain contact time. The concentrations of metal ions in the aqueous phase were measured via atomic absorption spectrophotometry (AAS, SP3880, Shanghai Guangpu Co., Ltd., Shanghai, China) and sequential plasma spectrometry (ICP-AES, ICP-7510, Shimadzu Corporation, Kyoto, Japan). The distribution coefficients (K_d) and adsorption amount (Q_{eq}) of the metal ions were calculated by the following equations [28]:

$$K_d = \frac{C_0 - C_e}{C_e} \times \frac{V}{m}, \quad (1)$$

$$Q_{eq} = (C_0 - C_e) \times \frac{V}{m}, \quad (2)$$

where C_0 and C_e (mmol/L) represent the initial and equilibrium concentrations of metal ions in an aqueous phase, respectively. V (mL) and m (g) denote the volume of the aqueous phase and the weight of the adsorbent, respectively.

2.6. Chromatographic Partitioning of Re

The chromatographic partitioning of Re from simulated HLLW (SHLLW) containing typical fission and non-fission elements was conducted to evaluate the feasibility and practicability of the $N3/SiO_2$ in a column system. The $N3/SiO_2$ of 1 g was packed in the glass column (10 mm diameter × 50 mm long); the SHLLW with the nitric acid concentration of 1 M was then fed through the adsorbent-packed column at a flow rate of 0.5 mL/min. The concentrations of the target elements in the effluent were tested by ICP and AAS.

2.7. ^{99}Tc Removal Performance

The adsorption property for the real radioactive waste was further tested. A certain amount of radioactive liquid waste was obtained from a batch of nuclear facilities in China, and its nitric acid concentration was adjusted to 1 and 3 mol/L respectively. After adjustment, the gross radioactivity of technetium used in this experiment was approximately 1×10^3 Bq. The abovementioned liquid waste was mixed with $N3/SiO_2$ and shaken mechanically for 12 h at 25 °C. The ratio of the solid phase to the aqueous phase was 0.5 g to 5 cm^3. The amounts of ^{90}Sr and ^{99}Tc were determined by a liquid scintillation analyzer. An HPGe gamma spectrometer was used to measure the activities of ^{137}Cs and ^{241}Am.

3. Results and Discussion

3.1. Characterization

3.1.1. Particle Size and Pore Size Analysis

The particle-size distribution characteristics of the N3/SiO_2 obtained by the laser particle size analyzer are shown in Figure 2. The cumulative curve showed that 80% of the particles were mainly distributed between 60 and 140 microns, with an average particle size of 114.9 microns. The separation efficiency was improved by reducing the back-mixing when the fluid passed through the adsorbent bed filled with the adsorbent of uniform particles.

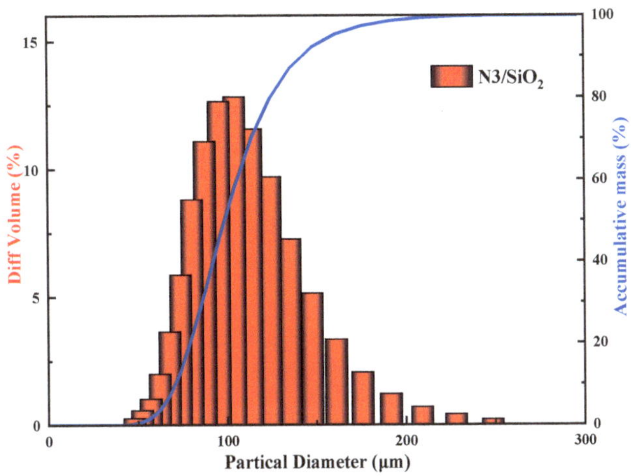

Figure 2. Particle size analysis curve of the N3/SiO_2 adsorbent.

The MIP method was used to analyze the porosity and pore size of the SiO_2 and N3/SiO_2. Figure 3 presents the pore size distribution of the SiO_2 and N3/SiO_2. The most probable pore size was located at the abscissa value of the peak. There was no significant difference in the pore-size distribution tendency between the two samples. The two peaks located at 10 to 50 μm were considered to arise from the interspace between the SiO_2 particles. The pore capacity increment of the silica carrier was concentrated in the pore size range of 26–62 nm and reached its maximum value at about 40 nm. The pore size distribution with good uniformity was conducive to the diffusion and impregnation of the adsorbate into the SiO_2 pores. The pore size of the SiO_2 was calculated to be 43.3 nm, and it decreased to 34.8 nm after the pyridine resin was loaded (Table 1). Similarly, the porosity of the N3/SiO_2 was evaluated at 70.7%, slightly lower than the SiO_2. The impregnation of the pyridyl resin into the SiO_2 pores resulted in a decrease in the pore size and porosity. Furthermore, a larger porosity was beneficial to improve the ion exchange rate in the adsorption process.

Table 1. The pore size and porosity of SiO_2 and N3/SiO_2.

Parameters	Average Pore Size (nm)	Porosity (%)
SiO_2	43.3	75.6
N3/SiO_2	34.8	70.7

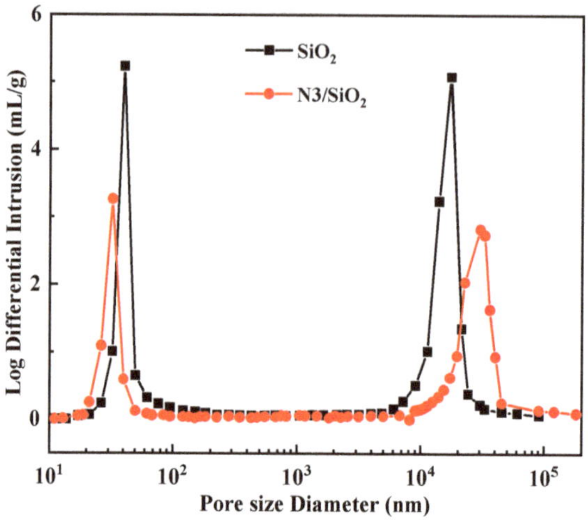

Figure 3. The pore size distribution of SiO_2 and N3/SiO_2.

3.1.2. Thermal Property

As shown in Figure 4, the thermal decomposition behavior of N3/SiO_2 was analyzed. The weight loss within 100 °C was 2.2 wt%, corresponding to the evaporation of water in the adsorbent. The continuous rapid weight loss starting at about 200 °C could be regarded as the decomposition of the resin polymer loaded into the SiO_2 carrier. There were two obvious exothermic peaks in the DTA curves at about 238 °C and 334 °C, respectively. The N3 resin obtained by the polymerization of 4-vinylpyridine and divinylbenzene as monomers mainly contained two crosslinking structures, a pyridine ring and a benzene ring, which were connected to the skeleton structure; the two main structures of the resin polymer were decomposed successively under these two temperature conditions. The weight loss calculated at around 404.8 °C was inferred to be the thermal desorption of the self-polymers of a few monomers with a corresponding endothermic peak. The residue after 650.6 °C was the stable silica carrier. In summary, the N3/SiO_2 lost approximately 28.54 wt% of its original weight at 30–900 °C, which indicated that 26.34 wt% of the organics were immobilized. All the results demonstrated the successful synthesis of the adsorbent, which exhibited good thermal stability under 100 °C.

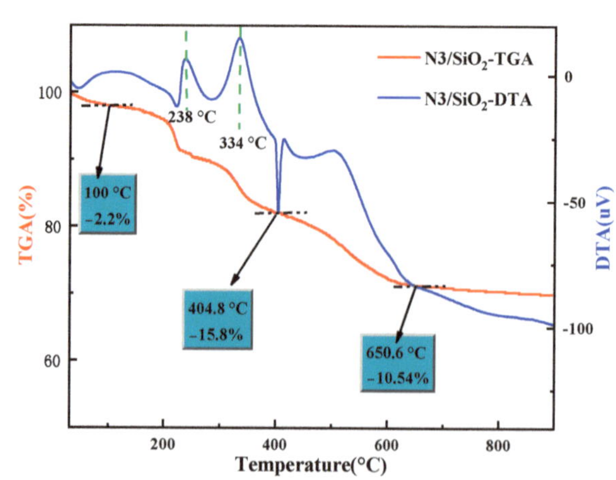

Figure 4. The TGA/DTA curves of the N3/SiO_2.

3.1.3. XPS

As shown in Figure 5, XPS was used to investigate the mechanism of the adsorption of N3/SiO_2 for Re. In Figure 5A, the peaks of O 1s, N 1s, C 1s, and Si 2p are observed in the full spectrum of the adsorbent before and after the Re adsorption. Moreover, new peaks at 264.62 eV and 45.37 eV were assigned to Re 4d and Re 4f, respectively, in the latter case, thereby confirming the successful adsorption of Re into the N3/SiO_2. The N 1s spectra of the absorbent before and after the adsorption of Re was further analyzed, as shown in Figure 5B. The components of the N 1s spectrum before and after adsorption could be deconvoluted into four peaks, indicating that no new valence of N was produced in the adsorption reaction. The lowest binding energy peak centered at 398.7 eV was ascribed to the deprotonated N atoms in the pyridine functional group, whereas the 399.3 and 401.2 eV contributions were due to the N atoms in different degrees of H acceptance and protonation in the pyridine functional group [29,30]. Since the atomic orbital arrangement of the N element was $1S^22S^22P^3$, there was a lone pair of electrons outside its nucleus, which easily combined with hydrogen ions to form a stable state. Due to the inevitable differences in the degree of protonation of the adsorbent in the two cases, the intensities of the contribution in both cases were also different [30]. From the above results, the adsorption mechanism of the N3/SiO_2 for Re mainly arose from the ion exchange reaction; that is, ReO_4^- combined with the protonated nitrogen element on the pyridine functional group.

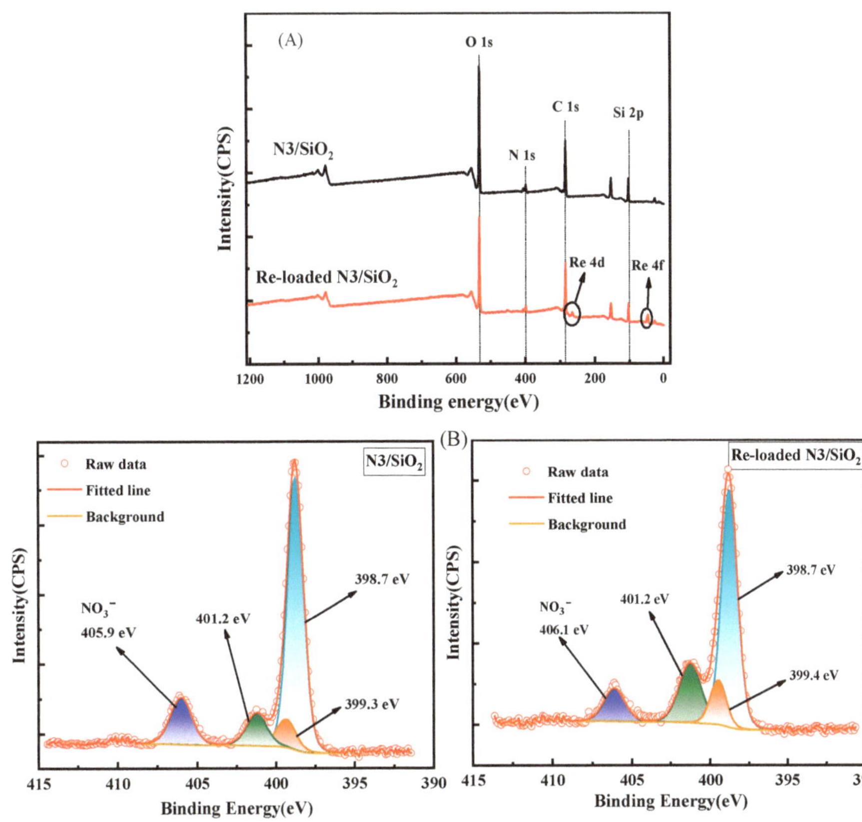

Figure 5. (**A**) Wide XPS spectra of N3/SiO_2 and Re-loaded N3/SiO_2; (**B**) High resolution XPS spectra of N 1s in N3/SiO_2 and Re-loaded N3/SiO_2.

The XPS spectrum of the N3/SiO_2 with an absorbed dose of 100 kGy was further analyzed. In Figure 6, the N 1s spectrum of irradiated N3/SiO_2 could be decomposed into four peaks, consistent with the fresh adsorbent. The areas of the two peaks near 399.2 eV and 401.4 changed greatly, which may be due to the gamma irradiation on the protonation process of the adsorbent.

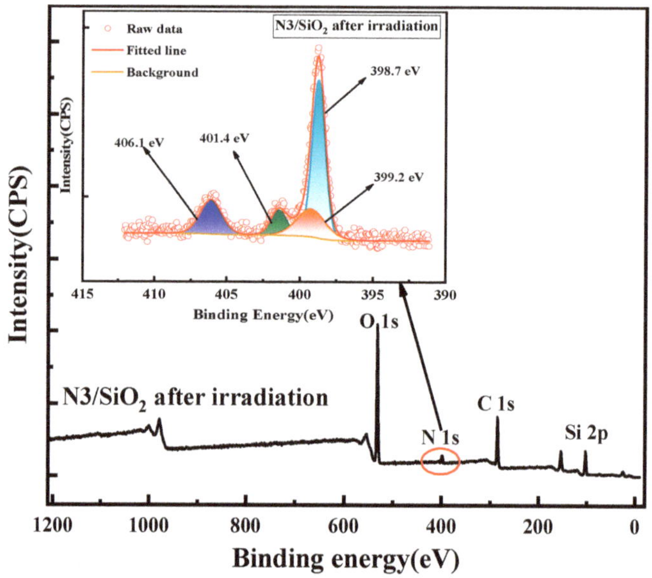

Figure 6. XPS spectra of the irradiated N3/SiO_2.

3.2. Effects of the γ-ray Irradiation on the Re Adsorption in Nitric Acid

The effect of the HNO_3 concentration on the K_d of the Re on the fresh and irradiated N3/SiO_2 was studied by varying the HNO_3 concentration in the range of 0.01–3 mol/L (Figure 7). At the absorbed dose of 50 and 100 kGy, the K_d of Re was consistent with the trend before irradiation; that is, it gradually decreased with the increase in the HNO_3 concentration. This phenomenon of the decrease in the K_d value with an increasing concentration of HNO_3 is explained by the competitive adsorption reaction of ReO_4^- and NO_3^- on the anion exchanger. Generally, the effect of nitric acid on the adsorption process in the adsorption environment depends on the strength of the two effects of salting-out and competition. As the concentration of HNO_3 increased, its competitive adsorption gradually dominated, which reduced the electrostatic attraction to ReO_4^-, resulting in the weakening of the adsorption capacity for Re under this experimental condition. In addition, with the increase in the radiation-absorbed dose, the K_d of Re by the adsorbent gradually decreased. This result might be attributed to the influence of the gamma irradiation on the protonation of the adsorbent, and the gamma irradiation might also lead to structural changes in the resins. However, at the 100 kGy absorbed dose, the K_d value only decreased in the range of 5.6–18.4% when the concentration of HNO_3 increased from 0.01 to 1 M, indicating an acceptable radiation-absorbed dose under 100 kGy. Considering the actual situation of the HLLW and the adsorption performance of the N3/SiO_2 for Re, 1 M HNO_3 was selected as the nitric acid condition for the later adsorption experiments.

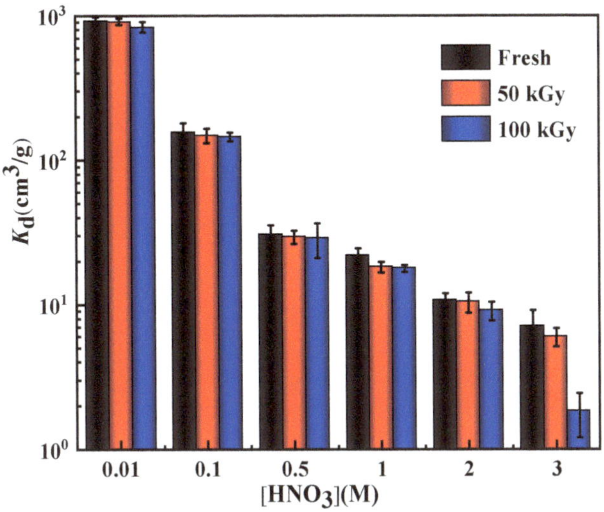

Figure 7. The effects of the γ-ray irradiation on different concentrations of nitric acid, with the following adsorption conditions: V/m = 100 cm^3/g, [Re] =1 mM, T = 25 °C, and contact time = 12 h.

3.3. Adsorption Isotherms of N3/SiO_2 before and after Irradiation

The adsorption capacity is a key parameter to evaluate the adsorption performance of the adsorbents. The experiments were carried out by changing the initial concentration of Re in the HNO_3. As seen in Figure 8, the adsorption capacity increased gradually and then tended to be stable with the increase in the initial Re concentration. The adsorption isotherm can predict the interaction between the target ions and the active sites of the adsorbent. The adsorption behavior of the adsorbent before and after irradiation for Re was further analyzed using the classical Langmuir and Freundlich adsorption isotherm models. The Langmuir model assumes that the adsorption between the adsorbent and the target nuclide is a monolayer adsorption of finite sites. The equation is defined as follows [31,32]:

$$Q_e = \frac{Q_m K_L C_e}{1 + K_L C_e}, \tag{3}$$

where Q_m is the maximum adsorption capacity (mg/g); Q_e indicates the equilibrium adsorption capacity (mg/g); C_e indicates the metal ion concentration in the aqueous phase (mmol/L); and K_L (L/g) represents the adsorption equilibrium constant.

The Freundlich adsorption isotherm model is usually applied to multilayer physical and chemical adsorption processes. The Freundlich adsorption isotherm equation is defined as follows [33–35]:

$$Q_e = K_F C_e^{\frac{1}{n}}, \tag{4}$$

where K_F is the adsorption capacity constant, and n is the adsorption intensity constant.

The considered adsorption isotherm models were used to fit the experimental data by a nonlinear regression method. Figure 8 shows the fitting curves, and Table 2 shows the fitting parameters of the adsorption isotherm model. As shown in Figure 8A,B, in 1 M HNO_3 and a mixture of 1 M $NaNO_3$ and 1 M HNO_3, the maximum adsorption capacity for fresh N3/SiO_2 obtained by the Langmuir isotherm model was 35.8 and 22.2 mg/g, respectively, which matched well with the experimental data. Further, the correlation constants of the Langmuir model in the two media (1 M HNO_3 and a mixture of 1 M $NaNO_3$ and 1 M HNO_3) were greater than 0.98. The better fitting with the Langmuir model indicated that the homogeneous monolayer adsorption was the governing adsorption process. The difference in the adsorption capacity of the adsorbent in the two solutions was attributed to the competitive effect of the increased NO_3^- in the solution on the adsorption of the ReO_4^- by the $N3/SiO_2$.

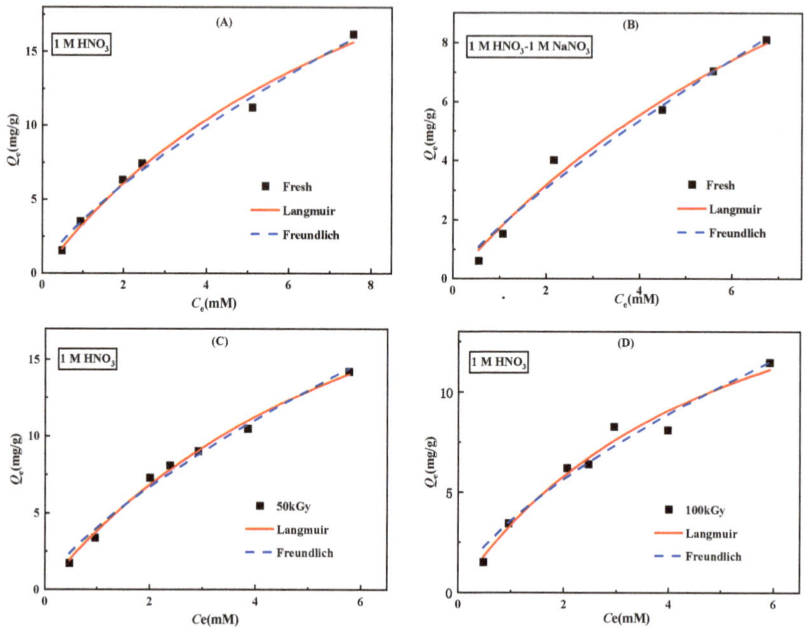

Figure 8. The adsorption isotherms of (**A**) the fresh $N3/SiO_2$ in 1 M HNO_3, (**B**) the fresh $N3/SiO_2$ in a mixture of 1 M HNO_3 and 1 M $NaNO_3$, (**C**) the irradiated $N3/SiO_2$ (50 kGy) in 1 M HNO_3, and (**D**) the irradiated $N3/SiO_2$ (100 kGy) in 1 M HNO_3, with the following adsorption conditions: $V/m = 100$ cm^3/g, T = 25 °C, and contact time = 12 h.

Table 2. The estimated isotherm parameters of the adsorption isotherms.

Models	Parameters	Fresh (1 M HNO_3)	Fresh (1 M HNO_3–1 M $NaNO_3$)	50 kGy (1 M HNO_3)	100 kGy (1 M HNO_3)
Langmuir	Q_m (mg/g)	35.8	22.2	31.9	20.9
	K_L (L/mg)	0.1022	0.0834	0.1357	0.1918
	R^2	0.9879	0.9850	0.9936	0.9717
Freundlich	K_F (mg/g)	3.6	1.8	4.1	3.6
	$1/n$	0.7307	0.8054	0.7241	0.6548
	R^2	0.9913	0.9804	0.9870	0.9660

Figure 8C,D show the adsorption isotherm model fitting curves of $N3/SiO_2$ to Re at different absorbed doses. It can be observed that as the absorbed dose increased from 50 to 100 kGy, the adsorption capacity obtained by the Langmuir model decreased from 31.9 to 20.9 mg/g, consistent with the results of the acidity experiment. The adsorption capacity of the $N3/SiO_2$ was only 10.9% lower than that of fresh adsorbent at 50 kGy, indicating that the γ-ray irradiation had little effect on the absorption of Re at this condition.

3.4. Chromatographic Partitioning of Re

To evaluate the separation of Re from the SHLLW in the column system, a feed solution containing 5 mmol/L Re(VII), La(III), Nd(III), Sm(III), Gd(III), Dy(III), Sr(II), Zr(IV), Mo(VI), Ru(III), Cs(I), and Pd(II) was passed through the $N3/SiO_2$ packed column. Then, the given volumes of 0.1 M HNO_3, 0.01 M HNO_3–0.1 M Thiourea (TU), and 3 M NH_4Cl were subsequently pumped into the column. As shown in Figure 9, except for Re(VII) and Pd(II), all coexisting ions exhibited no adsorption onto N3/SiO_2 and flowed into the effluent along with the 0.1 mol/L HNO_3 at stage I. In accordance with the adsorption behavior of La(III), Nd(III), Sm(III), and Gd(III), it could be inferred that almost all the trivalent RE(III) contained in HLLW would have no adverse effect on the separation of Re(VII) in a 1 mol/L HNO_3 solution [36]. Unlike Re, the affinity of the adsorbent for Pd was a presumed complexation reaction; that is, the nitrogen atom with a lone pair of electrons on the pyridine functional group of the adsorbent could form a stable coordination bond with the empty orbit of the palladium atom [37,38]. The elution of Pd(II) by 0.01 M HNO_3–0.1 M TU was effective, and the elution band appearing in the elution curve of Pd(II) was sharp and narrow, without elution tailing. This was ascribed to the strong complexation affinity of thiourea for Pd(II) [39]. Moreover, the presence of Re in the eluate could hardly be detected during the Pd(II) elution process (stage II). As 3 M NH_4Cl was supplied to the column, the adsorbed Re(VII) on the N3/SiO_2 was eluted efficiently (stage III). A high concentration of nitric acid is generally used as the eluent of anion exchange resins. Using 3 M NH_4Cl as a substitute for concentrated nitric acid could not only improve the service life of the adsorbent but also reduce the corrosion of the equipment in practical applications. The calculated recovery percentage was approximately 86% for Re(VII) and 98% for Pd(II). The experimental results showed that the N3/SiO_2 packed column could realize the selective separation and recovery of Re(VII) in SHLLW.

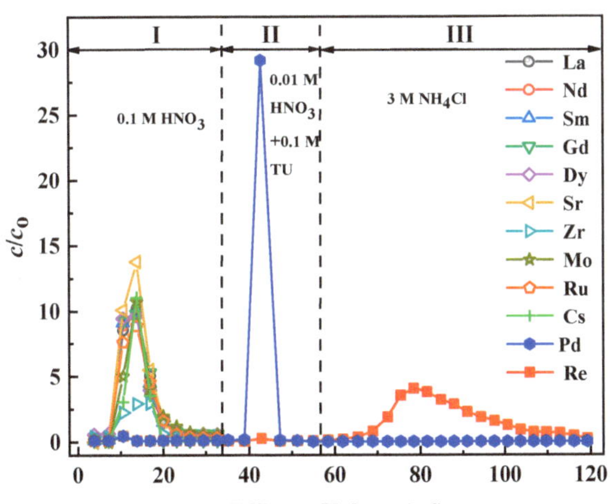

Figure 9. Chromatographic partitioning of Re from a simulated solution by the N3/SiO_2 packed column, with the following adsorption conditions: [M] = 5 mmol/L, [HNO_3] = 1 mol/L, and T = 25 °C.

3.5. ^{99}Tc Removal Performance

The decontamination efficiency (DE) of the adsorbent for ^{90}Sr, ^{99}Tc, ^{137}Cs, ^{241}Am, and U in radioactive liquid waste at 1 and 3 M HNO_3 is shown in Figure 10. In the radioactive liquid waste of 1 M HNO_3, the DE of ^{99}Tc by the adsorbent reached 64.5%, and 45.8% in 3 M HNO_3, much higher than those of the ^{90}Sr, ^{137}Cs, ^{241}Am, and U. The DE of the N3/SiO_2 for U was approximately 15% in both radioactive wastes. Uranyl nitrate is the main form of hexavalent uranium in nitric acid solution, and it is not easy to obtain ionic compounds using nitrite ions. Therefore, with the increase in the nitric acid concentration, the adsorption of U on N3/SiO_2 did not change significantly. The adsorbent had basically no adsorption effect on the other nuclides, and the small amount of ^{90}Sr, ^{137}Cs, ^{241}Am, and

U attached to the adsorbent could be eluted by the dilute nitric acid to achieve the selective separation and recovery of ^{99}Tc.

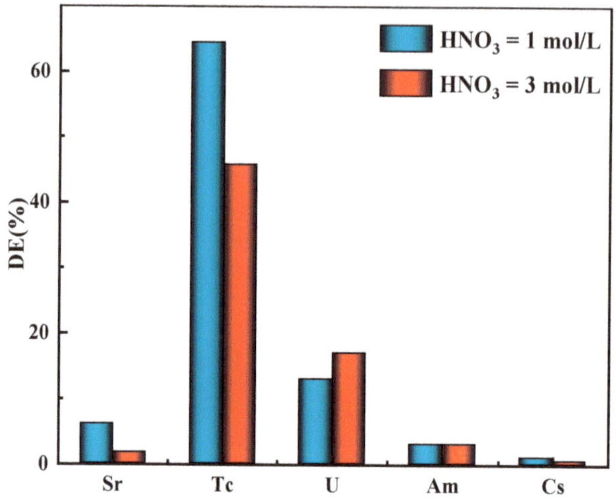

Figure 10. The decontamination efficiency of ^{90}Sr, ^{99}Tc, ^{137}Cs, ^{241}Am, and U for N3/SiO_2 in radioactive liquid waste, with the following adsorption conditions: V/m = 10 cm^3/g, T = 25 °C, and 12 h.

4. Conclusions

Re was selected as the analogue of radioactive ^{99}Tc to carry out the batch adsorption experiments, and XPS analysis demonstrated that the adsorption of Re by the adsorbent was the result of the ion-exchange reaction. The adsorption experiment showed that the competitive adsorption produced by the high concentration of NO_3^- inhibited the adsorption of ReO_4^-, resulting in the decrease in K_d with an increasing concentration of HNO_3. The adsorption capacity of Re decreased with the increase in the absorbed doses, and the maximum value in 1 M HNO_3 reached 20.9 mg/g, even with the absorbed dose up to 100 kGy. The decontamination efficiency of ^{99}Tc by N3/SiO_2 was estimated to be 64.5% after adsorption in radioactive waste at 1 M HNO_3. Hence, N3/SiO_2 is considered to be a competent and feasible adsorbent and has potential application prospects for the extraction of ^{99}Tc from HLLW.

Author Contributions: Conceptualization, Y.W. (Yan Wu), H.S., and Y.W. (Yuezhou Wei); methodology, Y.W. (Yan Wu) and H.S.; investigation, H.S. and C.M.; writing—original draft preparation, H.S.; writing—review and editing, Y.W. (Yan Wu); validation and formal analysis, H.S. and C.M.; supervision, Y.W. (Yan Wu) and Y.W. (Yuezhou Wei). All authors have read and agreed to the published version of the manuscript.

Funding: This research was funded by the National Natural Science Foundation of China (No. 12175143).

Institutional Review Board Statement: Not applicable.

Informed Consent Statement: Not applicable.

Data Availability Statement: Not applicable.

Conflicts of Interest: The authors declare no conflict of interest.

References

1. Wang, Y.; Wen, Y.; Mao, C.; Sang, H.; Wu, Y.; Li, H.; Wei, Y. Development of chromatographic process for the dynamic separation of ^{90}Sr from high level liquid waste through breakthrough curve simulation and thermal analysis. *Sep. Purif. Technol.* **2022**, *282*, 120103. [CrossRef]
2. Tagami, K. Technetium-99 Behavior in the Terrestrial Environment Field Observations and Radiotracer Experiments. *J. Nucl. Radiochem. Sci.* **2003**, *4*, A1–A8. [CrossRef]
3. Pepper, S.E.; Ogden, M.D. Perrhenate extraction studies by Cyphos 101-IL; screening for implementation in technetium removal. *Sep. Purif. Technol.* **2013**, *118*, 847–852. [CrossRef]
4. Lieser, K.H. Technetium in the Nuclear Fuel Cycle, in Medicine and in the Environment. *Radiochim. Acta* **1993**, *63*, 5–8. [CrossRef]
5. Wildung, R.E.; Li, S.W.; Murray, C.J.; Krupka, K.M.; Xie, Y.; Hess, N.J.; Roden, E.E. Technetium reduction in sediments of a shallow aquifer exhibiting dissimilatory iron reduction potential. *FEMS Microbiol. Ecol.* **2004**, *49*, 151–162. [CrossRef] [PubMed]
6. Bennett, R.; Willey, N. Soil availability, plant uptake and soil to plant transfer of ^{99}Tc—A review. *J. Environ. Radioact.* **2003**, *65*, 215–231. [CrossRef]
7. Darab, J.G.; Smith, P.A. Chemistry of Technetium and Rhenium Species during Low-Level Radioactive Waste Vitrification. *Chem. Mater.* **1996**, *8*, 1004–1021. [CrossRef]
8. Sheng, D.; Lin, Z.; Chao, X.; Xiao, C.; Wang, S. Efficient and Selective Uptake of TcO_4^- by a Cationic Metal–Organic Framework Material with Open Ag^+ Sites. *Environ. Sci. Technol.* **2017**, *51*, 3471. [CrossRef]
9. Kong, Y.R.; Jia, Y.F.; Li, H.B.; Wang, X.R.; Lou, H.L. Adsorption of Tc (VII) from nitric acid medium by silica-based anion exchanger. *J. Nucl. Radiochem.* **2013**, *35*, 359–363.
10. Hercigonja, R.V.; Maksin, D.D.; Nastasović, A.B.; Trifunović, S.S.; Glodić, P.B.; Onjia, A.E. Adsorptive removal of technetium-99 using macroporous poly(GMA-co-EGDMA) modified with diethylene triamine. *J. Appl. Polym. Sci.* **2011**, *123*, 1273–1282. [CrossRef]
11. Luo, W.; Kelly, S.D.; Kemner, K.M.; Watson, D.; Zhou, J.; Jardine, P.M.; Gu, B. Sequestering uranium and technetium through co-precipitation with aluminum in a contaminated acidic environment. *Environ. Sci. Technol.* **2009**, *43*, 7516–7522. [CrossRef] [PubMed]
12. Hölttä, P.; Hautojärvi, A.; Hakanen, M. Transport and Retardation of Non-Sorbing Radionuclides in Crystalline Rock Fractures. *Radiochim. Acta* **1992**, *58–59*, 285–290. [CrossRef]
13. Drout, R.J.; Otake, K.; Howarth, A.J.; Islamoglu, T.; Zhu, L.; Xiao, C.; Wang, S.; Farha, O.K. Efficient Capture of Perrhenate and Pertechnetate by a Mesoporous Zr Metal–Organic Framework and Examination of Anion Binding Motifs. *Chem. Mater.* **2018**, *30*, 1277–1284. [CrossRef]
14. George, K.; Masters, A.J.; Livens, F.R.; Sarsfield, M.J.; Taylor, R.J.; Sharrad, C.A. A review of technetium and zirconium extraction into tributyl phosphate in the PUREX process. *Hydrometallurgy* **2022**, *211*, 105892. [CrossRef]
15. Li, Y.; Fu, X.; Li, X.-X.; Zhang, F.; Wu, Q.; Wang, Y.; Yan, Z.-Y. Rapid and efficient column separation of Re(VII) as a surrogate for Tc(VII) with benzimidazole-based cross-linked poly(ionic liquid)s. *J. Radioanal. Nucl. Chem.* **2022**, *331*, 877–888. [CrossRef]
16. Wang, P.-Y.; Zu, J.-H.; Wei, Y.-Z. Synthesis and characterization of porous 4VP-based adsorbent for Re adsorption as analogue to ^{99}Tc. *Nucl. Sci. Tech.* **2017**, *28*, 30. [CrossRef]
17. Wu, Y.; Lee, C.-P.; Mimura, H.; Zhang, X.; Wei, Y. Stable solidification of silica-based ammonium molybdophosphate by allophane: Application to treatment of radioactive cesium in secondary solid wastes generated from fukushima. *J. Hazard. Mater.* **2018**, *341*, 46–54. [CrossRef]
18. Wang, S.; Alekseev, E.V.; Diwu, J.; Casey, W.H.; Phillips, B.L.; Depmeier, W.; Albrecht-Schmitt, T.E. NDTB-1: A Supertetrahedral Cationic Framework That Removes TcO_4^- from Solution. *Angew. Chem. Int. Ed.* **2010**, *49*, 1057–1060. [CrossRef]
19. Daniels, N.; Franzen, C.; Murphy, G.L.; Kvashnina, K.; Petrov, V.; Torapava, N.; Bukaemskiy, A.; Kowalski, P.; Si, H.; Ji, Y.; et al. Application of layered double hydroxides for ^{99}Tc remediation. *Appl. Clay Sci.* **2019**, *176*, 1–10. [CrossRef]
20. Neeway, J.J.; Asmussen, R.M.; Lawter, A.R.; Bowden, M.E.; Lukens, W.W.; Sarma, D.; Riley, B.J.; Kanatzidis, M.G.; Qafoku, N.P. Removal of TcO_4^- from Representative Nuclear Waste Streams with Layered Potassium Metal Sulfide Materials. *Chem. Mater.* **2016**, *28*, 3976–3983. [CrossRef]
21. Fu, L.; Zu, J.; He, L.; Gu, E.; Wang, H. An adsorption study of ^{99}Tc using nanoscale zero-valent iron supported on D001 resin. *Front. Energy* **2020**, *14*, 11–17. [CrossRef]
22. Suzuki, T.; Fujii, Y.; Yan, W.; Mimura, H.; Koyama, S.-i.; Ozawa, M. Adsorption behavior of VII group elements on tertiary pyridine resin in hydrochloric acid solution. *J. Radioanal. Nucl. Chem.* **2009**, *282*, 641. [CrossRef]
23. Sarri, S.; Misaelides, P.; Zamboulis, D.; Gaona, X.; Altmaier, M.; Geckeis, H. Rhenium(VII) and technetium(VII) separation from aqueous solutions using a polyethylenimine–epichlorohydrin resin. *J. Radioanal. Nucl. Chem.* **2016**, *307*, 681–689. [CrossRef]
24. Ashley, K.R.; Whitener, G.D.; Schroeder, N.C.; Ball, J.R.; Radzinski, S.D. Sorption behavior of pertechnetate ion on reillextm-hpq anion exchange resin from hanford and melton valley tank waste simulants and sodium hydroxide/sodium nitrate solutions. *Solvent Extr. Ion Exch.* **1998**, *16*, 843–859. [CrossRef]
25. Ashley, K.R.; Ball, J.R.; Pinkerton, A.B.; Abney, K.D.; Norman, C. Sorption behavior of pertechnetate on reillex(tm)-hpq anion exchange resin from nitric acid solution. *Solvent Extr. Ion Exch.* **1994**, *12*, 239–259. [CrossRef]
26. Chen, L.; Yin, X.; Yu, Q.; Siming, L.; Meng, F.; Ning, S.; Wang, X.; Wei, Y. Rapid and selective capture of perrhenate anion from simulated groundwater by a mesoporous silica-supported anion exchanger. *Microporous Mesoporous Mater.* **2019**, *274*, 155–162. [CrossRef]

27. Chen, Y.L.; Zhao, L.; Wei, Y.Z.; He, L.F.; Tang, F.D. Adsorption characteristics of thorium on silica-based anion exchange resins. *Nucl. Sci. Tech.* **2015**, *26*, 60303.
28. Holdsworth, A.F.; Eccles, H.; Rowbotham, D.; Bond, G.; Kavi, P.C.; Edge, R. The Effect of Gamma Irradiation on the Ion Exchange Properties of Caesium-Selective Ammonium Phosphomolybdate-Polyacrylonitrile (AMP-PAN) Composites under Spent Fuel Recycling Conditions. *Separations* **2019**, *6*, 23. [CrossRef]
29. Saha, S.; Dutta, B.; Ghosh, M.; Chowdhury, J. Adsorption of 4-Mercapto Pyridine with Gold Nanoparticles Embedded in the Langmuir-Blodgett Film Matrix of Stearic Acid: SERS, XPS Studies Aided by Born-Oppenheimer on the Fly Dynamics, Time-Resolved Wavelet Transform Theory, and DFT. *ACS Omega* **2022**, *7*, 27818–27830. [CrossRef]
30. Herrera, S.E.; Tasca, F.; Williams, F.J.; Calvo, E.J.; Carro, P.; Salvarezza, R.C. Surface Structure of 4-Mercaptopyridine on Au(111): A New Dense Phase. *Langmuir* **2017**, *33*, 9565–9572. [CrossRef]
31. Wu, Y.; Zhang, X.; Kim, S.-Y.; Wei, Y. Simultaneous separation and recovery of Cs(I) and Sr(II) using a hybrid macrocyclic compounds loaded adsorbent. Kinetic, equilibrium and dynamic adsorption studies. *J. Nucl. Sci. Technol.* **2016**, *53*, 1968–1977. [CrossRef]
32. Loc, N.X.; Tuyen, P.T.; Mai, L.C.; Phuong, D.T. Chitosan-Modified Biochar and Unmodified Biochar for Methyl Orange: Adsorption Characteristics and Mechanism Exploration. *Toxics* **2022**, *10*, 500. [CrossRef] [PubMed]
33. Richards, R.M.; Gómez, I.; Otazo-Sánchez, E.M.; Prieto, F.; Hernández-Ávila, J.; Linares, G.; González, C.A.; Gordillo, A.J.; Villagómez, R. High Mn^{2+} Uptake by a New Polyvinyl Alcohol Derivative: Isotherm Model Analysis. *Sep. Sci. Technol.* **2015**, *50*, 233–244. [CrossRef]
34. Tolner, L.; Füleky, G. Determination of the originally adsorbed soil phosphorus by modified Freundlich isotherm. *Commun. Soil Sci. Plant Anal.* **1995**, *26*, 1213–1231. [CrossRef]
35. Wang, Y.; Zheng, K.; Jiao, Z.; Zhan, W.; Ge, S.; Ning, S.; Fang, S.; Ruan, X. Simultaneous Removal of Cu^{2+}, Cd^{2+} and Pb^{2+} by Modified Wheat Straw Biochar from Aqueous Solution: Preparation, Characterization and Adsorption Mechanism. *Toxics* **2022**, *10*, 316. [CrossRef]
36. Zhang, A.; Li, J.; Dai, Y.; Xu, L. Development of a new simultaneous separation of cesium and strontium by extraction chromatograph utilization of a hybridized macroporous silica-based functional material. *Sep. Purif. Technol.* **2014**, *127*, 39–45. [CrossRef]
37. Zhang, A.; Wei, Y.; Arai, T.; Kumagai, M. Palladium Removal from the Simulated Nuclear Spent Fuel Solution Using a Silica-Based SiPyR-N3 Anion Exchanger. *Solvent Extr. Ion Exch.* **2006**, *24*, 447–462. [CrossRef]
38. Wei, Y.; Kumagai, M.; Takashima, Y.; Asou, M.; Namba, T.; Suzuki, K.; Maekawa, A.; Ohe, S. The Application of an Advanced Ion Exchange Process to Reprocessing Spent Nuclear Fuels, (I). Separation Behavior of Fission Products from Uranium. *J. Nucl. Sci. Technol.* **1998**, *35*, 357–364. [CrossRef]
39. Kumaresan, R.; Sabharwal, K.N.; Srinivasan, T.G.; Vasudeva Rao, P.R.; Dhekane, G. Studies on the Sorption of Palladium using Cross-Linked Poly (4-Vinylpyridine-Divinylbenzene) Resins in Nitric Acid Medium. *Solvent Extr. Ion Exch.* **2008**, *26*, 643–671. [CrossRef]

toxics MDPI

Article

Fabrication of a Ternary Nanocomposite g-C_3N_4/Cu@CdS with Superior Charge Separation for Removal of Organic Pollutants and Bacterial Disinfection from Wastewater under Sunlight Illumination

Malik Imran Afzal [1], Sammia Shahid [1], Sana Mansoor [1], Mohsin Javed [1], Shahid Iqbal [2,*], Othman Hakami [3], El Sayed Yousef [4,5], Foziah F. Al-Fawzan [6], Eslam B. Elkaeed [7], Rami Adel Pashameah [8], Eman Alzahrani [9] and Abd-ElAziem Farouk [10]

1 Department of Chemistry, School of Science, University of Management and Technology, Lahore 54770, Pakistan
2 Department of Chemistry, School of Natural Sciences (SNS), National University of Sciences and Technology (NUST), H-12, Islamabad 46000, Pakistan
3 Chemistry Department, Faculty of Science, Jazan University, Jazan, Saudi Arabia
4 Research Center for Advanced Materials Science (RCAMS), King Khalid University, P.O. Box 9004, Abha 61413, Saudi Arabia
5 Physics Department, Faculty of Science, King Khalid University, P.O. Box 9004, Abha 61413, Saudi Arabia
6 Department of Chemistry, College of Science, Princess Nourah bint Abdulrahman University, P.O. Box 84428, Riyadh 11671, Saudi Arabia
7 Department of Pharmaceutical Sciences, College of Pharmacy, AlMaarefa University, Riyadh 13713, Saudi Arabia
8 Department of Chemistry, Faculty of Applied Science, Umm Al-Qura University, Makkah 24230, Saudi Arabia
9 Department of Chemistry, College of Science, Taif University, P.O. Box 11099, Taif 21944, Saudi Arabia
10 Department of Biotechnology, College of Science, Taif University, P.O. Box 11099, Taif 21944, Saudi Arabia
* Correspondence: shahiducas@gmail.com

Citation: Afzal, M.I.; Shahid, S.; Mansoor, S.; Javed, M.; Iqbal, S.; Hakami, O.; Yousef, E.S.; Al-Fawzan, F.F.; Elkaeed, E.B.; Pashameah, R.A.; et al. Fabrication of a Ternary Nanocomposite g-C_3N_4/Cu@CdS with Superior Charge Separation for Removal of Organic Pollutants and Bacterial Disinfection from Wastewater under Sunlight Illumination. *Toxics* **2022**, *10*, 657. https://doi.org/10.3390/toxics10110657

Academic Editors: Yuezhou Wei, Yan Wu and Xinpeng Wang

Received: 28 September 2022
Accepted: 26 October 2022
Published: 29 October 2022

Publisher's Note: MDPI stays neutral with regard to jurisdictional claims in published maps and institutional affiliations.

Copyright: © 2022 by the authors. Licensee MDPI, Basel, Switzerland. This article is an open access article distributed under the terms and conditions of the Creative Commons Attribution (CC BY) license (https://creativecommons.org/licenses/by/4.0/).

Abstract: The synthesis of a photo-catalyst with a narrow bandgap and efficient capability to degrade contaminants in the presence of sunlight is currently challenging but exciting. In this work, an efficient photocatalytic ternary nanocomposite g-C_3N_4/Cu@CdS has been synthesized successfully by using the co-precipitation method. The synthesized composite was then characterized by SEM, XRD studies, EDX analysis, and ultra-violet-visible (UV-VIS) spectroscopy. The catalytic efficiency for the methylene blue (MB) dye and drug degradation (ciprofloxacin) was assessed by UV-visible absorption spectra. Gram-positive and Gram-negative bacteria were used to test the fabrication composite's antibacterial properties. Various compositions (1%, 3%, 5%, 7%, and 9%) of/Cu@CdS nanocomposite (NCs) and 20%, 30%, 40%, 50%, and 60% of g-C_3N_4 NCs were prepared. Results reveal that 5%Cu@CdS and 40%g-$C_3N_4$5%Cu@CdS showed maximum antibacterial activity and photocatalytic degradation of dye and drug. The X-ray pattern showed no remarkable change in doped and pristine CdS nanoparticles (NPs). The efficient photocatalytic degradation activity of the fabricated ternary nanocomposite against MB dye and ciprofloxacin an antibiotic drug makes it a viable contender for solving environmental problems.

Keywords: ternary nanocomposite; antibacterial activity; catalytic degradation; methylene blue; ciprofloxacin

1. Introduction

In recent years, semiconductor photo-catalysis has attained much attention because of its ability to use light energy to drive catalytic reactions. The application of photo-catalyst technology is needed to compete with the environmental pollution and energy shortage problems [1–4]. TiO_2 is considered to be the first and most investigated semiconductor, first studied by Fujishima. The drawback was that the TiO_2 semiconductor works efficiently only with UV (ultraviolet) radiation and thus cannot be used in solar-driven applications. For increasing the efficiency of TiO_2 semiconductors, TiO_2 was doped with various metallic and non-metallic elements. The resulting dopants, therefore, do not enhance the photocatalytic activity to a much greater extent [5–7]. The development of a photo-catalyst that is stable, abundant, and performs an efficient catalytic activity with visible light is still a challenge of great interest for researchers.

Nowadays, due to exponential population growth, the availability of fresh drinking water has become one of the most concerning worldwide problems [8]. Moreover, drastic industrial development has resulted in the release of untreated waterborne contaminants, including natural organic matter, antibiotics, heavy metal complexes, and microorganisms [9]. Over 7×10^5 to 1×10^6 tons and 100,000 types of various dyes are produced globally each year for use in industries such as leather and textile fabrics, paper printing, synthetic polymers, paints, and pigments [10]. About 10 to 15% of total dyes produced are discharged as waste, either into industrial effluents or the surrounding environment. These pollutants eventually reach freshwater sources, which cause various health problems including allergy, dermatitis, cancer, dysfunction of kidneys, and hormonal and reproductive malfunctions in humans [11]. Over the past few decades, researchers have made strenuous efforts to mitigate the deterioration of this indispensable natural resource around the world. Advanced photocatalytic oxidation processes employing semiconductors, such as CdS [12], TiO_2 [13–16], and ZnO [17], have attracted the attention of experts because of their remarkable effectiveness in eliminating water impurities by photo-degradation while producing no carcinogenic waste [18].

Graphitic carbon nitride(g-C_3N_4), a polymeric compound comprised of carbon, nitrogen, graphene and a minor amount of hydrogen content, has been introduced as a metal-free photo-catalyst that exhibits solar-driven applications [19] and water purification applications [20,21]. The use of this organic semiconductor is advantageous because of its economic feasibility, ease of availability, relatively high stability, and most important intrinsic visible light response [22–25]. Tian et al. reported different dimensional structures of g-C_3N_4 photo-catalyst. For example, 0D quantum dots, 1D nanorods, nanotubes, 2D nanosheets, and 3D nano-spheres and nano-flowers were designed to investigate photocatalytic efficiency [26]. Liao et al. investigated the g-C_3N_4-based composite photo-catalysts for HER application [27]. The photocatalytic response can be enhanced by various methods such as the formation of porous structures, using metal or non-metal elements as doping materials, and coupling with graphene [28]. In this work, graphene-coupled semiconducting material was fabricated and its photocatalytic efficiency was examined. Creating a stable photo-catalyst with a high optical absorption capacity and that exhibits the ability to improve the separation of photo-generated charges is still challenging [29].

In the past few years, various semiconductor NPSs (CdS, CdSe, ZnSe, CuO, ZnO) were prepared by using different fabrication techniques [30–35]. Among all of the above semiconductor nanoparticles, CdS and NPs are of great importance because of their distinctive physical and chemical properties [36]. Cadmium sulfide (CdS) exhibits a wide band gap, with a band gap energy of 2.42 eV [37]. CdS nanoparticles show high photocatalytic efficiency for the wastewater treatment and catalytic degradation of dyes (MB). The heterojunction semiconductor g-C_3N_4/Cu@CdS is used as a photocatalyst to investigate the efficiency of fabricated nanocomposites for dye and drug degradation. Bacterial infections are one of the most serious health problems that humans face. The growth of infectious diseases caused by pathogenic strains, the emergence of bacterial antibiotic resistance, and the development of new bacterial mutations have piqued the interest of researchers who are looking for new ways to combat these organisms. Another concern connected with bacteria is biofilm formation, which can lead to serious medical and industrial issues [38].

Herein, we have fabricated a novel cadmium sulfide (CdS)-based heterojunction photo catalyst composite by co-precipitation and provided a panorama of its photocatalytic mechanism, photocatalytic and optical properties under different morphologies. To the best of our knowledge, a ternary composite with this composition has not been reported before. Additionally, we have especially emphasized the relationship between the doping concentrations and photo and antibacterial activity. The percentage composition of the g-C_3N_4 substrate in the final composite was also given special attention regarding its synthesis and properties. SEM, XRD, EDX, and UV-visible spectroscopy were used to analyze the produced nanocomposite. The novel catalyst series synthesized in this study may be potentially applied for the treatment of textiles and other industrial wastewater sources which discharge toxic organic dyes and antibiotic drugs in the effluents.

2. Experimental Work

2.1. Chemicals

Without additional purification, analytical-grade compounds of all types were utilized. In deionized water, the solutions were made. ($Cd(Ac)_2.2H_2O$, 98%) and copper sulfate ($CuSO_4.5H_2O$) were obtained from Sigma Aldrich. Urea $(NH_2)_2CO$, and thiourea $(NH_2)_2CS$) were obtained from AppliChem, Germany while sodium hydroxide (NaOH) was acquired from Omicron, China.

2.2. Preparation of the Photo-Catalyst

2.2.1. Synthesis of Graphitic Carbon Nitride g-C_3N_4

g-C_3N_4 was synthesized by calcination of urea. About 8–9 g of urea was heated in a muffle furnace at about 5 °C per minute temperature rise until 550 °C, and kept at this temperature for 2 h [39]. The crucible was left in the furnace for cooling to ambient temperature. An off-white fluffy solid was obtained which was pulverized into a fine powder and preserved for further use.

2.2.2. Synthesis of Cadmium Sulfide Nanoparticles

The co-precipitation process, which was modified somewhat from that used by Xu et al., was employed to create CdS NPs [40]. Cadmium acetate dehydrates and thiourea was used as precursors for Cd and S, respectively. A total of 25 mL of 0.25 M solutions of $CdAc_2$ were stirred for 20 min with a 0.2 g Polyvinylpyrrolidone (PVP) capping agent for 30 min at 80 °C. pH was adjusted at 11–12 using a 4 M NaOH solution. A total of 30 mL of 0.25 M thiourea solution was added dropwise while stirring strongly at 80 °C. Yellow-colored CdS precipitates were produced, and they were filtered before being repeatedly washed with deionized water, absolute ethanol, and ethanol. Precipitates were then dried at 80 °C for 2 h. The solid obtained was pulverized to a fine powder.

2.2.3. Synthesis of Copper-Doped Cadmium Sulfide

Chemical doping of CdS was done by adding dopants before the precipitates were formed. A total of 0.05 M $CuSO_4$ solution was used for doping. The method given in Section 2.2.2 was repeated by adding 2.5, 7.5, 12.5, 17.5 and 22.5 mL of 0.05 M $CuSO_4$ solution after PVP addition for preparing 1, 3, 5, 7 and 9% of Cu-doped photo-catalyst.

2.2.4. Synthesis of Ternary g-C_3N_4/Cu@Cds Nanocomposite Photo-Catalyst

g-C_3N_4, already prepared in 2.2.1, was used in this section. Heterogeneous ternary composites were synthesized by adding 0.144, 0.246, 0.383, 0.575, 0. 8625, and 1.342 g of g-C_3N_4 to obtain the composite with 20, 30, 40, 50, 60, and 70% of g-C_3N_4, respectively. To create a fine suspension, these quantities were dispersed individually in deionized water and agitated for an hour. Method 2.2.3 was followed subsequently. A total of 25 mL of 0.25 M solutions of $CdAc_2$ were added to this suspension with a 0.2 g Polyvinylpyrrolidone (PVP) capping agent. A total of 12.5 mL of 0.05 M $CuSO_4$ solution was added, followed by stirring for 30 min at 80 °C. pH was adjusted at 11–12 using a 4 M NaOH solution. 30 mL of 0.25 M thiourea solution was added dropwise. Yellow-colored precipitates of CdS were obtained which were filtered and washed. The formation of g-C_3N_4/Cu@Cds nanocomposite is shown in (Figure 1).

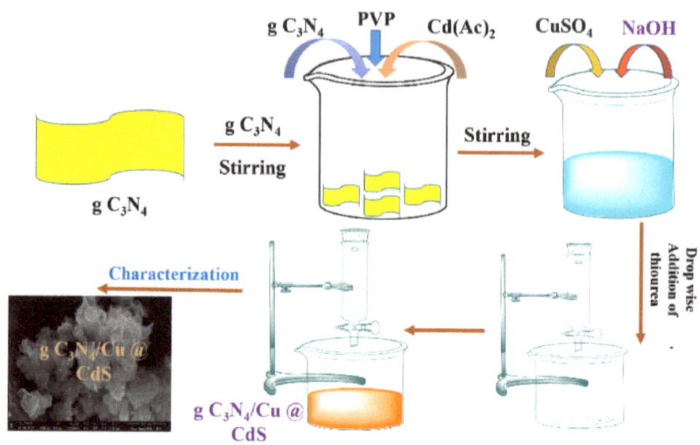

Figure 1. Schematic illustration of synthesis of g-C_3N_4/Cu@Cds nanocomposite.

3. Characterization

Crystallinity and the crystalline phase composition were analyzed by XRD using an X-ray diffractometer (XRD, Bruker D2-Phaser, Dublin, Ireland). The morphology of the fabricated samples was determined by scanning electron microscopy (SEM) and elemental proportions by energy-dispersive spectroscopy (EDS) using FEI Nova 450 NanoSEM. The UV-visible spectrophotometer was used to monitor and record dye degradation, which was used to gauge the photocatalytic performance of the catalysts (Cary 60 UV-Vis, Agilent, Santa Clara, CA, USA).

4. Photo-Degradation Activity

4.1. Degradation of Methylene Blue

To study the efficiency of the catalysts, photodegradation of the model pollutant methylene blue (MB) was studied with and without a catalyst. CdS NPs and Cu-doped (1%, 3%, 5%, 7% and 9%) CdS NPs were individually applied as described in a later section. Among the composites, (20, 30, 40, 50, 60, 70%) g-C_3N_4 with 5%Cu@CdS NPs were evaluated for their degradation activity. To order to achieve adsorption/desorption equilibrium before exposing to the light, 80 mL of MB solution (20 mg/L) containing 50 mg of each catalyst was mixed vigorously for 20 min in the dark. Every 20 min, 5 mL of the sample was taken, centrifuged for 10 min to remove the catalyst, and then a UV-visible spectrum from 200 nm to 800 nm was acquired.

4.2. Degradation of Ciprofloxacin

Ciprofloxacin (CIP) was used as a model drug to study the photocatalytic degradation of drug pollutants with and without catalysts. CdS NPs, Cu-doped (1%, 3%, 5%, 7% and 9%) CdS NPs and composites (20, 30, 40, 50, 60, 70%) g-C_3N_4-5%Cu@CdS NPs) were evaluated for their photocatalytic activity. A total of 80 mL of CIP solution (25 mgL^{-1}) was mixed with 30 mg of each of the catalysts and stirred in darkness for 20 min to reach adsorption/desorption equilibrium between the pollutant and the catalyst. The mixture was then transferred to Petri dishes before exposure to solar radiation [41]. Every 15 min, 5 mL of each sample was taken, centrifuged for 10 min to remove the catalyst, and then a UV-visible spectrum from 200 nm to 800 nm was acquired.

5. Antibacterial Activity

The antibacterial activity of the photo-catalysts (CdS, Cu@CdS NPs, g-C_3N_4, and g-C_3N_4-Cu@CdS composite) for *Staphylococcus aureus* (*S. aureus*) and *Escherichia coli* (*E. coli*) bacteria was carried out using the agar well diffusion approach. The agar well diffusion method was used to test the antibacterial activity of the photo-catalysts (CdS, Cu@CdS NPs, g-C_3N_4, and g-C_3N_4-Cu@CdS composite) for Gram-positive and Gram-negative bacteria [42]. *Staphylococcus aureus* (Gram-positive) and *Escherichia coli* were the bacterial strains employed to test the antibacterial capabilities (Gram-negative). The bacterial strains came from the PCSIR laboratory facility in Lahore, Pakistan. For the positive control, ciprofloxacin was utilized as a typical antibiotic, while distilled water served as the negative control.

6. Results and Discussion

6.1. Scanning Electron Microscope (SEM) Analysis

(Figure 2a–d) represents the SEM images of CdS NPs, Cu@Cds, gC_3N_4, and $g\text{-}C_3N_4$/ Cu@Cds nanocomposite. The nanocomposite was synthesized by the co-precipitation method. The morphology and size of the fabricated nanocomposite were investigated by SEM (Nova Nano SEM-LUMS). The CdS NPs possess a spherical shape and nanoparticles were agglomerated. The surface of Cu-doped CdS NPs appears to be smooth and has a small grain-like shape with irregular growth that may be due to Ostwald ripening. The graphitic carbon nitride $g\text{-}C_3N_4$ exhibits a flaky texture and the $g\text{-}C_3N_4$-Cu@Cds nanocomposite have a flaky texture.

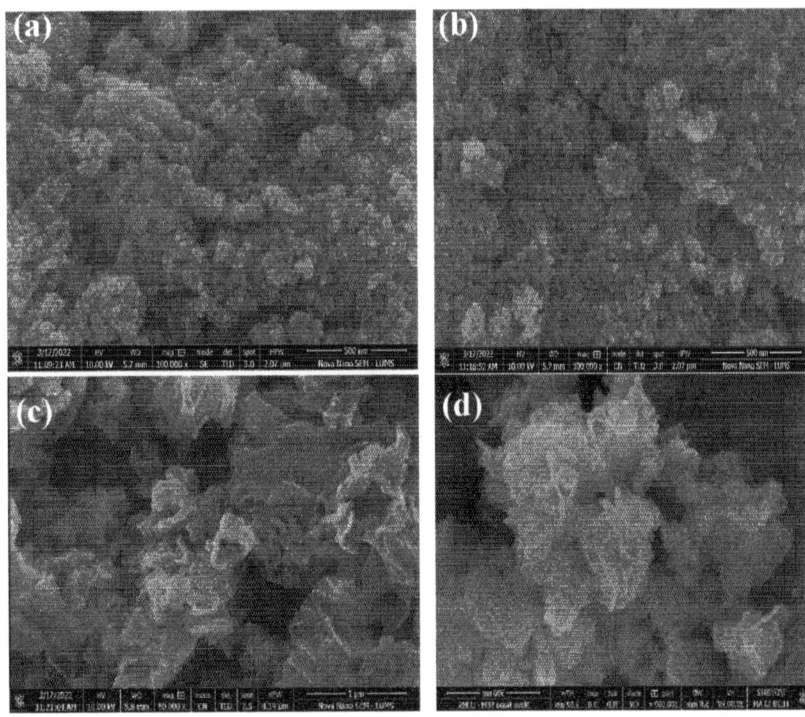

Figure 2. SEM analysis of (**a**) CdS NPs (**b**) Cu@Cds (**c**) gC_3N_4 (**d**) $g\text{-}C_3N_4$/Cu@CdS.

6.2. EDX Spectroscopy

EDX is an analytical technique that provides information about the elemental composition of synthesized materials. When a material is hit by electromagnetic radiation, it emits X-rays, which were then analyzed by using the EDX technique. The sample is targeted by a high-energy laser in an EDX instrument. (Figure 3a–c) shows the EDX spectra of gC_3N_4 CdS NPs, and 40%$g\text{-}C_3N_4$-5%Cu@Cds nanocomposite. In (Figure 3a) the appearance of cadmium (Cd) and sulfur (S) peaks show the presence of these elements in CuS nanoparticles. (Figure 3b) indicates the presence of carbon (C) and nitrogen (N) present in graphitic carbon nitride. (Figure 3c) displays the EDX spectra of/g C_3N_4/Cu@Cds nanocomposite.

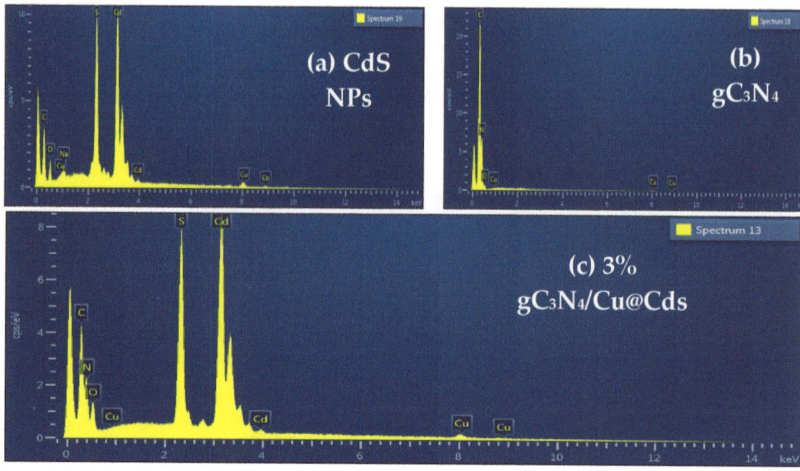

Figure 3. EDX pattern of (**a**) CdS NPs, (**b**) gC_3N_4 and (**c**) 3% g-C_3N_4/Cu@Cds nanocomposite.

6.3. XRD and XPS Analyses

All materials' phases and crystalline structures were determined using XRD analysis. The virgin and Cu-doped CdS nanoparticles corresponded with the wurtzite structure, according to the XRD patterns in Figure 4 (cod ref code 96-900-8863) [43]. The almost complete absence of peaks demonstrated that the 5%Cu@CdS nanocomposite structures were well matched with the hexagonal phase. According to an analysis of the XRD peaks, the peaks at 2θ = 24.92°, 26.55°, 27.85°, 36.55°, 43.90°, and 52.05° were produced, respectively, by the contributions of the (100), (002), (101), (102), (110) and (112) reflection planes. The CdS photo-diffraction catalyst's peaks at 2 = 24.860, 26.540, 28.240, 43.890, 47.940, and 51.910 were attributed to contributions from the reflection planes (100), (002), (101), (110), (103), and (112), respectively. The XRD peaks associated with the (100), (002), (101), and (112) reflection planes of CdS were identified in the ZnO/CdS photocatalyst at 2θ = 25.17°, 26.65°, 28.23°, and 52.23°, respectively. Additionally, the effective integration of the Cu dopant was shown by two diffraction peaks at 2θ = 44.280 and 48.130. The peak intensity of the doped CdS phase is noticeably stronger as compared to CdS alone, as a few peaks in pristine CdS are missing. The peak 2θ = 10.63 in the final composite was detected due to impurity. As seen in Figures S1 and S2, 5% g-C_3N_4/Cu@Cd nanocomposite formation and the electronic states of each of its component parts were determined using XPS analysis. The XPS examination further confirmed the results of the SEM and EDX that the 5% g-C_3N_4/Cu@CdS included g-C_3N_4, CdS, and Cu.

6.4. Degradation

6.4.1. Photocatalytic Degradation of Dye (MB)

Figure 5a–c shows how the synthetic Cu-doped CdS nanoparticles combined with g-C3N4 degraded the standard pollutant methylene blue (MB) under sunshine, respectively. The highest absorption was seen at 662 nm in all spectral graphs. The intensity of blue dye was decreased initially by adsorption on the catalyst's surface and further decreased by photo-degradation [44,45]. A decrease in the intensity of peaks showed that MB was oxidized to secondary compounds when exposed to sunlight in presence of a synthesized catalyst. All the samples were exposed to the sun for 120 min, measured to be an average 90–100 kLux using a Lux light meter, during March and April at the University of Management and Technology, Lahore (31.4835° N, 74.4121° E), Punjab, Pakistan.

Figure 5a shows that 5%Cu@CdS nanoparticles proved to be the most efficient catalyst, which decomposed a net 80% of the dye after adsorption. Furthermore, efficiency was increased by fixing the doped nanoparticles on g-C_3N_4 to fabricate the composite

(Figures S3 and S4). The composite 40%g-C_3N_4/5%Cu@CdS showed 95% MB degradation. This increase in efficiency is because of a decrease in band gap by adding Cu and making it a Z-scheme catalyst by merging with g-C_3N_4. The mechanism of photocatalytic degradation is shown in (Figure 6).

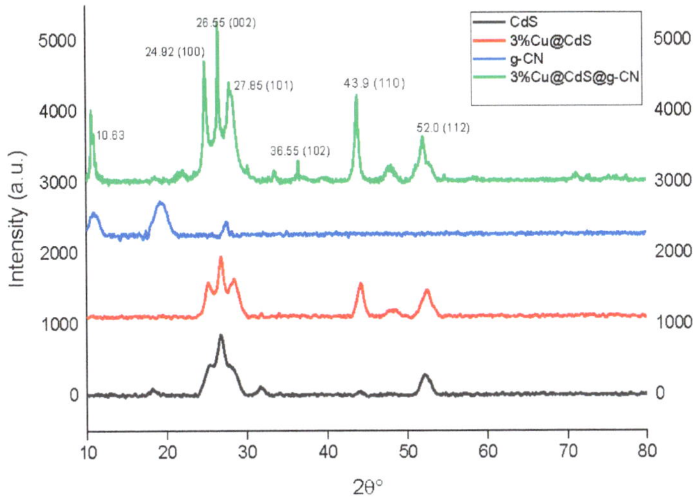

Figure 4. XRD pattern of CdS NPs, gC_3N_4, and 5% g-C_3N_4/Cu@Cds nanocomposite.

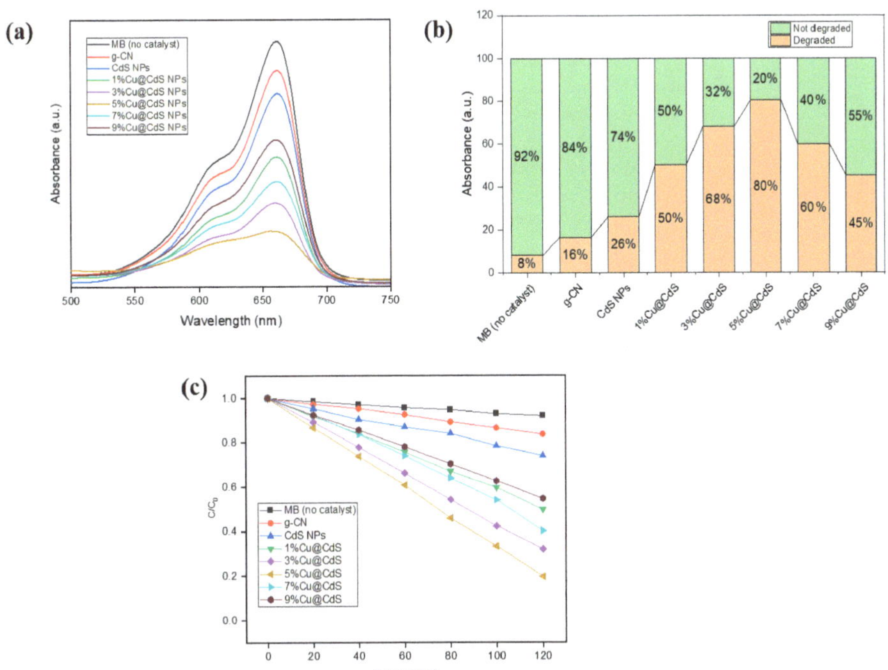

Figure 5. UV-visible spectra of photocatalytic degradation of methylene blue with g-C_3N_4, CdS NPs, Cu-doped CdS NPs: (**a**) absorption curves, (**b**) percentage dye degradation, (**c**) relative degradation curves C/C_0.

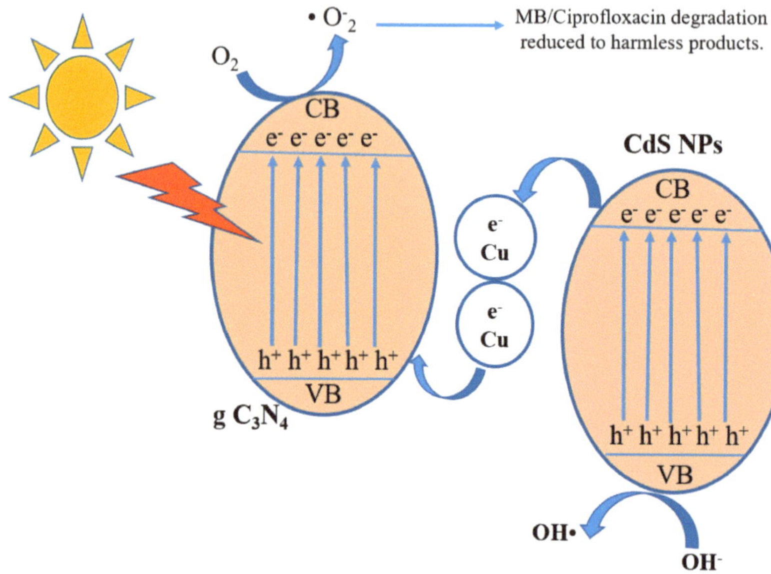

Figure 6. Mechanism of photocatalytic activity.

6.4.2. Photocatalytic Degradation of the Drug (CIP)

All the samples were exposed to solar radiation for 90 min, measured to be an average of 80 k Lux using the Lux light meter, during the months of February–March at the University of Management and Technology, Lahore (31.4835° N, 74.4121° E), Punjab, Pakistan. (Figure 7a–c) illustrates the degradation of CIP by the synthesized Cu-doped CdS nanoparticles and their composites with g-C_3N_4, respectively. Maximum absorption was shown at λ_{max} = 270 nm. The concentration of CIP decreased by incident light and 5%Cu@CdS NPs proved to be most efficient among Cu-doped CdS NPs series (59% in 90 min) and 40%g-CN/5%Cu@CdS among composites (76% in 90 min). A decrease in the intensity of peaks showed that CIP successfully degraded secondary compounds when exposed to sunlight in presence of a synthesized catalyst.

6.5. Scavenging Activity

For studying the photo-degradation mechanism of MB by the g-C_3N_4/Cu@CdS catalyst, the role of the species $^{\bullet}OH$, H^+, and, $^{\bullet}O^{2-}$ responsible for the degradation of MB in the photo-degradation reaction were examined, as given in (Figure 8). To establish the role of the leading active species, benzoquinone (BQ), isopropanol (IPA), and ammonium oxalate (AO) was added to remove superoxide ($^{\bullet}O^{2-}$), the hydroxyl radical ($^{\bullet}OH$) and holes (h^+), respectively. With the addition of AO to the MB dye solution, the degradation activity of the 40%g-C_3N_4/3%Cu@CdS catalyst was reduced to 82% (to 76.8% without scavenger), whereas the addition of IPA and BQ showed comparatively little reduction (12% and 16%, respectively) in degradation of MB under the same conditions of light, temperature and pH of dye solution. This confirmed that the generation of holes (h^+) plays a key role in the degradation of MB. The photo-generated electrons and superoxide radicals are not as significant as photo-generated holes [46].

Two mechanisms—the "type II-heterojunction" and the "straight Z-scheme"—were proposed in light of the aforementioned findings (Figure 8). The presence of copper supported by g-C_3N_4 has enhanced the charge separation increasing the electron-hole recombination time. The differences in the standard potential of conduction bands of CdS, Cu, and g-C_3N_4 enable the photo-generated electrons to be accommodated in Cu^{2+} ions.

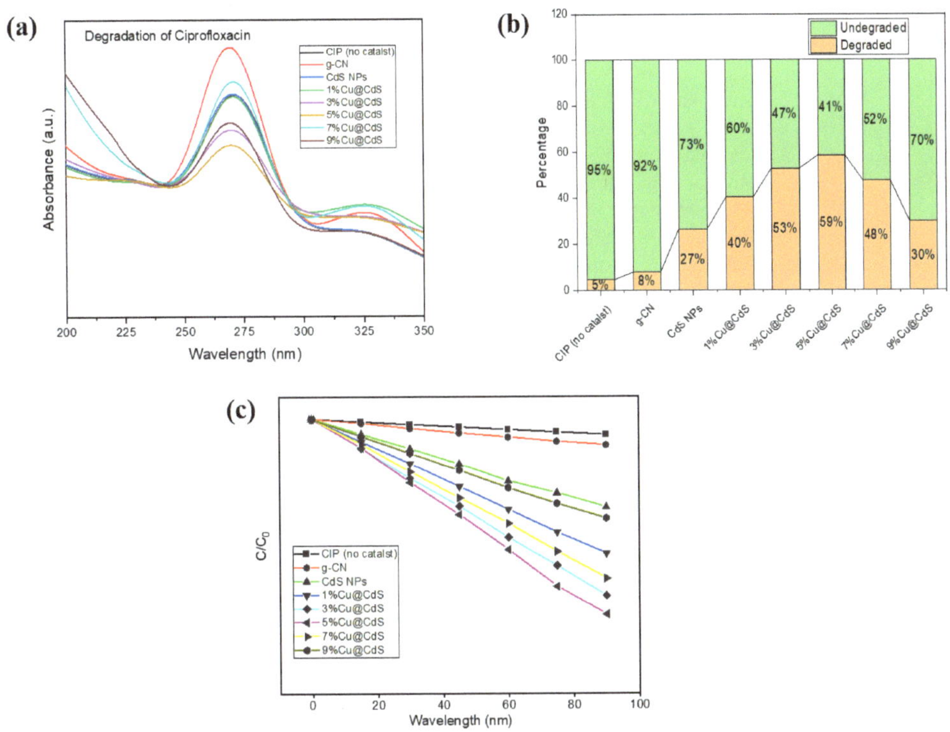

Figure 7. UV-visible spectra of photocatalytic degradation of Ciprofloxacin with g-C_3N_4, CdS NPs, Cu-doped CdS NPs: (**a**) absorption curves, (**b**) percentage dye degradation, (**c**) relative degradation curves C/C_0.

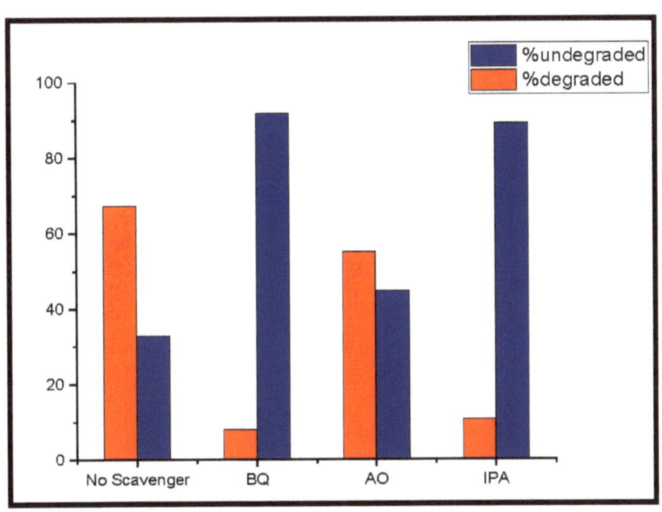

Figure 8. Influence of different scavengers on the photo-degradation of MB with 40%g-C_3N_4/5%Cu@CdS.

6.6. Antibacterial Activity

Stock suspensions of each of CdS, Cu@CdS NPs, g-C_3N_4, g-C_3N_4/CdS, and g-CN/Cu@CdS NCs were prepared in distilled water by ultra-sonication to yield a final concentration of 50 mg mL^{-1}, 100 mg mL^{-1}, 150 mg mL^{-1}, and 200 mg mL^{-1}. Suspensions were stored at room temperature and sonicated once again for 20 min just before adding to the assay. The reference antibiotic Ciprofloxacin concentration was 1 mg mL^{-1}. The antibacterial activities were studied using agar well diffusion method. Muller–Hinton agar plates were prepared for testing antibacterial activity. The prepared inoculum of *S. aureus* (Gram-positive) and *E. coli* (Gram-negative) bacteria was spread on plates. Wells were made using sterilized steel boring and filled using micropipette with 20 µL of CdS nps in a concentration of 200 mg/mL, 150 mg/mL 100 mg/mL and 50 mg/mL each. The method was repeated using Cu@CdS NPs, g-C_3N_4, g-C_3N_4/CdS, and g-CN/Cu@CdS NCs. Each MHA plate was incubated at 37 °C for 24 h. After incubation, the zone of inhibition was measured as shown in Figure 9.

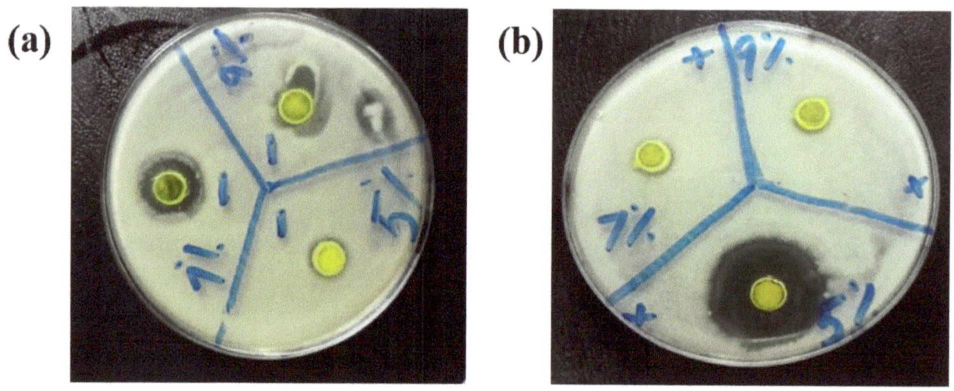

Figure 9. (**a**). The antimicrobial activity of; a. 7%Cu@CdS against S. aureus (**b**). 5%Cu@CdS against *E. coli*.

The 5%Cu@CdS showed the greatest antibacterial activity against *S. aureus* and 7%Cu@CdS was found to be most effective against *E. coli bacteria*. g-CN/Cu@CdS NCs showed a weaker action against both G-positive and G-negative bacteria.

7. Conclusions

In summary, g-C_3N_4/Cu@CdS NCs were successfully synthesized by the chemical co-precipitation method which showed photocatalytic properties in the visible region of the light spectrum. The interval of electron-hole pairs recombination of CdS np was remarkably increased when doped with Cu. Serving as a perfect electron carrier, g-C_3N_4 gave stability to the doped CdS. The photocatalytic efficiency was maximized, with 95% degradation of MB in 120 min and 75% of CIP in 90 min, with 40% g-C_3N_4 contents by weight. The antibacterial activity of synthesized nanocomposite against S. aureus and E. coli was studied. The antibacterial study showed that 40%g-C_3N_4-5%Cu@CdS NCs and 40%g-C_3N_4-7%Cu@CdS NCs have significant antibacterial activity against Gram-positive and Gram-negative bacteria, respectively. The fabricated ternary nanocomposite 40%gC_3N_4/5%Cu@CdS NCs exhibited superior photocatalytic activity over CdS NPs, g-C_3N_4, and 5%Cu@CdS nanomaterials against dye degradation of methylene blue (MB). The 40%g-C_3N_4/5%Cu@Cds showed 95% degradation of MB as compared to 80% degradation by 5%Cu@Cds and 26% by pristine CdS NPs in 90 min under the sun. Similar efficiency was observed in drug degradation (Ciprofloxacin). 40%g-C_3N_4/5%Cu@Cds reduced the drug concentration by 76% whereas 5%Cu@Cds and CdS NPs showed up to 59% and 27% degradation, respectively.

Supplementary Materials: The following supporting information can be downloaded at: https://www.mdpi.com/article/10.3390/toxics10110657/s1. Figure S1. High-resolution XPS spectra of 5% g-C3N4/Cu@CdS nanocomposite; (a) Cd 3d, (b) S 2p, (c) Cu 2p and (d) C 1s. Figure S2. High-resolution N 1s XPS spectra of 5% g-C3N4/Cu@CdS nanocomposite. Figure S3. UV-Visible Spectra of Photocatalytic degradation of Methylene Blue with g-C3N4, CdS NPs, Cu doped CdS NPs (a) Absorption curves (b) Percentage dye degradation (c) Relative degradation curves C/C0. Figure S4. UV-Visible Spectra of Photocatalytic degradation of Ciprofloxacin with g-C3N4, CdS NPs, Cu doped CdS NPs (**a**) Absorption curves (b) Percentage dye degradation (c) Relative degradation curves C/C0.

Author Contributions: Conceptualization, writing-original draft Preparation, M.I.A. and S.S.; methodology, S.M.; software, M.J.; validation, project administration, S.I., O.H. and E.S.Y.; writing review and editing, funding, F.F.A.-F. and E.B.E.; investigation, resources, R.A.P.; data curation, E.A.; writing-original draft preparation, A.-E.F. All authors have read and agreed to the published version of the manuscript.

Funding: The authors would like to thank the Deanship of Scientific Research at Umm Al-Qura University for supporting this work under Grant Code: (22UQU4320141DSR26). The authors extend their appreciation to the Deanship of Scientific Research at King Khalid University for funding this work through research groups program under grant number (RGP.2/193/42). This research was funded by Princess Nourah bint Abdulrahman University Researchers Supporting Project number (PNURSP2022R156), Princess Nourah bint Abdulrahman University, Riyadh, Saudi Arabia.

Institutional Review Board Statement: Not applicable.

Informed Consent Statement: Not applicable.

Data Availability Statement: The data will be available on request.

Conflicts of Interest: The authors declare no conflict of interest.

References

1. Costa, J.C.S.; Franco, N.; Soares, T.A.S.; Saraiva, N.A.M.; Garcia, M.A.S.; Gonzalez, J.R.; Machado, G. TiO_2 nanotubes decorated with Au nanoparticles for Photocatalytic Hydrogen Generation under UV-Visible and Visible Light Irradiations. *An. Da Acad. Bras. De Ciências* **2020**, *92*, e20200504. [CrossRef] [PubMed]
2. Wang, L.; Zhao, J.; Liu, H.; Huang, J. Design, modification and application of semiconductor photocatalysts. *J. Taiwan Inst. Chem. Eng.* **2018**, *93*, 590–602. [CrossRef]
3. Kudo, A.; Miseki, Y. Heterogeneous photocatalyst materials for water splitting. *Chem. Soc. Rev.* **2008**, *38*, 253–278. [CrossRef] [PubMed]
4. Chen, X.; Shen, S.; Guo, L.; Mao, S.S. Semiconductor-based Photocatalytic Hydrogen Generation. *Chem. Rev.* **2010**, *110*, 6503–6570. [CrossRef] [PubMed]
5. Kamat, P.V. TiO_2 Nanostructures: Recent Physical Chemistry Advances. *J. Phys. Chem. C* **2012**, *116*, 11849–11851. [CrossRef]
6. Teoh, W.Y.; Scott, J.A.; Amal, R. Progress in Heterogeneous Photocatalysis: From Classical Radical Chemistry to Engineering Nanomaterials and Solar Reactors. *J. Phys. Chem. Lett.* **2012**, *3*, 629–639. [CrossRef] [PubMed]
7. Kamat, P.V. Manipulation of Charge Transfer Across Semiconductor Interface. A Criterion That Cannot Be Ignored in Photocatalyst Design. *J. Phys. Chem. Lett.* **2012**, *3*, 663–672. [CrossRef]
8. Elimelech, M.; Phillip, W.A. The Future of Seawater Desalination: Energy, Technology, and the Environment. *Science* **2011**, *333*, 712–717. [CrossRef]
9. Shannon, M.A.; Bohn, P.W.; Elimelech, M.; Georgiadis, J.G.; Mariñas, B.J.; Mayes, A.M. Science and technology for water purification in the coming decades. *Nature* **2008**, *452*, 301–310. [CrossRef]
10. Shaban, M.; Ashraf, A.M.; Abukhadra, M.R. TiO_2 Nanoribbons/Carbon Nanotubes Composite with Enhanced Photocatalytic Activity; Fabrication, Characterization, and Application. *Sci. Rep.* **2018**, *8*, 781. [CrossRef]
11. Guo, Y.; Wang, J.; Tao, Z.; Dong, F.; Wang, K.; Ma, X.; Yang, L. Facile synthesis of mesoporous CdS nanospheres and their application in photocatalytic degradation and adsorption of organic dyes. *CrystEngComm* **2012**, *14*, 1185–1188. [CrossRef]
12. Sun, L.; Qin, Y.; Cao, Q.; Hu, B.; Huang, Z.; Ye, L.; Tang, X. Novel photocatalytic antibacterial activity of TiO_2 microspheres exposing 100% reactive {111} facets. *Chem. Commun.* **2011**, *47*, 12628–12630. [CrossRef] [PubMed]
13. Chen, J.S.; Liu, J.; Qiao, S.Z.; Xu, R.; Lou, X.W. Formation of large 2D nanosheets via PVP-assisted assembly of anatase TiO_2 nanomosaics. *Chem. Commun.* **2011**, *47*, 10443–10445. [CrossRef] [PubMed]
14. Zhang, X.; Pan, J.H.; Du, A.J.; Fu, W.; Sun, D.D.; Leckie, J.O. Combination of one-dimensional TiO_2 nanowire photocatalytic oxidation with microfiltration for water treatment. *Water Res.* **2009**, *43*, 1179–1186. [CrossRef]
15. Kim, J.; Monllor-Satoca, D.; Choi, W. Simultaneous production of hydrogen with the degradation of organic pollutants using TiO_2 photocatalyst modified with dual surface components. *Energy Environ. Sci.* **2012**, *5*, 7647–7656. [CrossRef]

16. Lee, J.S.; Kwon, O.S.; Jang, J. Facile synthesis of SnO_2 nanofibers decorated with N-doped ZnO nanonodules for visible light photocatalysts using single-nozzle co-electrospinning. *J. Mater. Chem.* **2012**, *22*, 14565–14572. [CrossRef]
17. Han, Z.; Liao, L.; Wu, Y.; Pan, H.; Shen, S.; Chen, J. Synthesis and photocatalytic application of oriented hierarchical ZnO flower-rod architectures. *J. Hazard. Mater.* **2012**, *217*, 100–106. [CrossRef]
18. Chong, M.N.; Jin, B.; Chow, C.W.K.; Saint, C. Recent developments in photocatalytic water treatment technology: A review. *Water Res.* **2010**, *44*, 2997–3027. [CrossRef]
19. Wu, W.; Xuan, Y.; Jin, Y.; Liang, X.; Meng, S.; Chen, M. Construction of Novel CdS/$SnNb_2$ O_6 Heterojunctions with Enhanced Photocatalytic Degradation Activity Under Visible Light. *Eur. J. Inorg. Chem.* **2018**, *2018*, 4812–4818. [CrossRef]
20. Zhang, S.; Gu, P.; Ma, R.; Luo, C.; Wen, T.; Zhao, G.; Cheng, W.; Wang, X. Recent developments in fabrication and structure regulation of visible-light-driven g-C3N4-based photocatalysts towards water purification: A critical review. *Catal. Today* **2018**, *335*, 65–77. [CrossRef]
21. Wu, K.; Cui, J.; Yi, J.; Liu, X.; Ning, F.; Liu, Y.; Zhang, J. Biodegradable Gel Electrolyte Suppressing Water-Induced Issues for Long-Life Zinc Metal Anodes. *ACS Appl. Mater. Interfaces* **2022**, *14*, 34612–34619. [CrossRef]
22. Li, X.H.; Wang, X.; Antonietti, M. Solvent-free and metal-free oxidation of toluene using O_2 and g-C_3N_4 with nanopores: Nanostructure boosts the catalytic selectivity. *ACS Catal.* **2012**, *2*, 2082–2086. [CrossRef]
23. Zhang, P.; Wang, Y.; Li, H.; Antonietti, M. Metal-free oxidation of sulfides by carbon nitride with visible light illumination at room temperature. *Green Chem.* **2012**, *14*, 1904–1908. [CrossRef]
24. Serpone, N.A.V.E.; Emeline, A.V. Semiconductor Photocatalysis- Past, Present, and Future Outlook. *J. Phys. Chem. Lett.* **2012**, *3*, 673–677. [CrossRef] [PubMed]
25. Hu, Z.; Liu, N.; He, P.; Bai, H.; Hao, L.; Min, J.; Fan, Z.; Chen, B.; Niu, R.; Gong, J. Green Synthesis of Carbon Nitride-Based Conjugated Copolymer for Efficient Photocatalytic Degradation of Tetracycline. *Macromol. Rapid Commun.* **2022**, *43*, 2200043. [CrossRef]
26. Tian, N.; Huang, H.; Du, X.; Dong, F.; Zhang, Y. Rational nanostructure design of graphitic carbon nitride for photocatalytic applications. *J. Mater. Chem. A* **2019**, *7*, 11584–11612. [CrossRef]
27. Liao, G.; Gong, Y.; Zhang, L.; Gao, H.; Yang, G.-J.; Fang, B. Semiconductor polymeric graphitic carbon nitride photocatalysts: The "holy grail" for the photocatalytic hydrogen evolution reaction under visible light. *Energy Environ. Sci.* **2019**, *12*, 2080–2147. [CrossRef]
28. Yu, X.; Liu, Z.; Wang, Y.; Luo, H.; Tang, X. Fabrication of corncob-derived biomass charcoal decorated g-C_3N_4photocatalysts for removing 2-mercaptobenzothiazole. *New J. Chem.* **2020**, *44*, 15908–15918. [CrossRef]
29. Fu, J.; Chang, B.; Tian, Y.; Xi, F.; Dong, X. Novel C_3N_4–CdS composite photocatalysts with organic-inorganic heterojunctions: In situ synthesis, exceptional activity, high stability, and photocatalytic mechanism. *J. Mater. Chem. A.* **2013**, *1*, 3083–3090. [CrossRef]
30. Salem, A.; Saion, E.; Al-Hada, N.M.; Kamari, H.M.; Shaari, A.H.; Abdullah, C.A.C.; Radiman, S. Synthesis and characterization of CdSe nanoparticles via thermal treatment technique. *Results Phys.* **2017**, *7*, 1556–1562. [CrossRef]
31. Han, J.; Dai, F.; Liu, Y.; Zhao, R.; Wang, L.; Feng, S. Synthesis of CdSe/SrTiO3 nanocomposites with enhanced photocatalytic hydrogen production activity. *Appl. Surf. Sci.* **2018**, *467*, 1033–1039. [CrossRef]
32. Zheng, Z.; Mounsamy, M.; Viguerie, N.L.-D.; Coppel, Y.; Harrisson, S.; Destarac, M.; Mingotaud, C.; Kahn, M.L.; Marty, J.-D. Luminescent zinc oxide nanoparticles: From stabilization to slow digestion depending on the nature of polymer coating. *Polym. Chem.* **2018**, *10*, 145–154. [CrossRef]
33. Mali, S.C.; Dhaka, A.; Githala, C.K.; Trivedi, R. Green synthesis of copper nanoparticles using Celastrus paniculatus Wild leaf extract and their photocatalytic and antifungal properties. *Biotechnol. Rep.* **2020**, *27*, e00518. [CrossRef]
34. Beena, V.; Ajitha, S.; Rayar, S.L.; Parvathiraja, C.; Kannan, K.; Palani, G. Enhanced Photocatalytic and Antibacterial Activities of ZnSe Nanoparticles. *J. Inorg. Organomet. Polym. Mater.* **2021**, *31*, 4390–4401. [CrossRef]
35. Rafiq, A.; Imran, M.; Aqeel, M.; Naz, M.; Ikram, M.; Ali, S. Study of Transition Metal Ion Doped CdS Nanoparticles for Removal of Dye from Textile Wastewater. *J. Inorg. Organomet. Polym. Mater.* **2019**, *30*, 1915–1923. [CrossRef]
36. Ayodhya, D.; Veerabhadram, G. Facile fabrication, characterization, and efficient photocatalytic activity of surfactant-free ZnS, CdS, and CuS nanoparticles. *J. Sci. Adv. Mater. Dev.* **2019**, *4*, 381–391. [CrossRef]
37. Devi, R.A.; Latha, M.; Velumani, S.; Oza, G.; Reyes-Figueroa, P.; Rohini, M.; Becerril-Juarez, I.G.; Lee, J.-H.; Yi, J. Synthesis and Characterization of Cadmium Sulfide Nanoparticles by Chemical Precipitation Method. *J. Nanosci. Nanotechnol.* **2015**, *15*, 8434–8439. [CrossRef]
38. Liu, J.; Zhang, T.; Wang, Z.; Dawson, G.; Chen, W. Simple pyrolysis of urea into graphitic carbon nitride with recyclable adsorption and photocatalytic activity. *J. Mater. Chem.* **2011**, *21*, 14398–14401. [CrossRef]
39. Xu, F.; Yuan, Y.; Han, H.; Wu, D.; Gao, Z.; Jiang, K. Synthesis of ZnO/CdS hierarchical heterostructure with enhanced photocatalytic efficiency under nature sunlight. *CrystEngComm* **2012**, *14*, 3615–3622. [CrossRef]
40. Kaur, N.; Verma, A.; Thakur, I.; Basu, S. In-situ dual effect of Ag-Fe-TiO2 composite for the photocatalytic degradation of Ciprofloxacin in aqueous solution. *Chemosphere* **2021**, *276*, 130180–130188. [CrossRef]
41. Qamar, M.A.; Shahid, S.; Khan, S.A.; Zaman, S.; Sarwar, M.N. Synthesis characterization, optical and antibacterial studies of Co-doped SnO2 nanoparticles. *Dig J. Nanomater Biostruct.* **2017**, *12*, 1127–1135.

42. Xu, J.; Li, X.; Niu, J.; Chen, M.; Yue, J. Synthesis of direct Z-Scheme Bi_3TaO_7/CdS composite photocatalysts with enhanced photocatalytic performance for ciprofloxacin degradation under visible light irradiation. *J. Alloys Compd.* **2020**, *834*, 155061. [CrossRef]
43. Senasu, T.; Chankhanittha, T.; Hemavibool, K.; Nanan, S. Visible-light-responsive photocatalyst based on ZnO/CdS nanocomposite for photodegradation of reactive red azo dye and ofloxacin antibiotic. *Mater. Sci. Semicond. Process.* **2020**, *123*, 105558. [CrossRef]
44. Pathania, D.; Katwal, R.; Sharma, G.; Naushad, M.; Khan, M.R.; HAla'a, H. Novel guar gum/Al2O3 nanocomposite as an effective photocatalyst for the degradation of malachite green dye. *Int. J. Biol. Macromol.* **2016**, *87*, 366–374. [CrossRef]
45. Qamar, M.A.; Shahid, S.; Javed, M. Synthesis of dynamic g-C3N4/Fe@ ZnO nanocomposites for environmental remediation applications. *Ceram. Int.* **2020**, *46*, 22171–22180. [CrossRef]
46. Mirsalari, S.A.; Nezamzadeh-Ejhieh, A. Focus on the photocatalytic pathway of the CdS-AgBr nano-catalyst by using the scavenging agents. *Sep. Purif. Technol.* **2020**, *250*, 117235–117246. [CrossRef]

toxics

MDPI

Article

Separation of Minor Actinides from High-Level Liquid Waste Using Novel Silica-Based Butyl-BTP Adsorbents

Tianjiao Jiang [1], Shunyan Ning [1], Tao Yu [1], Ji Wang [1,*], Yuezhou Wei [1,2], Yan Wu [2], Hui He [3], Fangqiang Chen [4,*] and Qingsong Wang [4]

1 School of Nuclear Science and Technology, University of South China, Hengyang 421001, China
2 School of Nuclear Science and Engineering, Shanghai Jiao Tong University, Shanghai 200240, China
3 Department of Radiochemistry, China Institute of Atomic Energy, Beijing 102413, China
4 Nuclear and Radiation Safety Center, MEE, Beijing 102400, China
* Correspondence: wangji@usc.edu.cn (J.W.); frankcfq@126.com (F.C.)

Abstract: To separate the long-lived minor actinides (MA = Am, Cm) from high-level liquid waste (HLLW), we have been studying an advanced separation process via selective adsorption that uses minimal amounts of organic solvent and compact equipment. The process consists of two separation columns packed with a CMPO (octyl(phenyl)-N,N-diisobutylcarbamoyl-methyl phosphine oxide) adsorbent for elemental group separation and a soft-donor named the R-BTP (2,6-bis-(5,6-dialkyl-1,2,4-triazine-3-yl) pyridine) adsorbent for the isolation of MA from lanthanides (Ln). In this work, the effects of nitrate ion (NO_3^-) on the adsorption behavior of Am(III) and a typical fission product Ln(III) onto the *iso*Bu-BTP/SiO_2-P adsorbent were studied experimentally. Then, the desorption properties of the adsorbed element were examined using different eluting agents. A hot test for the separation of MA from the fission product Ln in a genuine MA containing effluent from the irradiated MOX-fuel treatment process was carried out using a *n*Bu-BTP/SiO_2-P packed column. It was found that the separation factor between Am(III) and Ln(III)-FP is over 100 in the measured 0.5–4 M NO_3^-. The adsorbed elements could be effectively eluted off using a complexing agent such as DTPA or pure water. Complete separation between MA and Ln was achieved in the column results, indicating that the proposed MA separation process is feasible in principle.

Keywords: high-level liquid waste; minor actinides; separation; R-BTP adsorbent; hot test

Citation: Jiang, T.; Ning, S.; Yu, T.; Wang, J.; Wei, Y.; Wu, Y.; He, H.; Chen, F.; Wang, Q. Separation of Minor Actinides from High-Level Liquid Waste Using Novel Silica-Based Butyl-BTP Adsorbents. *Toxics* **2022**, *10*, 741. https://doi.org/10.3390/toxics10120741

Academic Editor: Alexei Konoplev

Received: 2 November 2022
Accepted: 28 November 2022
Published: 30 November 2022

Publisher's Note: MDPI stays neutral with regard to jurisdictional claims in published maps and institutional affiliations.

Copyright: © 2022 by the authors. Licensee MDPI, Basel, Switzerland. This article is an open access article distributed under the terms and conditions of the Creative Commons Attribution (CC BY) license (https://creativecommons.org/licenses/by/4.0/).

1. Introduction

As a final disposal method for the high-level liquid waste (HLLW) generated by spent fuel reprocessing, as with the PUREX process, the geological disposal of vitrified HLLW has been proposed and investigated worldwide. Minor actinides (MAs; Am, Cm, etc.), which are alpha emitters with long half-lives (e.g., $^{241}Am_{t1/2}$ = 432.2y, $^{243}Am_{t1/2}$ = 7.38 × 103 y), contribute most of the long-term radiotoxicity of high-level liquid waste [1,2]. The efficient separation of MAs is of great significance for the final geological disposal of the HLLW and is also an essential component of the partition and transmutation (P&T) strategy to transmute the long-lived nuclides into short-lived or stable nuclides [3–5]. The fission products lanthanides (Lns) are much more abundant than MAs in HLLW and are neutron poisons, so they need to be separated before MA transmutation. Since Ln(III) possesses very similar chemical properties to MA(III) in solution, the selective separation of MA(III) from Ln(III) has become one of the most challenging technical problems in HLLW partitioning [3–7]. From the viewpoints of minimizing the long-term radiotoxic risk to the environment and facilitating the P&T strategy, the separation of the long-lived minor actinides (Am, Cm) from HLLW is an important task. For this purpose, we have been studying an advanced separation process involving a selective adsorption technology that uses a minimal organic solvent and compact equipment [2,3,5,8–10]. The schematic flowsheet of this process is shown in Figure 1 [11]. The process is constituted by two separation columns packed with CMPO (octyl(phenyl)-N,N-diisobutylcarbamoyl-methyl phosphine oxide) and R-BTP (2,6-bis-(5,6-dialkyl-1,2,4-triazine-3-yl)pyridine, R = alkyl group) adsorbents, respectively. In the first column, the elements contained in the HLLW can be partitioned into (a) non-adsorptive fission products (FP), (b) MA-Ln, and (c) Zr-Mo, according to their different adsorption and elution properties with CMPO and the eluents. The MA-Ln effluent is then supplied to the second column packed with the R-BTP adsorbent, where the MA is expected to be selectively isolated from the Ln fission products, since the trivalent actinides have much stronger adsorbability onto R-BTP than the trivalent lanthanides [4–7]. For this process, we synthesized some new silica-based adsorbents by impregnating CMPO or R-BTP into a styrene–divinylbenzene copolymer, which was immobilized in porous silica particles with a mean diameter of 60 μm (SiO_2-P). Compared to conventional polymeric matrix adsorbents, these highly porous silica-based adsorbents with a small particle size have faster diffusion kinetics with a relatively lower pressure drop in packed columns, making them suitable for efficient chromatographic separation [8,9]. In our previous work, it was demonstrated that Am together with the fission product Ln can be successfully separated from other FPs contained in a genuine HLLW solution, which was generated from an irradiated MOX fuel treatment process using a CMPO/SiO_2-P packed column [12].

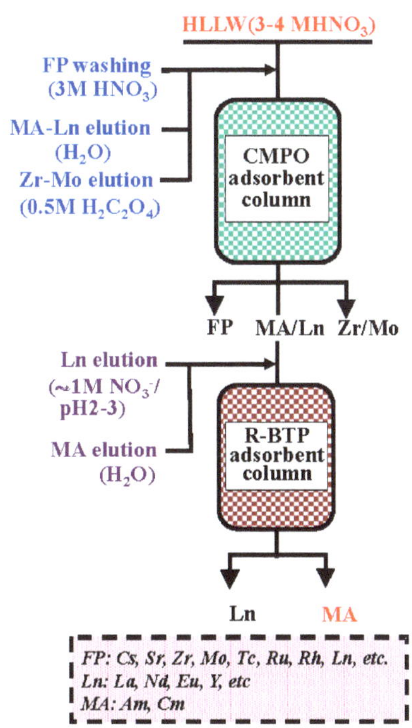

Figure 1. Schematic flowsheet of the MA separation process from HLLW using the adsorption method [11].

It is well known that the mutual separation of the trivalent actinides with 5f-orbital electrons and the lanthanides with 4f-orbital electrons is one of the most difficult tasks in the nuclear fuel cycle due to their very closed ionic size and similar chemical properties. At the end of last century, Kolarik et al. reported that a multidentate nitrogen donor ligand, 2,6-bistriazinylpyridine (BTP), exhibits much higher extraction selectivity for Am(III) than Ln(III) [6,7]. Since 2000, we have synthesized some new R-BTP ligands with different alkyl groups (C = 1–8) and studied their separation behavior for MA and Ln and their chemical and radiation stability in nitric acid solutions [13–15]. As a result, the adsorption affinity and stability strongly depend on the structure of the alkyl group in 2,6-bis-(5,6-dibutyl-1,2,4-triazine-3-yl)pyridine adsorbent. It was found that Bu-BTP/SiO_2-P showed significantly higher adsorption ability and selectivity for Am(III) than Ln(III). The separation experiment results for a simulated HLLW containing trace amounts of ^{243}Am(III) and typical Ln(III) using a *n*Bu-BTP/SiO_2-P packed column demonstrated that the Am(III) was completely separated from the Ln(III) [15].

Since Am(III) is present as a cationic ion in weak nitric acid solutions and R-BTP is a neutral compound, their coordination reaction requires anionic nitrate to form a stably neutral complex. In this work, we investigate the effects of the nitrate ion (NO_3^-) concentration on the adsorption behavior of Am(III) and the typical fission product Ln(III) on the Bu-BTP/SiO_2-P adsorbent. Then, the desorption(elution) properties of the adsorbed element are examined using different eluting agents. A small-scale column separation experiment using a simulated high-level waste solution and a hot test for a genuine MA-containing effluent obtained from an irradiated MOX-fuel treatment process are carried out to verify the feasibility of the proposed separation process.

2. Materials and Experiment Methods

2.1. Reagents

^{241}Am(III) and ^{152}Eu(III) were obtained from a stock solution from the China Institute of Atomic Energy (CIAE). Commercially available $Ln(NO_3)_3$•nH_2O at 99.9% purity was used to prepare the adsorption solution. The other agents used were of analytical grade. The aqueous solutions were prepared with deionized water.

2.2. Synthesis and Characterization of Materials

As the support for the *n*Bu-BTP, the silica/polymer composite support (SiO_2-P) developed in our previous work was utilized. It contains a macroreticular styrene–divinylbenzene copolymer that is immobilized in porous silica particles with a pore size of 0.6 μm and mean diameter of 60 μm. The *n*Bu-BTP was impregnated into the SiO_2-P support at 33.3 wt% of Bu-BTP and the preparation procedure was described in our previous works [8,9]. The chemical structure of the *n*Bu-BTP and an SEM image of *n*Bu-BTP/SiO_2-P adsorbent are illustrated in Figure 2. The *Iso*Bu-BTP/SiO_2-P adsorbent was prepared using the vacuum impregnation method [2]. Typically, 5 g *iso*Bu-BTP dissolved in CH_2Cl_2 was impregnated into 10 g of SiO_2-P under vacuum conditions using a rotary evaporator with a pressure controller.

Here, 2,6-bis-(5,6-dibutyl-1,2,4-triazine-3-yl)pyridine (*n*Bu-BTP or *iso*Bu-BTP) was synthesized in our laboratory with a purity level of 97% according to the high-performance liquid chromatography (HPLC) procedure (Prominence Plus, Shimadzu, Kyoto, Japan). The morphology and microstructure of the *n*Bu-BTP and *iso*Bu-BTP/SiO_2-P adsorbents were characterized using scanning electron microscopy (SEM-EDS) (SU5000, HITACHI, Kyoto, Japan). The concentrations of stable Ln elements in the aqueous phase were determined using inductively coupled plasma atomic emission spectrometry (ICP-AES) (ICPS-7510, Shimadzu, Kyoto, Japan). The ^{241}Am was measured using a high-purity germanium multichannel gamma spectrometer (GEM70P-PLUS, ORTEC, Oak Ridge, TN, USA), while ^{244}Cm was determined using an alpha spectrometer (ALPHA-ENSEMBLE, ORTEC, Oak Ridge, TN, USA).

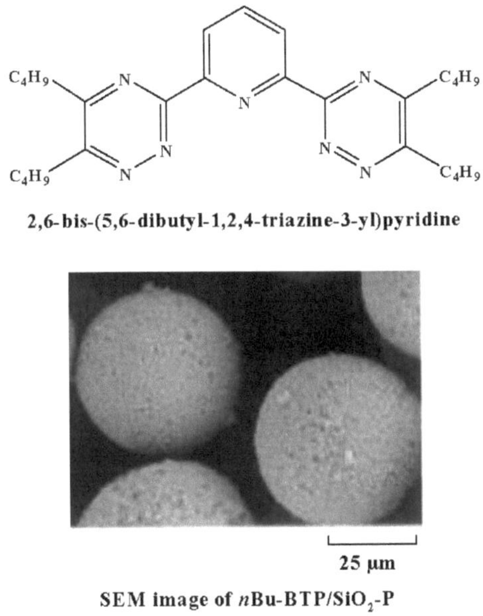

Figure 2. Structure and photograph of the *n*Bu-BTP/SiO_2-P adsorbent.

2.3. Batch Adsorption Experiment

Batch adsorption experiments were conducted to evaluate the adsorbability of Am(III) and some trivalent Ln fission products towards the *iso*Bu-BTP adsorbent. A determined volume (typically 5 cm^3) of a sample solution containing 10 mM of each Ln(III) was spiked with the radioactive nuclides ^{243}Am and ^{152}Eu. A weighed amount of adsorbent (typically 0.25 g) was combined in a glass vial with a Teflon stopper with the sample solution. The resultant mixture was maintained in a water bath shaker kept at 298 K and mechanically shaken at 120 rpm until reaching an equilibrium state. After adsorption, the aqueous phase was filtrated through a membrane filter with 4.5 μm pores and sampled for quantitative measurements. The gamma activities of ^{243}Am and ^{152}Eu were detected using a γ-ray spectrometer with a high-purity germanium detector (Seiko EG&G). The average count rate was taken from three measured values to determine the activity. The concentration of stable Ln elements in the aqueous phase was analyzed using an ICP-AES instrument. The distribution coefficient (K_d) was calculated using Equation (1). Typical experiments were performed in triplicate. The relative error rates of the adsorption experiments were all less than 5%.

$$K_d = \frac{C_0 - C_S}{C_S} \times \frac{V_S}{W_R} = \frac{A_0 - A_S}{A_S} \times \frac{V_S}{W_R} \left(\mathrm{dm}^3/\mathrm{kg}\right) \quad (1)$$

where C_0 and C_S denote the element concentration, while A_0 and A_S present the nuclide activity in the aqueous phase before and after adsorption, respectively. W_R indicates the weight of the dry adsorbent (kg) and vs. the volume of the aqueous phase (dm^3).

2.4. Hot Column Separation Experiment

The HLLW sample solution containing 6 M of HNO_3 was generated from the anion exchange process that had separated the U, Pu, Pd, Ru, and Tc from a spent LWR fuel solution [12,16]. Prior to the separation of MA/Ln, the HLLW sample was fed to a CMPO/SiO_2-P packed column for the elemental group partitioning according to the procedure shown in Figure 1. From this column, a genuine MA-Ln effluent containing Am, Cm, and almost all the FP lanthanides was obtained [12]. The solid-state $NaNO_3$ was added into the MA-Ln effluent to the concentration of 0.3 M (M = mol/dm^3). The pH value of the effluent was estimated to be around 1. Table 1 illustrates the analytic results of the representative nuclides contained in the sample solution.

Table 1. Analytic results of the genuine MA effluent used for the separation experiment.

Nuclide	Concentration (μg/g-soln)	Analysis Method
^{95}Mo	0.2	ICP-MS
^{141}Pr	0.7	ICP-MS
^{145}Nd	10.9	ICP-MS
Nd-total	62.9	ICP-MS
^{238}U	6.7	ICP-MS
^{239}Pu	0.2	ICP-MS
^{241}Am	84.4	γ spectrometry
^{244}Cm	0.1	α spectrometry

In this work, we conducted a separation experiment for the genuine MA-Ln effluent using a Pyrex glass column with a 10 mm inner diameter and 300 mm length. The *n*Bu-BTP/SiO_2-P adsorbent was filled to the column in a slurry state under 2–3 bars of nitrogen gas pressure. The temperature of the column and the throughput liquid phase were maintained constantly at 298 K. Before the chromatographic operation, the *n*Bu-BTP/SiO_2-P packed in the column was conditioned by pumping 100 cm^3 of 1 M $NaNO_3$–0.01 M HNO_3 solution through the column. Next, 30 cm^3 of the sample solution was fed into the column at a constant flow rate of 1.0 cm^3/min (0.76 m/h) using an HPLC metering pump. Subsequently, 70 cm^3 of 0.3 M $NaNO_3$ and 110 cm^3 of H_2O were supplied to the column as eluents, respectively. The outlet effluents from the column were collected using an auto-fractional collector in 10 cm^3 aliquots. The concentrations and activities of the nuclides in each aliquot were quantitatively measured using ICP-MS and alfa/gamma spectrometry.

3. Results and Discussion

3.1. Adsorption and Desorption Behavior of Am(III) and Ln(III)

Figure 3 presents the distribution coefficients (K_d) of ^{241}Am(III) and ^{152}Eu(III) towards the *iso*Bu-BTP/SiO_2-P adsorbent with the change in the initial HNO_3 concentration in the solution. The adsorption affinity of Am(III) increased drastically with the increase in HNO_3 concentration and the K_d reached near 5×10^3 cm^3/g in the range of 1–4 M HNO_3. The K_d of Eu(III) also increased with the increasing HNO_3 concentration, but the adsorbability of Eu(III) was much lower than that of Am(III). The separation factor ($SF_{Am/Ln}$ = ratio of K_d-Am to K_d-Eu) between Am(III) and Eu(III) is about 100 in the measured 0.5–4.0 M HNO_3 range, which would be high enough for their efficient separation by means of the chromatography method.

Since Bu-BTP is a neutral ligand, the adsorption of trivalent metal cations from nitric acid or nitrate solution towards the adsorbent is supposed to result from the following complexing reaction [2]:

$$M^{3+}{}_{aq} + 3NO_3{}^-{}_{aq} + 3(\text{Bu-BTP})_{ads} = [M(NO_3)_3(\text{Bu-BTP})_3]_{ads} \quad (2)$$

Because the stoichiometric ratio of M:NO_3^- is 1:3, the adsorption ability of trivalent MA strongly depends on the concentration of nitrate ion (NO_3^-) surrounding the nitrogen ligand of the adsorbent (Figure 2). Nitric acid is a very strong acid, which almost completely liberates into a nitrate ion (NO_3^-) and proton (H^+) in aqueous solutions. Therefore, the increase in adsorbability with HNO_3 corresponds to the increase in NO_3^- concentration. Notably, since the adsorption of Am(III) and Eu(III) drastically decreased when reducing the concentration of nitrate, and as they showed almost no adsorption in very dilute HNO_3 solution, the adsorbed elements are expected to be eluted out when decreasing the NO_3^- concentration in the chromatographic separation process. On the other hand, the mechanism for the soft donor ligands containing nitrogen or sulfur to exhibit a significantly higher adsorption affinity for trivalent actinides over lanthanides has not yet been clarified.

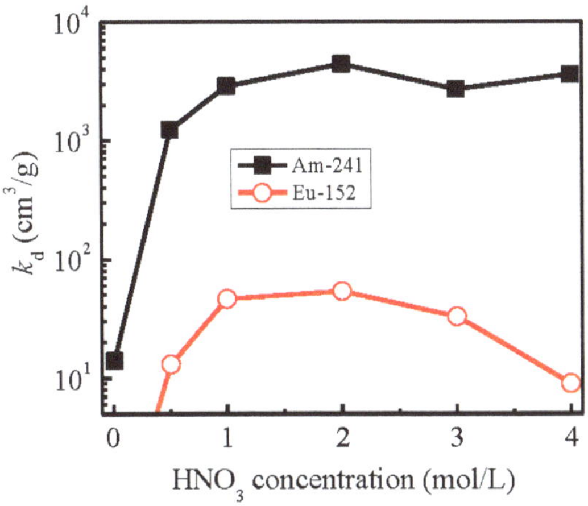

Figure 3. Effect of nitric acid concentration on the adsorption of ^{241}Am(III) and ^{152}Eu(III) onto *iso*Bu-BTP/SiO_2-P (pH 2, 298 K, 24 h).

The effects of the nitrate ($NaNO_3$) concentration on the K_d values of Am(III) and some trivalent lanthanides are shown in Figure 4. As can be seen, the adsorption ability of *iso*Bu-BTP/SiO_2-P towards both ^{241}Am(III) and Ln(III) increased with the NO_3^- concentration, but the adsorbability of Am(III) is much higher than that of Ln(III). The separation factors between Am(III) and the Ln(III) are over 100 at a wide range of NO_3^- concentrations, except $SF_{Am/Dy}$. In addition, the adsorption of Ln(III) increased regularly with the atomic number and Dy(III) showed the strongest adsorption among the tested Ln(III) elements. In our previous work, we studied the coordination chemistry for the adsorption of some Ln(III) ions towards *iso*Bu-BTP/SiO_2-P by means of an EXAFS analysis using the Ln(III)-adsorbed *iso*Bu-BTP/SiO_2-P samples, e.g., Sm(III), Eu(III), Gd(III), and Dy(III) [2,5]. It was found that the [Ln(*iso*Butyl-BTP/SiO_2-P)$_3$]$^{3+}$ complex is the dominant species and the Ln(III)-N bond length in the first coordination layer become shorter as the atomic number of Ln(III) increases, which indicates the enhanced interaction between Ln(III) and the *iso*Bu-BTP/SiO_2-P adsorbent as the atomic number increases. Therefore, the relationship between adsorption affinity of *iso*Bu-BTP/SiO_2-P and the different Ln(III) shown in Figure 4 can be reasonably explained by the EXAFS analysis results.

The lanthanide fission products resulting from the ^{235}U fission reaction mostly include the light lanthanides from La to Tb, while the yield of Dy is almost negligible [17]. Since Dy(III) exhibits the strongest adsorbability among the lanthanides onto the R-BTP adsorbents and its adsorption behavior is similar to Am(III), it can be used as a simulated element of Am in some cases [18–20].

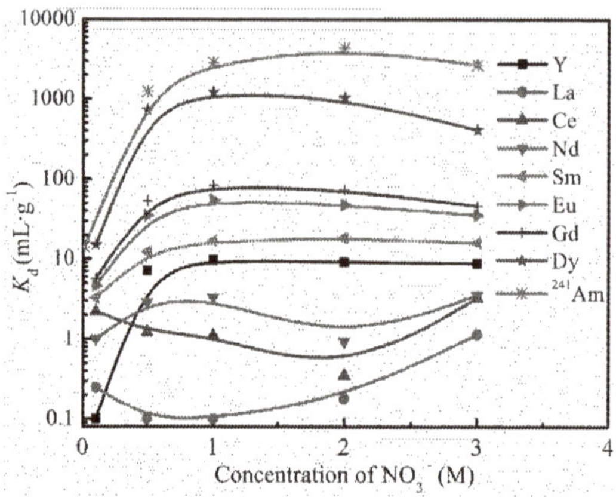

Figure 4. Relationship of adsorbability of ^{241}Am(III) and Ln(III) towards *iso*Bu-BTP/SiO_2-P with the nitrate concentration ($NaNO_3$, pH 2, 298 K, 24 h).

The SEM-EDS analysis result for the Dy-loaded isoBu-BTP/SiO_2-P is shown in Figure 5. After the adsorption experiment, the Dy-loaded isoBu-BTP/SiO_2-P adsorbent was inlaid into an epoxy resin. The particle size of the *iso*BTP/SiO_2-P adsorbent is around 50 μm. As can be seen, the Dy(III) was adsorbed from the surface into the inside center of the particle and then distributed across the whole cross-section of the adsorbent. The images prove that the Dy(III) was evenly adsorbed inside the *iso*Bu-BTP/SiO_2-P particles, not just on the surface.

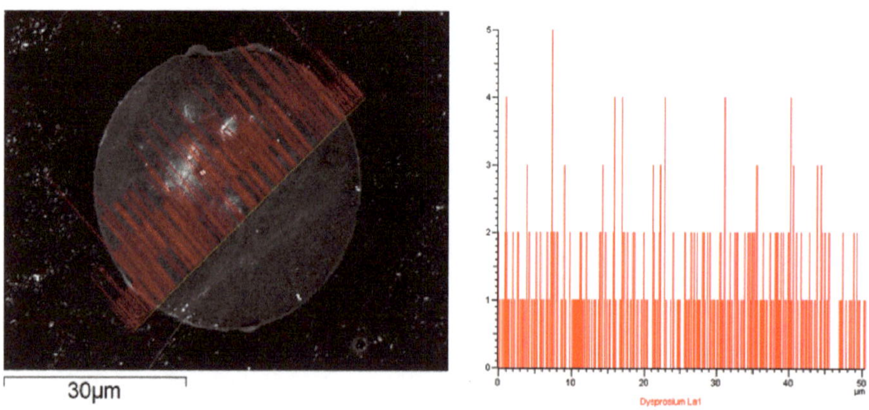

Figure 5. SEM-EDS of Dy(III)-loaded *iso*BTP/SiO_2-P after batch adsorption for 1 h.

To examine the desorption behavior of the trivalent metal ions from *iso*Bu-BTP/SiO_2-P, batch desorption experiments were conducted using different eluting solutions. Figure 6 shows the effect of the eluting agent on the desorption rate of the Dy(III) from the loaded *iso*Bu-BTP/SiO_2-P adsorbent. As can be seen, DTPA(diethylenetriaminepentaacetic acid) exhibited highest desorption rate of the Dy(III) ions. DTPA is an aqueous soluble nitrogen bearing a complexing agent and can coordinate with Dy(III), resulting in its desorption from the adsorbent. On the other hand, the adsorbed Dy(III) was quite effectively eluted out by the water due to a drastic decrease in the nitrate concentration in the adsorbent.

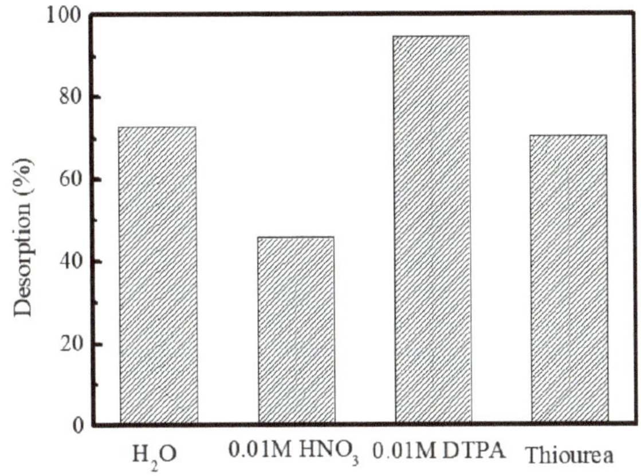

Figure 6. Effect of the eluting agent on the desorption rate of Dy(III) from the loaded *iso*Bu-BTP/SiO_2-P (adsorption conditions: 298 K, phase ratio: 0.1 g/5 cm^3, 120 rpm, 24 h).

3.2. Chromatographic Separation of Am(III) from Ln(III)

Prior to the hot test, a column dynamic adsorption–elution experiment using the *iso*Bu-BTP/SiO_2-P adsorbent was carried out for a mixed Ln(III) sample solution with Dy(III) as an analogue of Am(III). The feed sample contained 1 mM of Y(III), La(III), Ce(III), Nd(III), Sm(III), Eu(III), Gd(III), and Dy(III) in a 0.01 M HNO_3–0.49 M $NaNO_3$ solution, respectively. The elution curves of the Ln(III) ions are shown in Figure 7. As can be seen, the light Ln(III) ions, such as La(III), Ce(III), Nd(III), and Y(III), showed almost no adsorption and flowed out from the column with the feed sample and the rinsing solution. The Dy(III) was completely adsorbed and was effectively eluted off by 0.01 M DTPA as eluent. Small portions of Sm(III), Eu(III), and Gd(III) were adsorbed and were eluted together with Dy(III). It was found that the column dynamic adsorption–elution results show good agreement with the batch adsorption–desorption experiments.

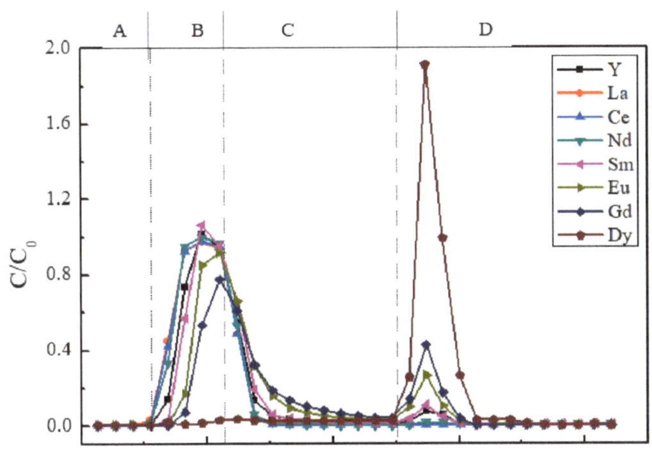

Figure 7. Column dynamic adsorption–elution experiment results for a Ln(III) mixed sample solution using *iso*Bu-BTP/SiO_2-P adsorbent (column: ø10 mm × *h*300 mm; flow rate: 0.3 mL/min; temperature: 298 K): (**A**) washing solution (0.01 M HNO_3–0.49 M $NaNO_3$); (**B**) feed sample solution; (**C**) rinsing solution (0.01 M HNO_3–0.49 M $NaNO_3$); (**D**) 0.01 M DTPA.

In our previous work, a hot test for an actual high-level liquid waste (HLLW) generated from an irradiated MOX fuel treatment process was carried out using a CMPO/SiO_2-P adsorbent packed column. As a result, Am and Cm together with the fission product Ln were successfully separated from other FPs contained in the HLLW, resulting in an MA-Ln effluent with a pH of around 1 [12]. In this work, this weakly acidic MA-Ln effluent was adjusted to 0.3 M of $NaNO_3$ and then subjected to a *n*Bu-BTP/SiO_2-P packed column for Am/Ln separation. Figure 8 shows the results of the separation experiment for the genuine MA-containing effluent with the nuclide composition, as shown in Table 1. As can be seen, the Ln(III) represented by Nd(III) and Pr(III) showed almost no adsorption. They flowed out with the sample solution and the 0.3 M $NaNO_3$ rinsing solution. On the other hand, Am(III) and Cm(III) were strongly adsorbed by the *n*Bu-BTP/SiO_2-P adsorbent, and then were effectively eluted off using pure water as an eluent. A complete separation between MA (Am and Cm) and the Ln was achieved.

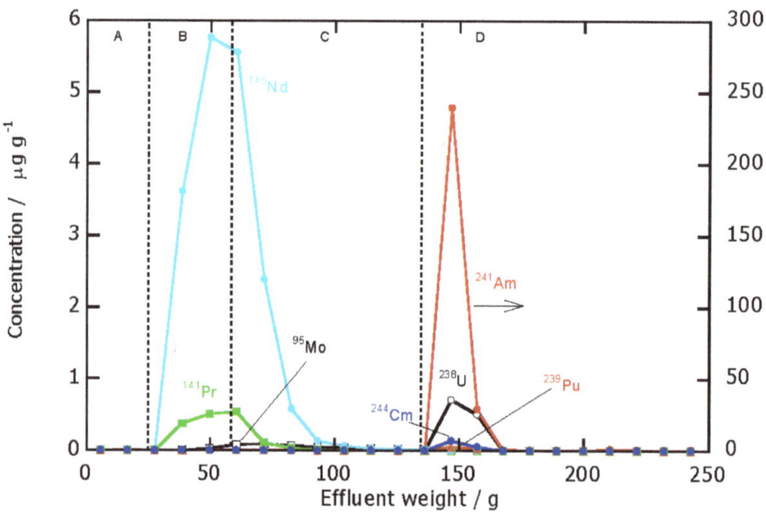

Figure 8. Separation experiment results for the genuine MA-Ln effluent by *n*Bu-BTP/SiO_2-P packed column: (**A**) dead volume; (**B**) sample solution; (**C**) 0.3 M $NaNO_3$; (**D**) H_2O.

In the sample solution, small amounts of U, Pu, and Mo were contained. The elution profiles of these nuclides together with the MA and Ln are illustrated in Figure 8. It was found that U and Pu were also adsorbed by the adsorbent and efficiently eluted off by H_2O together with the MA. Since the spent fuel solution was once reduced by electro-reduction prior to the anion exchange separation process, the valence representations of U and Pu in the MA-Ln effluent were considered to be U(IV) and Pu(IV) or Pu(III), respectively [16]. From this experimental result, we can conclude that the U and Pu that remained in MA-Ln effluent could be effectively recovered together with the MA. On the other hand, the Mo showed very weak adsorption and behaved similarly to the non-adsorptive Ln(III).

These experimental results demonstrate that the proposed MA separation process shown in Figure 1 is feasible in principle. For future practical applications, further investigations including improvements of the radiation stability of the adsorbents and scale-up testing are required.

4. Conclusions

The adsorption and desorption behavior of Am(III) and Ln(III) onto Bu-BTP/SiO_2-P adsorbent were investigated. Column separation experiments using simulated and genuine MA effluent were carried out. The *iso*Bu-BTP/SiO_2-P exhibited much stronger adsorbability for ^{241}Am(III) than ^{152}Eu(III) and all of the Ln(III) ions in the acidic nitrate solution. The separation factor between Am(III) and Ln(III)-FP was over 100 in the measured 0.5–4 M $NaNO_3$ solution with a pH of 2. The adsorbed Am(III) and Ln(III) ions were effectively eluted out using DTPA or pure water as the eluents. The column experiment results demonstrated that the Ln(III)-FP showed almost no adsorption, while the Am(III) and Cm(III), as well as the residual U and Pu, were completely adsorbed by the adsorbent and eluted off by using water as an eluent. A complete separation between the MA and Ln was achieved, indicating that the proposed MA separation process is feasible in principle. Further investigations including improvements of the radiation stability of the adsorbents and scale-up testing are required.

Author Contributions: Conceptualization, Y.W. (Yuezhou Wei) and T.Y.; methodology, J.W. and H.H.; writing-review and editing, F.C. and Y.W. (Yan Wu); writing—original draft preparation, T.J. and S.N.; investigation and data curation, S.N., T.J., Q.W. and J.W.; supervision, T.Y. and Y.W. (Yuezhou Wei); project administration, Y.W. (Yuezhou Wei) and S.N. All authors have read and agreed to the published version of the manuscript.

Funding: This research was financially supported by the National Natural Science Foundation of China, grant numbers U1967218, 11975082, and 22066005.

Institutional Review Board Statement: Not applicable.

Informed Consent Statement: Not applicable.

Data Availability Statement: Not applicable.

Conflicts of Interest: The authors declare no conflict of interest.

References

1. Zhang, W.; He, X.; Ye, G.; Yi, R.; Chen, J. Americium(III) capture using phosphonic acid-functionalized silicas with different mesoporous morphologies: Adsorption behavior study and mechanism investigation by EXAFS/XPS. *Environ. Sci. Technol.* **2014**, *48*, 6874–6881. [CrossRef] [PubMed]
2. Ning, S.Y.; Zhang, W.; Yu, S.Q.; Zhang, S.C.; Zhou, J.; Wang, X.P.; Wei, Y.Z. Selective separation of MA(III) from Ln(III) by highly stable silica-polymer based N-donor IsoBu-BTP/SiO_2-P adsorbent. *Solvent Extr. Ion Exch.* **2019**, *37*, 126–139. [CrossRef]
3. Madic, C. Overview of the hydrometallurgical and pyro-metallurgical processes studied world-wide for the partitioning of high active nuclear wastes. In Proceedings of the NEA/OECD 6th Information Exchange Meeting on Actinide and Fission Product Partitioning and Transmutation, Madrid, Spain, 11–13 December 2000; pp. 53–64.
4. Ning, S.Y.; Zhang, S.C.; Zhou, J.; Zhang, W.; Wei, Y.Z. Salt-free separation of ^{241}Am(III) from lanthanides by highly stable macroporous silica-polymer based Me_2-CA-BTP/SiO_2-P adsorbent. *J. Radioanal. Nucl. Chem.* **2019**, *322*, 1023–1030. [CrossRef]
5. Ning, S.Y.; Zou, Q.; Wang, X.P.; Liu, R.Q.; Wei, Y.-Z. Adsorption mechanism of silica/polymer-based 2,6-bis(5,6-diisohexyl-1,2,4-triazin-3-yl)pyridine adsorbent towards Ln(III) from nitric acid solution. *J. Nucl. Sci. Technol.* **2016**, *53*, 1417–1425. [CrossRef]
6. Kolarik, Z.; Müllich, U.; Gassner, F. Selective extraction of Am(III) over Eu(III) by 2,6-ditriazolyl-and 2,6-ditriazolylpyridines. *Solvent Extr. Ion Exch.* **1999**, *17*, 23–32. [CrossRef]
7. Kolarik, Z.; Müllich, U.; Gassner, F. Extraction of Am(III) and Eu(III) nitrates by 2,6-Di-(5,6-dipropyl-1,2,4-triazine-3-yl). *Solvent Extr. Ion Exch.* **1999**, *17*, 1155–1170. [CrossRef]
8. Wei, Y.-Z.; Kumagai, M.; Takashima, Y.; Modolo, G.; Odoj, R. Studies on the separation of minor actinides from high-level wastes by extraction chromatography using novel silica-based extraction resins. *Nucl. Technol.* **2000**, *132*, 413–423. [CrossRef]
9. Wei, Y.-Z.; Kumagai, M.; Takashima, Y.; Yokoi, H.; Hoshikawa, T.; Kawamura, F. Development of silica-based chelating exchangers and their separation behavior for lanthanides and americium by column chromatography. In Proceedings of the RECOD'98, Nice Acropolis, France, 25–28 October 1998; pp. 701–708.
10. Zhang, H.; Li, C.M.; Chen, X.J.; Fu, H.; Chen, Y.L.; Ning, S.Y.; Fujita, T.; Wei, Y.Z.; Wang, X.P. Layered ammonium vanadate nanobelt as efficient adsorbents for removal of Sr^{2+} and Cs^{+} from contaminated water. *J. Colloid Interf. Sci.* **2022**, *615*, 110–123. [CrossRef]
11. Wei, Y.; Ning, S.; Wang, Q.; Chen, Z.; Wu, Y.; Liu, R.; Mimura, H. Adsorption materials development for the separation of actinides and specific fission products from high level waste. *Adv. Sci. Technol.* **2014**, *94*, 103–110. [CrossRef]
12. Wei, Y.-Z.; Hoshi, H.; Kumagai, M.; Goethals, P.; Bruggeman, A. A hot test on minor actinides separation from high-level-waste by CMPO/SiO_2-P extraction resin. In *Recent Advances in Actinide Science*; Alvares, R., Bryan, N., May, I., Eds.; RSC: London, UK, 2006; pp. 647–649.

13. Wei, Y.-Z.; Sabharwal, K.N.; Kumagai, M.; Asakura, T.; Uchiyama, G.; Fujine, S. Preparation of novel silica-based nitrogen donor extraction resins and their adsorption performance for trivalent americium and lanthanides. *J. Nucl. Sci. Technol.* **2000**, *37*, 1108–1110. [CrossRef]
14. Wei, Y.-Z.; Hoshi, H.; Kumagai, M.; Asakura, T.; Uchiyama, G. Preparation of novel silica-based R-BTP extraction-resins and their application to trivalent actinides and lanthanides separation. *J. Nucl. Sci. Technol.* **2002**, *39* (Suppl. S3), 761–764. [CrossRef]
15. Wei, Y.-Z.; Hoshi, H.; Kumagai, M.; Asakura, T.; Morita, Y. Separation of Am(III) and Cm(III) from trivalent lanthanides by 2,6-bistriazinylpyridine extraction chromatography for radioactive waste management. *J. Alloys Compd.* **2004**, *374*, 447–450. [CrossRef]
16. Wei, Y.-Z.; Arai, T.; Hoshi, H.; Kumagai, M.; Bruggeman, A.; Goethals, P. Development of a new aqueous process for nuclear fuel reprocessing: Hot tests on the recovery of U and Pu from a nitric acid solution of spent LWR fuel. *Nucl. Technol.* **2005**, *149*, 217–231. [CrossRef]
17. Benedict, M.; Pigford, T.H.; Levi, H.W. *Nuclear Chemical Engineering*, 2nd ed.; McGraw-Hill: New York, NY, USA, 1981; Chapter 10.
18. Geist, A.; Hill, C.; Modolo, G.; Foreman, M.R.S.J.; Weigl, M.; Gompper, K.; Hudson, M.J. 6,6′-Bis(5,5,8,8-tetramethyl-5,6,7,8-tetrahydro-benzo [1,2,4]triazin-3-yl) [2.;2′]bipyridine, an effective extracting agent for the separation of americium(III) and curium(III) from the lanthanides. *Solv. Extr. Ion Exch.* **2006**, *24*, 463–482. [CrossRef]
19. Trumm, S.; Geist, A.; Panak, P.J.; Fanghänel, T. An improved hydrolytically-stable bis-triazinyl-pyridine (BTP) for selective actinide extraction. *Solv. Extr. Ion Exch.* **2011**, *29*, 213–229. [CrossRef]
20. Usuda, S.; Wei, Y.-Z.; Xu, Y.; Li, Z.; Liu, R.; Kim, S.-Y.; Wakui, Y.; Hayashi, H.; Yamazaki, H. Development of a simplified separation process of trivalent minor actinides from fission products using novel R-BTP/SiO_2-P adsorbents. *J. Nucl. Sci. Technol.* **2012**, *49*, 334–342. [CrossRef]

toxics

MDPI

Article

Evaluation of Heavy Metal Removal of Nanoparticles Based Adsorbent Using *Danio rerio* as Model

Antony V. Samrot [1,*], Muthiah Bavanilatha [2,*], Sivasuriyan Krithika Shree [2], Mahendran Sathiyasree [2], Jayaram Vanjinathan [3], Nagarajan Shobana [2], Rajendran Thirugnanasambandam [4], Chandrasekaran Kumar [4], Samraj Wilson [5], Deenadhayalan Rajalakshmi [2], Lawrence Xavier Noel Richard Prakash [2] and Ram Singh Sanjay Preeth [2]

1 School of Bioscience, Faculty of Medicine, Bioscience and Nursing, MAHSA University, Jalan SP2, Bandar Saujana Putra, Jenjarom 42610, Malaysia
2 Department of Biotechnology, School of Bio and Chemical Engineering Sathyabama Institute of Science and Technology, Chennai 600119, Tamil Nadu, India
3 Department of Civil Engineering, Sathyabama Institute of Science and Technology, School of Building and Environment, Chennai 600119, Tamil Nadu, India
4 Centre for Ocean Research (DST—FIST Sponsored Centre), MoES—Earth Science & Technology Cell, Sathyabama Institute of Science and Technology, Chennai 600119, Tamil Nadu, India
5 Department of Botany, St. John's College, Tirunelveli 627002, Tamil Nadu, India
* Correspondence: antonysamrot@gmail.com (A.V.S.); bavagold@gmail.com (M.B.)

Abstract: Nanoparticles are potential candidates for wastewater treatment especially for the removal of heavy metals due to their strong affinity. Many biopolymers are used as adsorbents and encapsulation of nanoparticle onto them can increase their efficiency. In this study, SPIONs, alginate, and SPIONs incorporated on alginate beads have been synthesized and characterized both microscopically and spectroscopically. These were then used for the removal of chromium metal and the percentage of removal was evaluated using a batch adsorption study. The percent removal of chromium using SPIONs, alginate and alginate–SPIONs beads were recorded to be 93%, 91% and 94%, respectively. The adsorption of chromium using SPIONs and alginate–SPIONs beads followed the Tempkin isotherm, whereas adsorption of chromium metal by alginate beads was found to be homogeneous in nature and followed the Langmuir isotherm with an R^2 value of 0.9784. An *in-vivo* study using *Danio rerio* as a model organism was done to examine the toxicity and the removal efficiency of the samples. It was observed that chromium water treated with alginate–SPIONs beads, which were removed after water treatment showed less damage to the fishes when compared to SPIONs and alginate beads treated with chromium water where the SPIONs and alginate beads were not removed after the treatment period.

Keywords: nanosorbent; alginate beads; SPIONs; nanotoxicity

Citation: Samrot, A.V.; Bavanilatha, M.; Krithika Shree, S.; Sathiyasree, M.; Vanjinathan, J.; Shobana, N.; Thirugnanasambandam, R.; Kumar, C.; Wilson, S.; Rajalakshmi, D.; et al. Evaluation of Heavy Metal Removal of Nanoparticles Based Adsorbent Using *Danio rerio* as Model. *Toxics* **2022**, *10*, 742. https://doi.org/10.3390/toxics10120742

Academic Editors: Yuezhou Wei, Yan Wu and Xinpeng Wang

Received: 30 September 2022
Accepted: 18 November 2022
Published: 30 November 2022

Publisher's Note: MDPI stays neutral with regard to jurisdictional claims in published maps and institutional affiliations.

Copyright: © 2022 by the authors. Licensee MDPI, Basel, Switzerland. This article is an open access article distributed under the terms and conditions of the Creative Commons Attribution (CC BY) license (https://creativecommons.org/licenses/by/4.0/).

1. Introduction

Water is one of the world's natural resources, and its availability in its purest form is critical for all living things [1]. Effluent discharge from various industries pollutes water bodies, resulting in water pollution which could be easily monitored continuously using simple—to—use analytical technology on—site using composite sensors [2–4]. Effective wastewater treatment is a key requirement for a rising global economy [5]. Adsorption materials such as membranes, nanocatalysts, nanofilm—functionalized surfaces and coatings are examples of novel advanced materials that can be developed using nanotechnology for efficient wastewater cleanup procedures. Nanosorbents are one of the most ideal candidates for wastewater treatment due to the ease in utilization, high sorption capacity, etc. [6]. The use of nanoparticles in water treatment applications is restricted due to environmental contamination, potential toxicity, and safety risks. Hence, these nanoparticles can be encapsulated within a matrix material, which contributes to the treatment of water without any addition of impurities. The most commonly used biopolymers for encapsulation include polysaccharide—like alginate, cellulose, carrageenan, chitosan, starch, etc. [7–10]. Alginate beads are commonly used as a support material in numerous technologies, and they have lately been used in the removal of pollutants due to their high affinity towards heavy metal ions [11,12]. The encapsulation of alginate beads with magnetic nanoparticles, such as magnetic nanosorbents or magsorbents, helps in increasing their biosorption ability [12–15]. The beads play an important function in water treatment and are a one—of—a—kind technology for removing various organic contaminants from water [16].

These nano—adsorbent beads are good nanocomposite materials for removing heavy metals from wastewater in a continuous treatment process and are proposed to be non-toxic in nature, less expensive, to have high hydrophilicity and high efficiency [12,17]. In contrast to conventional adsorbents, nanoadsorbents are superior in ranking, status, and quality and are useful in a variety of fields, which enhances their applicability [6]. Magnetic—based adsorbents offer the key advantages of simple separation from treated water using an external magnetic field and of being reusable. For example, in the removal of phenol and p—cresol present in wastewater from the pulp and paper industry, a novel iron oxide—hydrotalcite modified with dodecylsulfate and cyclodextrin magnetic adsorbent demonstrated maximum adsorption and removed about 216.08 mg/g and 272.48 mg/g, respectively, which is greater than was reported for certain activated carbon—type and activated char adsorbents [18]. Among magnetic nanoparticles in particular, superparamagnetic Iron Oxide Nanoparticles (SPIONs) have been proposed to be an effective candidate in removal of heavy metals and also in drug delivery and imaging [19–22]. In this study, SPIONs, alginate beads, SPIONs—loaded alginate beads were prepared. These were subjected to characterization by, for example, UV—Visible Spectroscopy, Fourier Transform Infrared Spectroscopy, X-ray Crystallography, Scanning Electron Microscopy and Energy Dispersive X-ray Analysis, Vibrating Sample Magnetometer, and Zeta Potential analysis. They were then tested for removal of chromium, and the percentage of removal was examined using batch adsorption studies. Further adsorption patterns of adsorbate upon adsorbent were studied using different isotherm models. These adsorbents were used for treating water with chromium and then tested for toxicity against zebrafishes. To understand the toxicity of nanosorbent/sorbent itself, we allowed the adsorbents to be present in the water even after the treatment period in some groups.

2. Materials and Methods

2.1. Synthesis of SPIONs

SPIONs were synthesized using a chemical co—precipitation method prepared using ferric chloride ($FeCl_3$) and ferrous chloride ($FeCl_2$) as precursor agents: 5.4 g of ferric chloride ($FeCl_3$) and 1.9 g of ferrous chloride ($FeCl_2$) were mixed in 1 L of nitrogen—purged water. Then the solution was reduced by adding 40 mL of 2 M sodium hydroxide (NaOH) to it. The mixture was stirred vigorously until a black precipitate appeared. It was then centrifuged at 5000 rpm, and the pellets were collected to obtain SPIONs, which were later lyophilized [23].

2.2. Synthesis of Alginate Beads

25 mL of 1.2% sodium alginate ($NaC_6H_7O_6$) solution was prepared and added dropwise to a mixture containing 10% of calcium chloride ($CaCl_2$) and 1 mL of Tween 20 to form beads. Sodium alginate was transformed into beads after interacting with calcium ions. These beads were then filtered from the solution using muslin cloth [24].

2.3. Synthesis of Alginate–SPIONs Beads

10 mL of 1.2% sodium alginate solution was prepared, and 0.03 g of hydroxypropyl methylcellulose (HPMC) was added to it as a stabilizing agent. 125 mg of SPIONs was then added to the mixture and stirred for 40 min until the SPIONs were completely dispersed in the solution. Then the solution was added dropwise to the solution containing 10% of

calcium chloride and 1 mL of Tween 20 to form beads. The beads were then filtered from the solution using muslin cloth [25].

2.4. Characterization of SPIONs

The UV–Visible absorption spectra of SPIONs were taken at room temperature using UV 3600 PLUS (Shimadzu, Kyoto, Japan) with a variable wavelength between 200 and 800 nm. Fourier Transform Infrared Spectroscopic experiments were carried out to check the functional groups using a IRTRACER 100 (Shimadzu, Kyoto, Japan). Scanning Electron Microscopic micrographs and elemental composition were recorded using a JSM—IT 200 (JEOL, Tokyo, Japan). The crystalline structure of the sample was determined by X-ray crystallography (PANalytical Corporation, X'Pert, Almelo, The Netherlands). The determination of the hysteresis loop was performed by Vibrating Sample Magnetometer (LakeShore Co. Ltd., Lake Shore 7400, Westerville, OH, USA) and its stability was checked using Zeta Potential analysis (ZETASIZER Nano Series ZSP, Malvern Instruments, Worcestershire, UK).

2.5. Characterization of Alginate Beads and Alginate–SPIONs Beads

The alginate beads and alginate–SPIONs beads were characterized using UV–Visible spectroscopy (Shimadzu, UV 3600 PLUS, Kyoto, Japan), Fourier Transform Infrared Spectroscopy (Shimadzu, IRTRACER 100, Kyoto, Japan), and crystallinity was observed using X-ray crystallography (PANalytical Corporation, X'Pert, Almelo, Netherlands).

2.6. Batch Adsorption Studies

The effective removal of chromium metal using SPIONs, alginate beads, alginate–SPIONs beads was done using batch adsorption studies by calculating triplicate OD values. The DPC method (1, 5—Diphenylcarbazide) was used to investigate the chromium ion adsorption [25,26]. The optimization of adsorbate concentration, adsorbent concentration, pH, and contact time was done. The adsorbate concentration (chromium metal) was initially adjusted between 1 ppm and 10 ppm, then incubated for 1 h at neutral pH. One mL of the supernatant was collected after the interaction between the samples and chromium and centrifuged for 15 min at 5000 rpm. The optimal adsorbate concentration fixed at the ppm which showed the maximum removal was observed. Then the adsorbent concentration (SPIONs, alginate beads, alginate—SPIONs beads) was altered from 0.01 g to 0.1 g/10 mL and the efficient removal was investigated under optimal adsorbate concentration. Under optimal adsorbate and adsorbent concentrations, pH was varied between 5 and 9 and checked for the removal efficiency. Similarly, effective removal of adsorbate by adsorbent was examined by altering the contact time between 30 min and 2 h 30 min under the optimized adsorbate, adsorbent and pH conditions. After optimization, 1 mL of supernatant was treated with 30 mL of 2N H_2SO_4, which produces Cr^{6+} dissociation, and further treated with 0.5% of diphenyl carbazide, which gives a reddish—violet color by binding to Cr^{6+}. It was then spectroscopically measured at 540 nm [26]. The formula used to calculate the adsorption removal is given below:

$$\% \text{ Removal} = [(\text{Initial concentration-Final concentration})/(\text{Initial concentration})] \times 100$$

2.7. Isotherm Studies

The adsorption pattern and the mechanism of adsorbate removal by adsorbent was examined using four types of isotherm models—Langmuir, Freundlich, Tempkin and Dubinin—Radushkevich (D—R) isotherm. A graph between 1/qe and 1/Ce was plotted for the Langmuir isotherm, ln qe vs. ln Ce for the Freundlich, the Tempkin isotherm was investigated by plotting qe vs. ln Ce, and D—R was determined by plotting between ε^2 and ln qe in order to determine the adsorption pattern, where Ce is the equilibrium concentration of adsorbate in mg/L, qe is the amount of adsorbate adsorbed upon the adsorbent, and ε^2 is the adsorption potential based on the Polanyi potential.

2.8. In-Vivo Study to Determine Removal of Chromium

2.8.1. Collection and Maintenance of Zebrafish

Male *Danio rerio* were purchased from Tarun fish farm, Chennai and kept in a glass aquarium filled with water with sufficient and continuous aeration. The fishes were introduced into the test environment after three to four days of acclimating to the new circumstances. The water was changed once a week and dry food fed twice a day. The initial pH of the water was at 7.2 ± 0.3 [27].

2.8.2. *In-Vivo* Study Using Zebrafish as Model

An *in-vivo* study for the effective removal of chromium using SPIONs, alginate beads, and alginate–SPIONs beads was done against *Danio rerio*. Test 1 with 1 ppm of chromium water was used to test the survivability of zebrafishes. In Test 2, 7 g of synthesized SPIONs were added to 5 L of water. In Test 3, 7 g of SPIONs were mixed with 1 ppm chromium water and was not removed. In Test 4, SPIONs—treated water was used, whereby 7 g of SPIONs were added to 1 ppm of chromium water and the SPIONs were withdrawn after 1 h of interaction before introducing the fishes. In Test 5, 7 g of synthesized SPIONs were put into the water, and after 1 h of interaction they were removed. In Test 6, SPIONs—encapsulated alginate beads were added to 1 ppm of chromium water and not removed. In Test 7, SPIONs—encapsulated alginate beads were put into chromium water and removed after 1 h of interaction. In Test 8, 7 g of alginate beads were introduced to 1 ppm of chromium water and were not removed. Ten fishes were allowed inside the aquarium which contained the aforementioned conditions. All the fishes were fed regularly, as mentioned.

2.8.3. Histology Studies

The damage that occurred in the fishes on exposure to the chromium water and SPIONs was analyzed using histology studies. The fishes were taken from each test on the 5th and 10th days and were put into 10% of formaldehyde solution for analysis. The organs of the fish—eyes and gills—were then subjected to Hematoxylin and Eosin (H&E) staining and examined using a fully inverted microscope (Leica DMI6000 B, Leica Microsystems, Stockach, Germany).

2.8.4. Inductively Coupled Plasma—Mass Spectrometry (ICP—MS)

The uptake of chromium and iron metals by the fish was examined on the 5th and 10th days using ICP—MS (Inductively Coupled Plasma Mass Spectrometry) through the microwave digestion method. The whole test fish was taken from the aquarium and immersed in 5 mL of concentrated nitric acid for 15 min. A few drops of concentrated hydrofluoric acid were then added and microwaved until fully digested. A clear solution was obtained by centrifugation which was then taken and kept at 4 °C and then analyzed using Agilent 7700x—ICP—MS (Agilent Technologies, Mulgrave, VIC, Australia).

3. Results and Discussion

3.1. Characterization of SPIONs, Alginate Beads and SPIONs—Coated Alginate Beads

3.1.1. UV—Visible Spectroscopy

The peak at the range between 250 and 300 nm shows the presence of iron oxide nanoparticles, and the peak between 200 to 400 nm shows the light absorption of the alginate beads (Figure 1a,b). A broad peak from 250 to 400 nm shows the presence of SPIONs indicating the encapsulation of SPIONs into the alginate beads (Figure 1c). Lam et al. [28] reported that absorption of light at 200 to 400 nm shows the presence of alginate beads. The maximum absorption at 250 to 300 nm indicates the characteristic peak for the SPIONs [29].

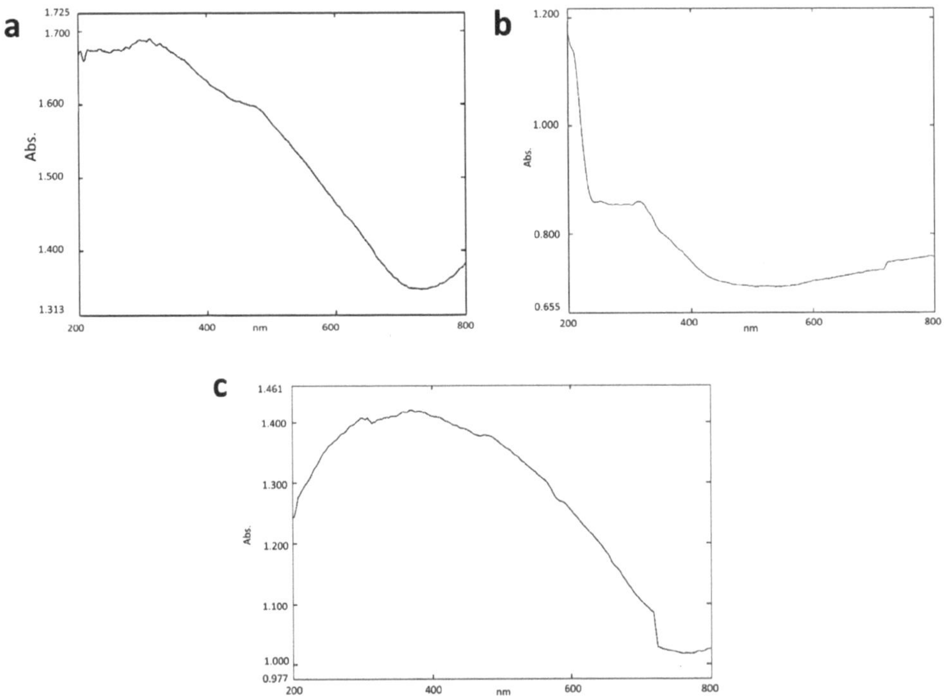

Figure 1. UV—Vis analysis of (**a**) SPIONs (**b**) Alginate beads (**c**) Alginate–SPIONs microspheres.

3.1.2. Fourier Transform Infrared Spectroscopy (FTIR)

The functional groups present in the SPIONs, alginate beads and SPIONs—coated alginate beads were analyzed using Fourier Transform Infrared Spectroscopy (FTIR) (Figure 2). The shift of the C=O bond due to the formation of an ionic bond between calcium chloride and sodium alginate was found at the peak at 1616 cm^{-1}. A peak around 3400 to 3300 cm^{-1} was due to hydroxyl groups and is seen in all the samples. A large peak seen in Figure 2b,c at 630–550 cm^{-1} depicts the presence of Fe–O vibrations of SPIONs [30]. Matos et al. [31] reported that peaks at 3500 cm^{-1}, 453 cm^{-1}, and 586 cm^{-1} showed the presence of an OH group on the surface of SPIONs and the presence of the Fe–O bond of SPIONs, respectively. In another study, the presence of Fe_3O_4 particles in the Alg—Ce magnetic bead was depicted by the presence of a novel absorption band at 561 cm^{-1} [32].

3.1.3. X-ray Diffraction (XRD)

The crystalline structure of the samples was examined through X-ray diffraction (XRD) (Figure 3). The crystal peaks of 2θ at 28.8°, 34.2°, 38.6°, 44.5°, 61.7° and 67.4° depict the crystallinity of synthesized SPIONs. The irregular peaks of alginate beads revealed that they were amorphous in nature. The encapsulation of SPIONs was revealed by its crystallite structure on alginate beads through the regular peaks at 9.04°, 13.78°, 38.93°, and 62.21°. The spinel structure of the SPIONs was depicted by the multiple peaks at 2θ of 18.25°, 30.06°, 35.63°, 43.48°, 53.78°, 57.33°, and 63.11° [33]. Lilhare et al. [34] reported that sharp peaks of calcium alginate entrapped on methionine—functionalized magnetic nanoparticles indicate that the particles are crystalline and tiny and correspond to the planes of a Fe_3O_4 face—centered cubic structure.

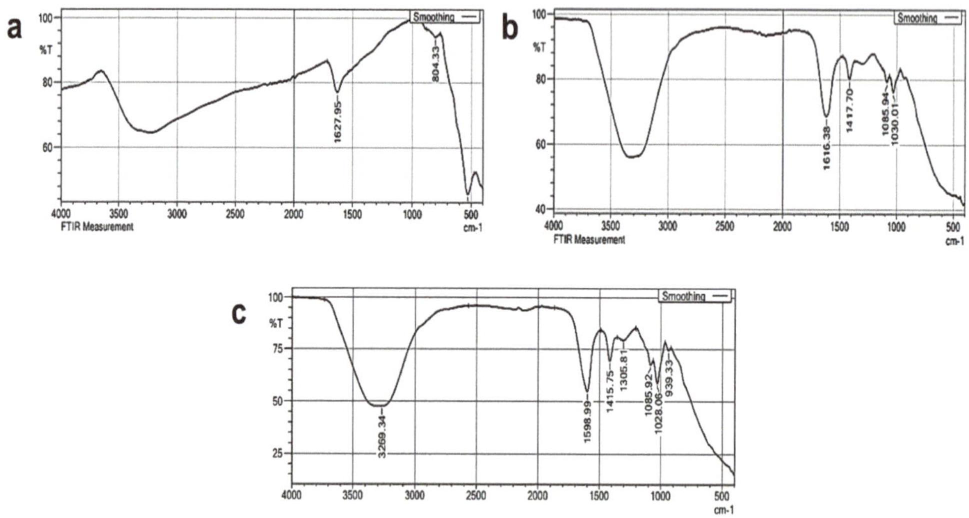

Figure 2. FTIR analysis of (**a**) SPIONs (**b**) Alginate beads (**c**) Alginate–SPIONs microspheres.

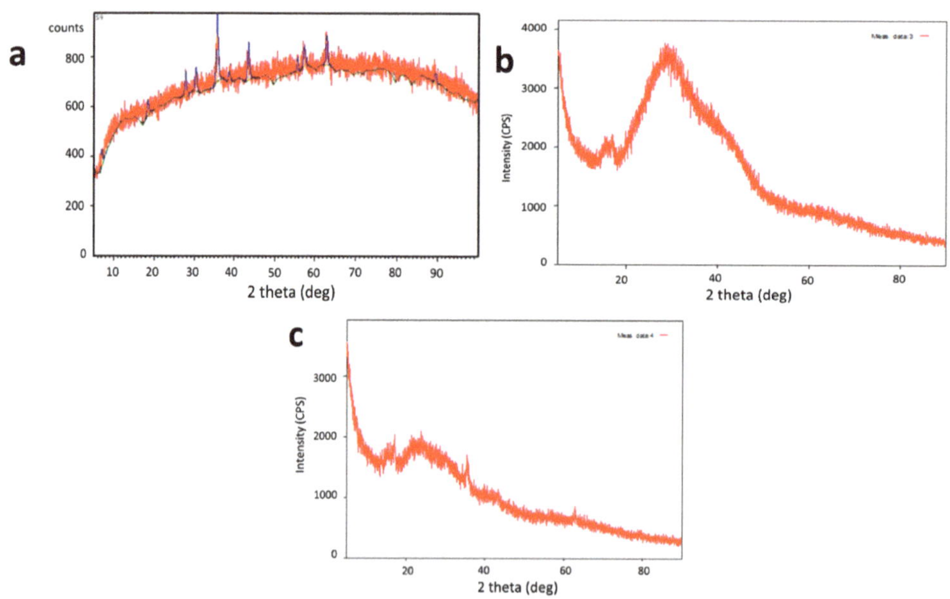

Figure 3. XRD analysis of (**a**) SPIONs (**b**) Alginate beads (**c**) Alginate–SPIONs microspheres.

3.1.4. Scanning Electron Microscopy and Energy Dispersive X-ray Analysis

The surface morphology and the elementary composition of the SPIONs were examined using Scanning Electron Microscopy (SEM) and Energy Dispersive X-ray Analysis (EDX) (Figure 4). The size of the synthesized SPIONs were 10–15 nm in size, and the major compounds like Fe, O and minor compounds such as Na and Cl were observed through EDX. Baykal et al. [35] reported sizes of SPIONs of around 12 nm. There are reports on SPIONs of size between 14.53 nm and 20.54 nm with the elemental composition of Fe and O [36].

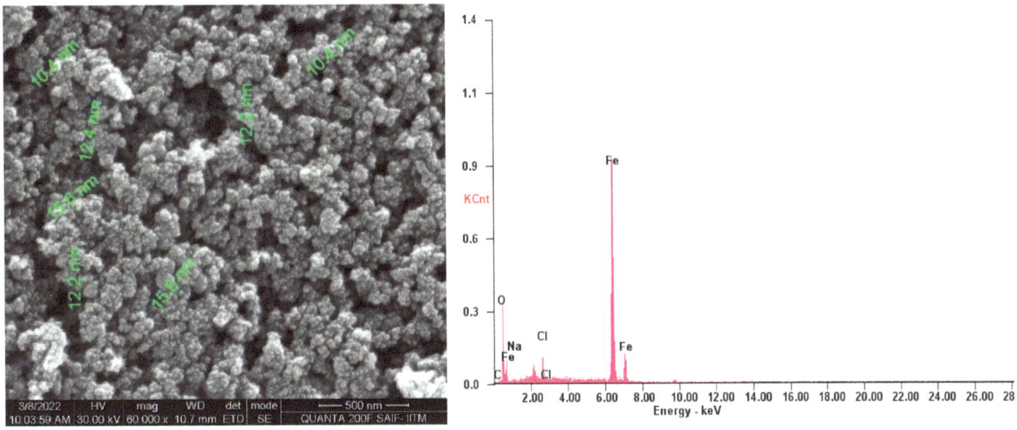

Figure 4. SEM and EDX analysis of SPIONs.

3.1.5. Zeta Potential Analysis

The charges of the SPIONs were observed using Zeta Potential analysis (Figure 5). The peaks of the analysis showed that the synthesized SPIONs carried negative charges. Unlike iron nanoparticles that exhibited a low surface charge, the synthesized SPIONs used in this study had a much higher Zeta Potential; Khatami et al. [37] also produced SPIONs using Stevia plant with a magnitude of—41.1 mV [37]. In contrast, Reczyńska et al. [38] also reported the charge of SPIONs to be 19.6 $\pm$ 0.8 mV.

3.1.6. Vibrating Sample Magnetometer

The magnetism of the SPIONs was examined using a vibrating sample magnetometer (VSM) (Figure 6). The hysteresis loop of the synthesized SPIONs showed the superparamagnetic nature of the SPIONs. Raees et al. [39] reported that saturation magnetization of SPIONs was found to be 71.82 emu/g, indicating a very high saturation magnetization and superparamagnetic behavior. The SPIONs produced in this study also showed a high saturation magnetization.

3.2. Batch Adsorption Studies

Batch adsorption experiments on the removal of chromium utilizing SPIONs, alginate beads, and alginate–SPIONs beads were studied using the DPC method. The conditions such as adsorbate concentration, adsorbent concentration, pH, and contact time were all optimized.

3.2.1. Removal of Chromium Using SPIONs

The optimized condition for the effective removal of chromium from water was studied (Figure 7). The maximum removal of chromium was observed at 6 ppm at a removal percentage of 88.19%. With the adsorbate concentration, the adsorbent concentration was optimized; 0.09 g/10 mL of SPIONs was observed to remove about 90.88%. At pH of 8, 6 ppm of chromium was removed efficiently, at 0.09 g/10 mL, with the increased removal of 1 h. The interaction time between the adsorbate and adsorbent plays a major role in the removal. The optimized contact time was examined to be 1 h with a removal efficiency of 93.8%. Farhana et al. [19] reported that SPIONs with impregnated activated carbon showed a removal percentage of nearly 99.7% with the optimized adsorbate concentration of 1 ppm and adsorbent concentration of 10 g/L. It was revealed that as the concentration of chromium increases, iron leaching was high, as there is no active site readily available for the adsorption reaction to take place. The electrostatic interaction between the adsorbent surface (SPIONs) and anion (chromates) is the driving force for the adsorption process [9,19].

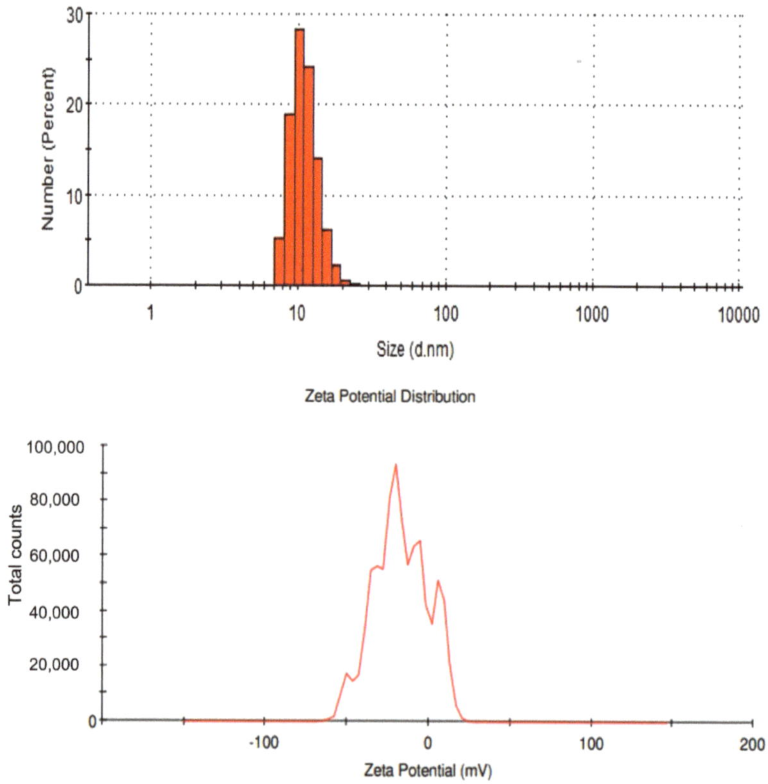

Figure 5. Zeta potential analysis of SPIONs.

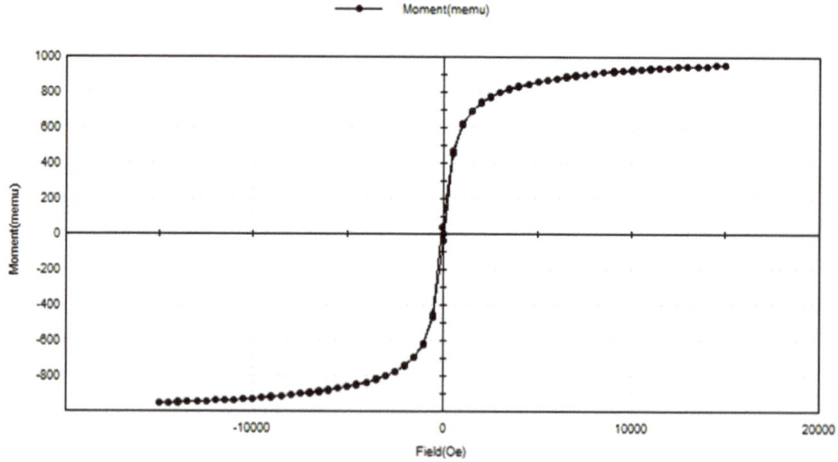

Figure 6. VSM analysis of SPIONs.

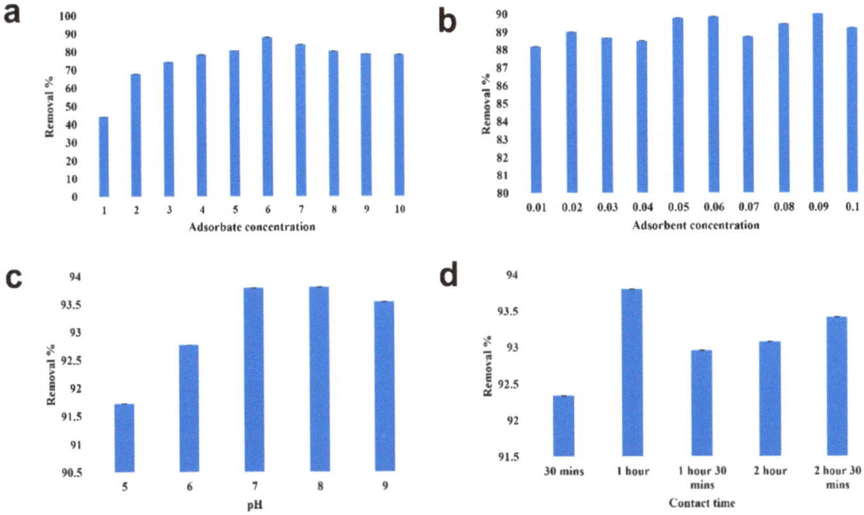

Figure 7. Batch adsorption study of SPIONs. (**a**) Optimization of adsorbate concentration, (**b**) Optimization of adsorbent concentration, (**c**) Optimization of pH, (**d**) Optimization of contact time.

3.2.2. Removal of Chromium Using Alginate Beads

The adsorbate concentration of 7 ppm was removed efficiently by alginate beads with a removal percentage of 81.61% (Figure 8). Under the optimized adsorbate concentration, the maximum removal of chromium was recorded at 0.07 g/10 mL of adsorbent concentration. Using 0.07 g/10 mL of alginate beads, 7 ppm of chromium was efficiently removed at pH 6 and the contact time was found to be 1 h. The overall removal percentage of chromium using alginate beads was recorded to be 90.98%. Babu et al. [40] reported that red mud doped in calcium alginate beads with hydrazine sulphate activation had the ability to remove 25 mg/L of lead with 100% efficiency and that the removal efficiency decreased with the increase in concentration. The adsorbent concentration of 2.0 g/100 mL removed nearly 91.5%. As lead is cationic in nature, pH 6 was found to be the optimized condition for the efficient removal of lead using alginate beads. As the pH increases, the removal percentage changes due to the imposition of a negative charge on the surface of the adsorbate. The optimized contact time was recorded to be 180 min and higher rate of adsorption was observed initially due to greater vacant site availability in the adsorbent.

3.2.3. Removal of Chromium Using Alginate–SPIONs Beads

The removal of chromium using alginate–SPIONs beads was found to be higher than that of SPIONs and alginate beads (Figure 9). 7 ppm of chromium were removed efficiently by the alginate–SPIONs beads with an efficiency of 85.03% due to the combined activity of alginate beads with that of SPIONs. The optimized adsorbent concentration was evaluated to be 0.08 g/10 mL with a removal percentage of 85.92%. The optimized pH and contact time under optimized adsorbate and adsorbent concentration was found to be 8 pH and 1 h, respectively. The maximum removal percentage of chromium by alginate–SPIONs beads was evaluated to be 94.84%. Idris et al. [41] revealed the magnetic alginate beds has the ability to removal heavy metals such as lead from water. The optimized adsorbent concentration was reported to be 50 mg/g with the adsorbate concentration of 400 mg/L. It was reported that the removal efficiency increased with the increase in pH until reaching neutral pH. The adsorption mechanism was observed to be of two stages: rapid adsorption on adsorbate—adsorbent contact, and lower adsorption until equilibrium was reached as the vacant spaces became fewer.

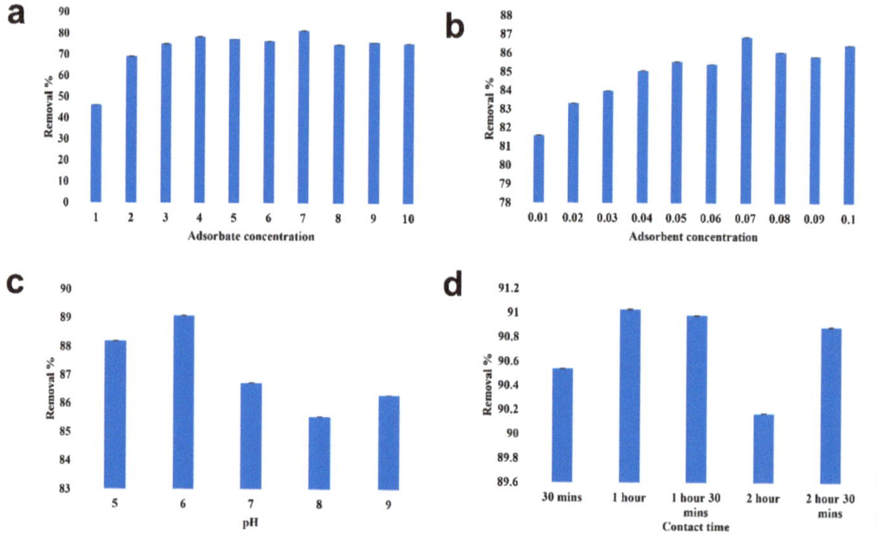

Figure 8. Batch adsorption study of alginate beads. (**a**) Optimization of adsorbate concentration, (**b**) Optimization of adsorbent concentration, (**c**) Optimization of pH, (**d**) Optimization of contact time.

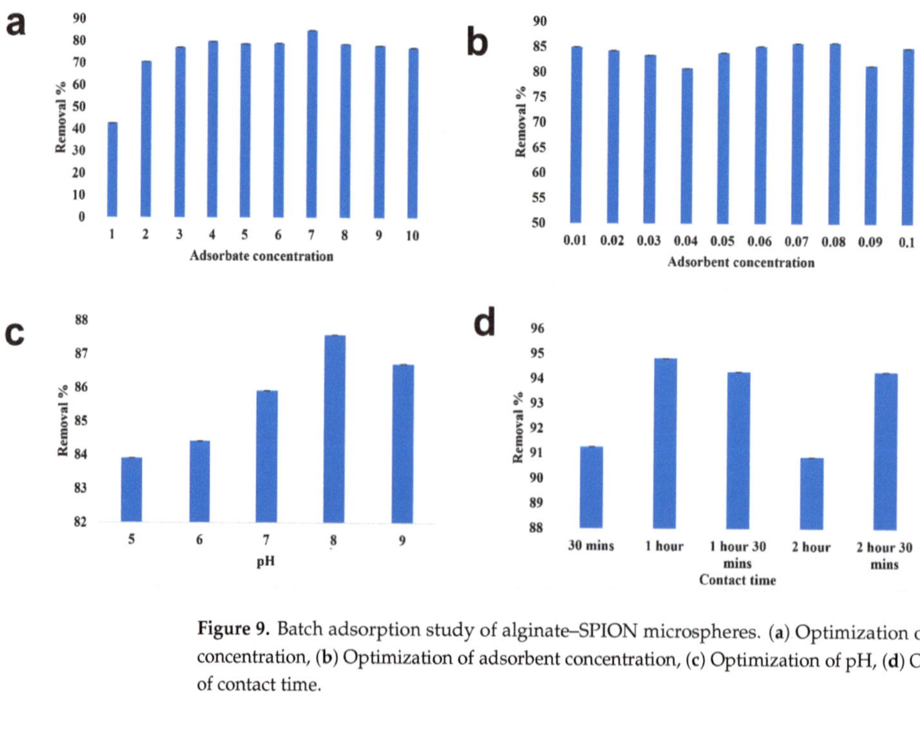

Figure 9. Batch adsorption study of alginate–SPION microspheres. (**a**) Optimization of adsorbate concentration, (**b**) Optimization of adsorbent concentration, (**c**) Optimization of pH, (**d**) Optimization of contact time.

3.2.4. Isotherm Studies

The pattern by which the adsorbent adsorbs the adsorbent is studied using four kinds of isotherm model: the Langmuir isotherm, the Freundlich isotherm, the Tempkin isotherm, and the D—R isotherm (Figures 10–12). The best—fit isotherm model was identified by the comparison of correlation coefficient (R^2) values of each isotherm (Highlighted in yellow). The adsorption of chromium using SPIONs and alginate–SPIONs beads was found to be ionic in nature, as it followed the Tempkin isotherm with an R^2 of 0.6003 and 0.7215. Homogeneous adsorption of chromium on alginate beads was recorded, as it followed the Langmuir isotherm with the highest R^2 value of 0.9784. The removal of chromium using SPIONs was found to follow the Freundlich isotherm, with the highest R^2 of 0.9228 [31]. Bilici et al. [42] reported that adsorption of oil contaminant from wastewater by calcium alginate beads functionalized with sodium dodecyl sulphate (SDS) was found to follow the Langmuir isotherm with a regression coefficient 0.9999. Further, the removal of lead using magnetic nano—sorbent was found to be heterogenous in nature, following the Freundlich isotherm [43].

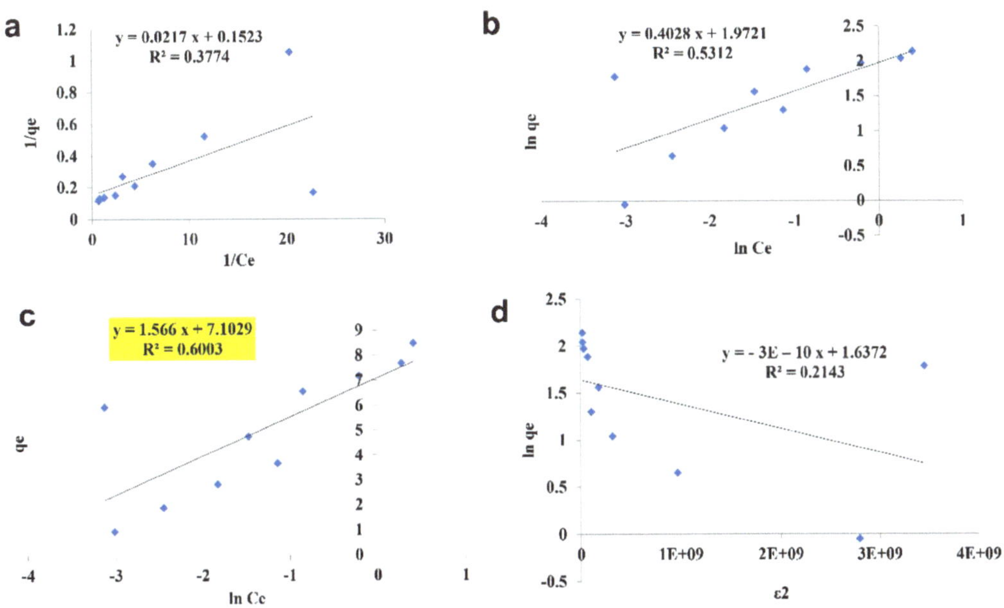

Figure 10. Isotherm study of SPIONs. (**a**) Langmuir Isotherm, (**b**) Freundlich Isotherm, (**c**) Tempkin Isotherm, (**d**) D—R Isotherm.

3.3. Toxicity Studies

3.3.1. Histology Studies

The damage caused by the exposure to the chromium and the adsorbents was evaluated by histology study. Histology images taken on the 5th day are shown in Figure 13a,b. On the 5th day of exposure, the direct interaction of fishes with the chromium water caused damage to the lens of the eyes, and complete distortion occurred in the vitreous region of the eyes. In the fishes exposed to the water containing SPIONs there were also major impacts. Huge accumulations of SPIONs in the pupil, iris and the lens region of the eyes were observed. Moreover, distortion of the lamellar epithelium in the gill region and the hepatic lobules of the liver were noted. However, less damage was observed in the SPIONs—treated chromium water, and adsorbent was removed. The damage in the retinal ganglion cell layer and deformation of the gill sacs and filaments were noted in Tests 5 and

7. In comparison to these tests, less damage to the fishes was observed in the chromium water treated with alginate beads/alginate–SPIONs beads, and these adsorbents were removed after 1 h.

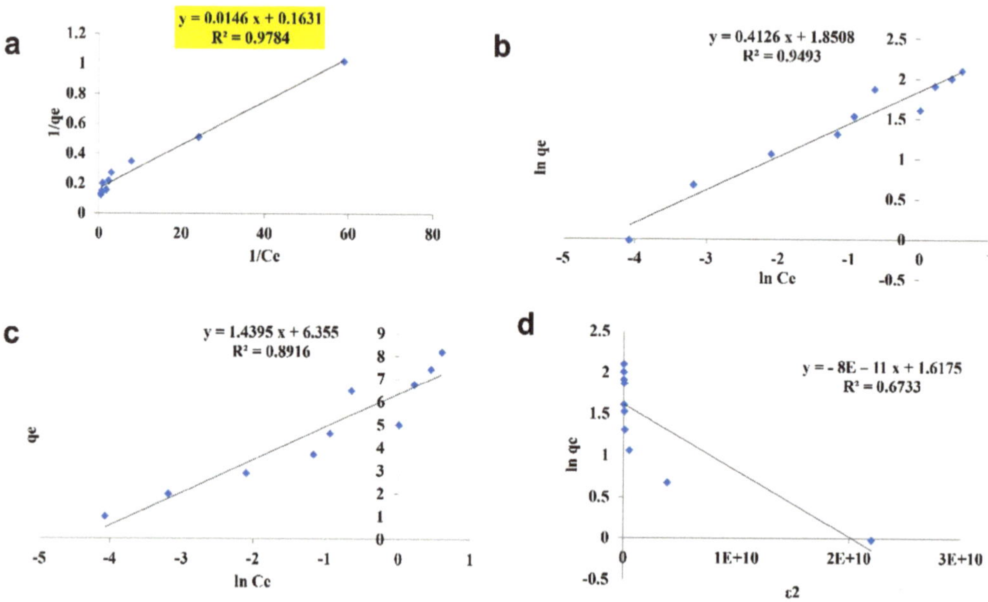

Figure 11. Isotherm study of alginate beads. (**a**) Langmuir Isotherm, (**b**) Freundlich Isotherm, (**c**) Tempkin Isotherm, (**d**) D—R Isotherm.

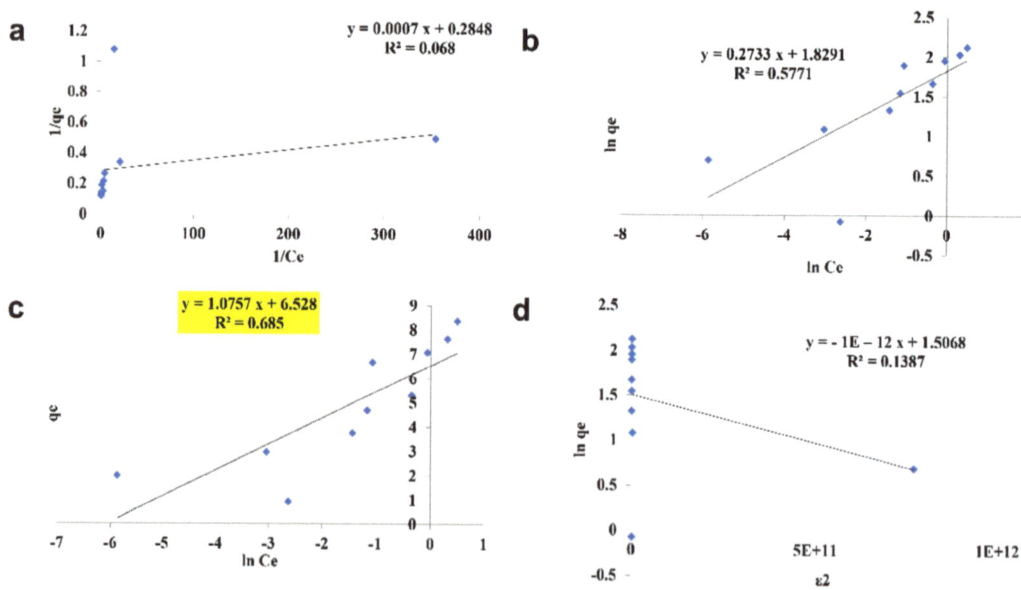

Figure 12. Isotherm study of alginate–SPIONs microspheres. (**a**) Langmuir Isotherm, (**b**) Freundlich Isotherm, (**c**) Tempkin Isotherm, (**d**) D—R Isotherm.

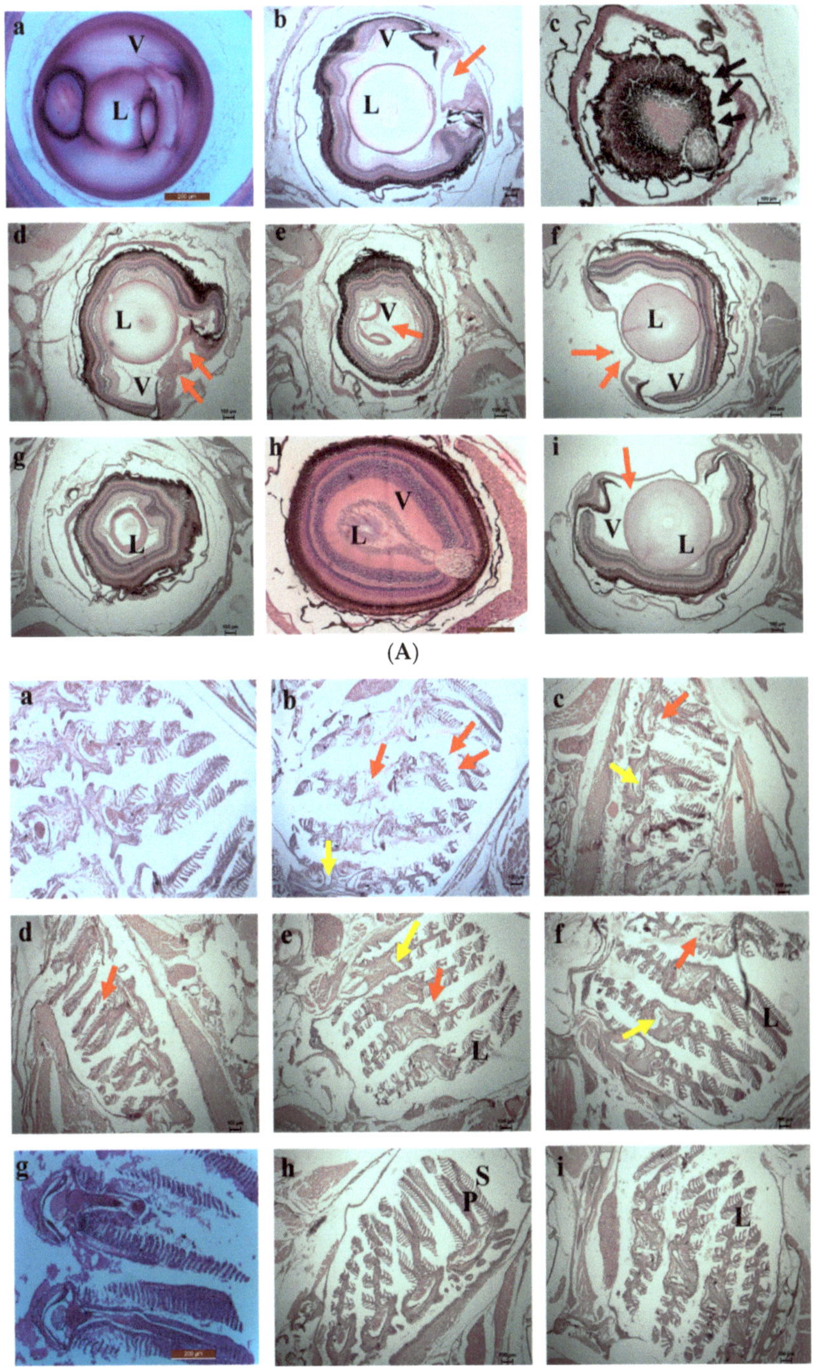

Figure 13. (**A**) 5th day. Histology images of eyes of zebrafish exposed to (**a**) Control (**b**) **Test 1**—Chromium metal in water (**c**) **Test 2**—SPIONs in water (**d**) **Test 3**—Chromium metal with wate in

SPIONs (**e**) **Test 4**—Chromium metal in water with SPIONs removed after interaction for 1 h (**f**) **Test 5**—Chromium metal in water with alginate–SPIONs microspheres (**g**) **Test 6**—Chromium metal in water with alginate–SPIONs microspheres removed after interaction for 1 h (**h**) **Test 7**—Chromium metal in water with alginate beads (**i**) **Test 8**—Chromium metal in water with alginate beads removed after interaction for 1 h. (Red arrow denotes erosion; L—lens, V—Vitreous). (**B**) 5th day. Histology images of gills of zebrafish exposed to (**a**) Control (**b**) **Test 1**—Chromium metal in water (**c**) **Test 2**—SPIONs in water (**d**) **Test 3**—Chromium metal with water in SPIONs (**e**) **Test 4**—Chromium metal in water with SPIONs removed after interaction for 1 h (**f**) **Test 5**—Chromium metal in water with alginate–SPIONs microspheres (**g**) **Test 6**—Chromium metal in water with alginate–SPIONs microspheres removed after interaction for 1 h (**h**) **Test 7**—Chromium metal in water with alginate beads (**i**) **Test 8**—Chromium metal in water with alginate beads removed after interaction for 1 h. (Red arrow denotes erosion, yellow arrow denotes the intestinal space; Black arrow denotes deposition (P—Primary Lamella, S—Secondary Lamella, L—Lamella).

On the 10th day of exposure, Test 1 showed increased damage in the eyes and gill regions (Figure 14a,b). Higher accumulation of SPIONs was observed in the lens of the eyes. The degradation of the lamella and the secondary lamella of the gills were observed in fishes exposed to chromium water treated with SPIONs (and SPIONs not removed). The fishes exposed to chromium water treated with alginate—SPIONs and then removed showed no damage in the eye lens, and lesser damage to the gill filament were recorded. Aldavood et al. [44] reported that the exposure of zebrafish embryos to heavy metals such as cadmium caused a decrease in myosin protein expression and affected the function and development of muscle. A study which investigated the removal efficiency of chromium using graphene oxide revealed damage in the skeletal muscles, misfolding of proteins and oxidative stress in zebrafishes and its embryos [45]. It was revealed that magnetite—based nanocomposites imposed a lower toxicity on zebrafish with less variation in shape, color and behavior [46,47].

3.3.2. Inductively Coupled Plasma—Mass Spectrometry (ICP—MS)

The accumulation of chromium and the iron in fish was evaluated using ICP—MS analysis (Figures 15 and 16). From the analysis, the highest accumulation of chromium was observed only in Test 1 at a concentration of 0.96–0.97 mg/kg, then in Test 3, whereas fishes in other tests were found to have negligible concentrations of chromium accumulation. The accumulation of iron was also estimated, and it was recorded in fishes in Tests 3 to 7. The highest concentration, of 0.71 mg/kg, was recorded in the fish grown in Test 3 and may have arisen from dispersion of SPIONs from alginate beads.

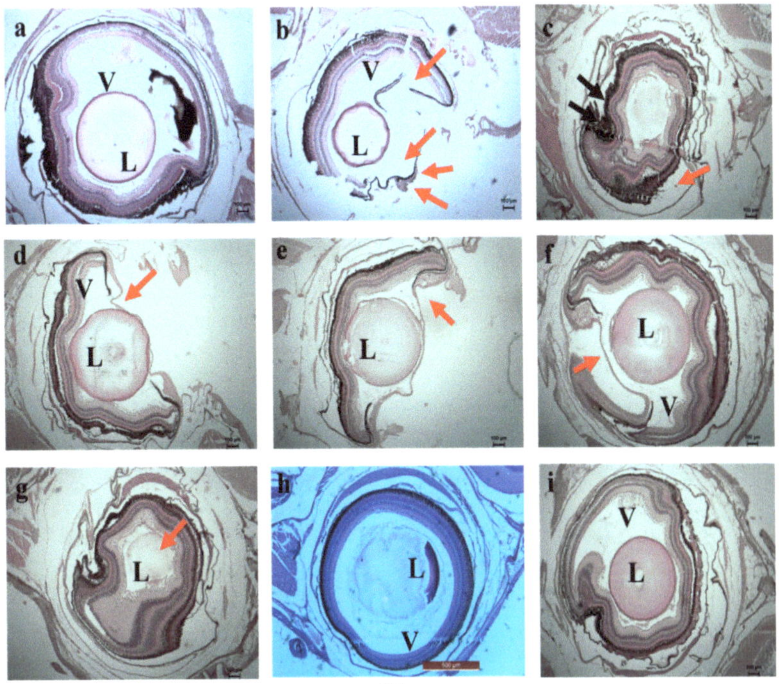

(A)

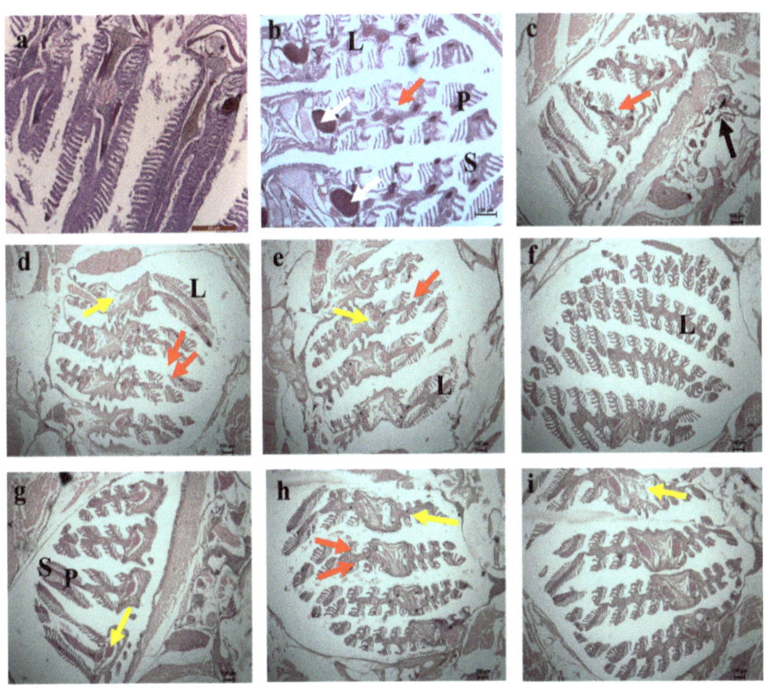

(B)

Figure 14. (**A**) 10th day. Histology images of eye of zebrafish exposed to (**a**) Control (**b**) **Test 1**—Chromium metal in water (**c**) **Test 2**—SPIONs in water (**d**) **Test 3**—Chromium metal with water in SPIONs (**e**) **Test 4**—Chromium metal in water with SPIONs removed after interaction for 1 h (**f**) **Test 5**—Chromium metal in water with alginate–SPIONs microspheres (**g**) **Test 6**—Chromium metal in water with alginate–SPIONs microspheres removed after interaction for 1 h (**h**) **Test 7**—Chromium metal in water with alginate beads (**i**) **Test 8**—Chromium metal in water with alginate beads removed after interaction for 1 h. (Red arrow denotes erosion; L—Lens, V—Vitreous). (**B**) 10th day. Histology images of gills of zebrafish exposed to (**a**) Control (**b**) **Test 1**—Chromium metal in water (**c**) **Test 2**—SPIONs in water (**d**) **Test 3**—Chromium metal with water in SPIONs (**e**) **Test 4**—Chromium metal in water with SPIONs removed after interaction for 1 h (**f**) **Test 5**—Chromium metal in water with alginate–SPIONs microspheres (**g**) **Test 6**—Chromium metal in water with alginate–SPIONs microspheres removed after interaction for 1 h (**h**) **Test 7**—Chromium metal in water with alginate beads (**i**) **Test 8**—Chromium metal in water with alginate beads removed after interaction for 1 h. (Red arrow denotes erosion, Yellow arrow denotes the intestinal space; Black arrow denotes deposition; White arrow denotes Lipofuscin (P—Primary Lamella, S—Secondary Lamella, L—Lamella).

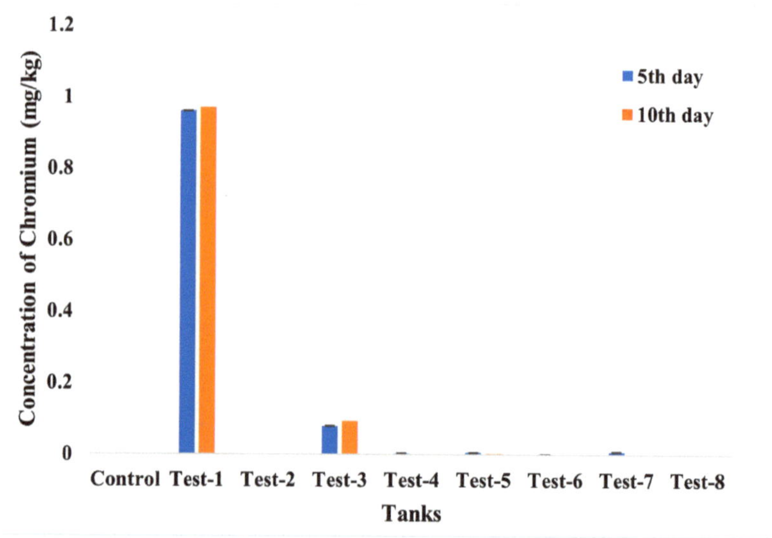

Figure 15. ICP—MS analysis for chromium accumulation analysis in control; **Test 1**—Chromium water; **Test 2**—SPIONs in water; **Test 3**—Chromium water with SPIONs; **Test 4**—Chromium water treated with SPIONs; **Test 5**—Chromium water with alginate–SPIONs microspheres; **Test 6**—Chromium water treated with alginate–SPIONs microspheres; **Test 7**—Chromium water with alginate beads; **Test 8**—Chromium water treated with alginate beads.

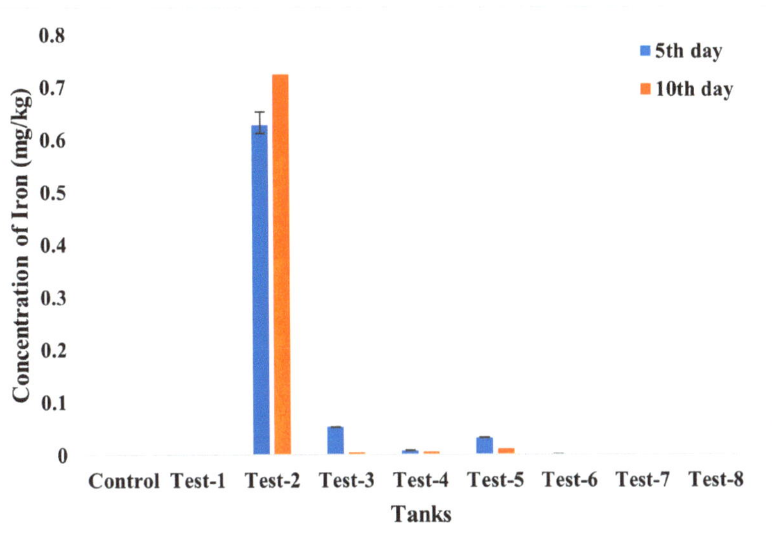

Figure 16. ICP—MS analysis for iron accumulation analysis in control; **Test 1**—Chromium water; **Test 2**—SPIONs in water; **Test 3**—Chromium water with SPIONs; **Test 4**—Chromium water treated with SPIONs; **Test 5**—Chromium water with alginate–SPIONs microspheres; **Test 6**—Chromium water treated with alginate–SPIONs microspheres; **Test 7**—Chromium water with alginate beads; **Test 8**—Chromium water treated with alginate beads.

4. Conclusions

SPIONs, alginate beads and alginate–SPIONs beads were successfully synthesized, characterized and evaluated for the removal of chromium from contaminated water. It was concluded that the removal of chromium by alginate–SPIONs beads is higher than that of SPIONs and alginate beads with the optimized adsorbate and adsorbent concentration of 7 ppm and 0.08 g/10 mL, respectively, at pH 8 with a contact time of 1 h. The toxicity study also revealed that chromium water treated with the alginate–SPIONs beads (where the adsorbents were removed) imposed less damage on zebrafish (*Danio rerio*), and therefore it may be one of the better and lower—cost biosorbents for removing heavy metals. The presence of absorbents was found to cause lethal impacts to the animals; therefore, it is suggested not to allow the adsorbents into the treatment system following the treatment process.

Author Contributions: A.V.S.: Conceptualization, Methodology, Writing—Original Draft, Editing, Visualization, Supervision; M.B.: Writing—Original Draft, Editing; S.K.S.: Methodology, Writing—Original Draft, Editing, Visualization; M.S.: Methodology, Writing—Original Draft, Editing, Visualization; J.V.: Writing—Original Draft, Editing; N.S.: Writing—Original Draft, Editing, Visualization; R.T.: Writing—Original Draft, Editing, Visualization; C.K.: Writing—Original Draft, Editing, Visualization; S.W.: Writing—Original Draft, Editing; D.R.: Writing—Original Draft, Editing, L.X.N.R.P.: Writing—Original Draft, Editing; R.S.S.P.: Writing—Original Draft, Editing. All authors have read and agreed to the published version of the manuscript.

Funding: This research received no external funding.

Institutional Review Board Statement: Not applicable.

Informed Consent Statement: Not applicable.

Data Availability Statement: The data used to support the findings of this study are included in the article. Should further data or information be required, these are available from the corresponding author upon request.

Acknowledgments: The authors thank the Sathyabama Institute of Science and Technology, India and MAHSA University, Malaysia for providing facilities and support to complete this research work.

Conflicts of Interest: The authors declare that there are no conflicts of interest regarding the publication of this paper.

References

1. Zhang, Y.; Wu, B.; Xu, H.; Liu, H.; Wang, M.; He, Y.; Pan, B. Nanomaterials-enabled water and wastewater treatment. *NanoImpact* **2016**, *3–4*, 22–39. [CrossRef]
2. Singh, U.; Singh, S.; Tiwari, R.K.; Pandey, R.S. Water Pollution due to Discharge of Industrial Effluents with special reference to Uttar Pradesh, India–A review. *Int. Arch. Appl. Sci. Technol.* **2018**, *9*, 111–121.
3. Salman, S.; Znad, H.; Hasan, N.; Hasan, M. Optimization of innovative composite sensor for Pb(II) detection and capturing from water samples. *Microchem. J.* **2021**, *160*, 105765. [CrossRef]
4. Shahat, A.; Kubra, K.T.; Salman, S.; Hasan, N.; Hasan, M. Novel solid-state sensor material for efficient cadmium(II) detection and capturing from wastewater. *Microchem. J.* **2021**, *164*, 105967. [CrossRef]
5. Adeleye, A.S.; Conway, J.R.; Garner, K.; Huang, Y.; Su, Y.; Keller, A.A. Engineered nanomaterials for water treatment and remediation: Costs, benefits, and applicability. *Chem. Eng. J.* **2016**, *286*, 640–662. [CrossRef]
6. El-Sayed, M.E. Nanoadsorbents for water and wastewater remediation. *Sci. Total. Environ.* **2020**, *739*, 139903. [CrossRef]
7. Motshekga, S.C.; Ray, S.S.; Maity, A. Synthesis and characterization of alginate beads encapsulated zinc oxide nanoparticles for bacteria disinfection in water. *J. Colloid Interface Sci.* **2018**, *512*, 686–692. [CrossRef]
8. Bée, A.; Talbot, D.; Abramson, S.; Dupuis, V. Magnetic alginate beads for Pb(II) ions removal from wastewater. *J. Colloid Interface Sci.* **2011**, *362*, 486–492. [CrossRef]
9. Samrot, A.V.; Shobana, N.; Sruthi, D.P.; Sahithya, C.S. Utilization of chitosan-coated superparamagnetic iron oxide nanoparticles for chromium removal. *Appl. Water Sci.* **2018**, *8*, 192. [CrossRef]
10. Chellapan, J.; Samrot, A.V.; Annamalai, A.A.; Bhattacharya, R.K.; Sathiyamoorthy, P.; Sahithya, C.S. Biopolymer Coated Coreshell Magnetite Nanoparticles for Rifampicin Delivery. *Orient. J. Chem.* **2018**, *34*, 2389–2396. [CrossRef]
11. Covarrubias, S.A.; De-Bashan, L.E.; Moreno, M.; Bashan, Y. Alginate beads provide a beneficial physical barrier against native microorganisms in wastewater treated with immobilized bacteria and microalgae. *Appl. Microbiol. Biotechnol.* **2011**, *93*, 2669–2680. [CrossRef] [PubMed]
12. Ferreira, I.L.D.M.; Bittencourt, R.F.; Júnior, C.S. Nanomagnetic Polymeric Absorbent Based on Alginate and Gamma-Maghemite Synthesized in Situ for Wastewater Treatment from Metallurgical Industry. In *Properties and Applications of Alginates*; Deniz, I., Imamoglu, E., Keskin-Gundogdu, T., Eds.; IntechOpen: London, UK, 2021. [CrossRef]
13. Asadi, S.; Eris, S.; Azizian, S. Alginate-Based Hydrogel Beads as a Biocompatible and Efficient Adsorbent for Dye Removal from Aqueous Solutions. *ACS Omega* **2018**, *3*, 15140–15148. [CrossRef] [PubMed]
14. Samrot, A.V.; Sahithya, C.S.; Selvarani, A.J.; Purayil, S.K.; Ponnaiah, P. A review on synthesis, characterization and potential biological applications of superparamagnetic iron oxide nanoparticles. *Curr. Res. Green Sustain. Chem.* **2021**, *4*, 100042. [CrossRef]
15. Samrot, A.V.; Sai Bhavya, K.; Sruthi, P.D.; Paulraj, P. Synthesis of SPIONs to deliver drug *in-vitro* and to use as contrasting agent. *Int. J. Adv. Res. Eng. Technol. (IJARET)* **2020**, *11*, 200–208.
16. Yaqoob, A.A.; Parveen, T.; Umar, K.; Mohamad Ibrahim, M.N. Role of nanomaterials in the treatment of wastewater: A review. *Water* **2020**, *12*, 495. [CrossRef]
17. Aziz, F.; El Achaby, M.; Lissaneddine, A.; Aziz, K.; Ouazzani, N.; Mamouni, R.; Mandi, L. Composites with alginate beads: A novel design of nano-adsorbents impregnation for large-scale continuous flow wastewater treatment pilots. *Saudi J. Biol. Sci.* **2020**, *27*, 2499–2508. [CrossRef] [PubMed]
18. Mudhoo, A.; Sillanpää, M. Magnetic nanoadsorbents for micropollutant removal in real water treatment: A review. *Environ. Chem. Lett.* **2021**, *19*, 4393–4413. [CrossRef]
19. Farhana, A.; Selvarani, A.J.; Samrot, A.V.; Alsrhani, A.; Raji, P.; Sahithya, C.S.; Cypriyana, P.J.J.; Senthilkumar, P.; Ling, M.P.; Yishak, S. Utilization of Superparamagnetic Iron Oxide Nanoparticles (SPIONs) Impregnated Activated Carbon for Removal of Hexavalent Chromium. *J. Nanomater.* **2022**, *2022*, 4326939. [CrossRef]
20. Samrot, A.; Bhavya, K.S.; Angalene, J.L.A.; Roshini, S.; Preethi, R.; Steffi, S.; Raji, P.; Kumar, S.S. Utilization of gum polysaccharide of *Araucaria heterophylla* and *Azadirachta indica* for encapsulation of cyfluthrin loaded super paramagnetic iron oxide nanoparticles for mosquito larvicidal activity. *Int. J. Biol. Macromol.* **2020**, *153*, 1024–1034. [CrossRef]
21. Samrot, A.V.; Saigeetha, S.; Mun, C.Y.; Abirami, S.; Purohit, K.; Cypriyana, P.J.J.; Dhas, T.S.; Inbathamizh, L.; Kumar, S.S. Utilization of *Carica papaya* latex on coating of SPIONs for dye removal and drug delivery. *Sci. Rep.* **2021**, *11*, 24511. [CrossRef]
22. Justin, C.; Samrot, A.V.; Sruthi, P.D.; Sahithya, C.S.; Bhavya, K.S.; Saipriya, C. Preparation, characterization and utilization of coreshell super paramagnetic iron oxide nanoparticles for curcumin delivery. *PLoS ONE* **2018**, *13*, e0200440. [CrossRef]

23. Singh, H.; Du, J.; Singh, P.; Mavlonov, G.T.; Yi, T.H. Development of superparamagnetic iron oxide nanoparticles via direct conjugation with ginsenosides and its *in-vitro* study. *J. Photochem. Photobiol. B Biol.* **2018**, *185*, 100–110. [CrossRef]
24. Alpdemir, Ş.; Vural, T.; Kara, G.; Bayram, C.; Haberal, E.; Denkbaş, E.B. Magnetically responsive, sorafenib loaded alginate microspheres for hepatocellular carcinoma treatment. *IET Nanobiotechnol.* **2020**, *14*, 617–622. [CrossRef]
25. Lace, A.; Ryan, D.; Bowkett, M.; Cleary, J. Chromium Monitoring in Water by Colorimetry Using Optimised 1,5-Diphenylcarbazide Method. *Int. J. Environ. Res. Public Health* **2019**, *16*, 1803. [CrossRef]
26. Samrot, A.V.; Sahithya, C.S.; Selvarani, A.J.; Pachiyappan, S.; Suresh Kumar, S. Surface-Engineered Super-Paramagnetic Iron Oxide Nanoparticles for Chromium Removal. *Int. J. Nanomed.* **2019**, *14*, 8105–8119. [CrossRef]
27. Avdesh, A.; Chen, M.; Martin-Iverson, M.T.; Mondal, A.; Ong, D.; Rainey-Smith, S.; Taddei, K.; Lardelli, M.; Groth, D.M.; Verdile, G.; et al. Regular Care and Maintenance of a Zebrafish (*Danio rerio*) Laboratory: An Introduction. *JoVE J. Vis. Exp.* **2012**, *69*, e4196. [CrossRef]
28. Lam, W.-H.; Chong, M.N.; Horri, B.A.; Tey, B.-T.; Chan, E.-S. Physicochemical stability of calcium alginate beads immobilizing TiO_2 nanoparticles for removal of cationic dye under UV irradiation. *J. Appl. Polym. Sci.* **2017**, *134*, 45002. [CrossRef]
29. Justin, C.; Philip, S.A.; Samrot, A.V. Synthesis and characterization of superparamagnetic iron-oxide nanoparticles (SPIONs) and utilization of SPIONs in X-ray imaging. *Appl. Nanosci.* **2017**, *7*, 463–475. [CrossRef]
30. Idris, M.I.; Zaloga, J.; Detsch, R.; Roether, J.A.; Unterweger, H.; Alexiou, C.; Boccaccini, A.R. Surface Modification of SPIONs in PHBV Microspheres for Biomedical Applications. *Sci. Rep.* **2018**, *8*, 7286. [CrossRef] [PubMed]
31. Matos, J.C.; Gonçalves, M.C.; Pereira, L.C.J.; Vieira, B.J.C.; Waerenborgh, J.C. SPIONs Prepared in Air through Improved Synthesis Methodology: The Influence of γ-Fe_2O_3/Fe_3O_4 Ratio and Coating Composition on Magnetic Properties. *Nanomaterials* **2019**, *9*, 943. [CrossRef]
32. Gopalakannan, V.; Viswanathan, N. Synthesis of magnetic alginate hybrid beads for efficient chromium (VI) removal. *Int. J. Biol. Macromol.* **2015**, *72*, 862–867. [CrossRef] [PubMed]
33. Khalkhali, M.; Rostamizadeh, K.; Sadighian, S.; Khoeini, F.; Naghibi, M.; Hamidi, M. The impact of polymer coatings on magnetite nanoparticles performance as MRI contrast agents: A comparative study. *DARU J. Pharm. Sci.* **2015**, *23*, 45. [CrossRef] [PubMed]
34. Lilhare, S.; Mathew, S.; Singh, A.; Carabineiro, S. Calcium Alginate Beads with Entrapped Iron Oxide Magnetic Nanoparticles Functionalized with Methionine—A Versatile Adsorbent for Arsenic Removal. *Nanomaterials* **2021**, *11*, 1345. [CrossRef] [PubMed]
35. Baykal, A.; Amir; Günerb, S.; Sözeri, H. Preparation and characterization of SPION functionalized via caffeic acid. *J. Magn. Magn. Mater.* **2015**, *395*, 199–204. [CrossRef]
36. Samrot, A.V.; Saipriya, C.; Selvarani, A.J.; Subbu, R.V.; Cypriyana, P.J.; Lavanya, Y.; Shehanaz, A.R.; Soundarya, P.; Priyanka, S.R.B.; Sangeetha, P.; et al. A study on influence of superparamagnetic iron oxide nanoparticles (SPIONs) on green gram (*Vigna radiata* L.) and earthworm (*Eudrilus eugeniae* L.). *Mater. Res. Express* **2020**, *7*, 055002. [CrossRef]
37. Khatami, M.; Alijani, H.Q.; Fakheri, B.; Mobasseri, M.M.; Heydarpour, M.; Farahani, Z.K.; Khan, A.U. Super-paramagnetic iron oxide nanoparticles (SPIONs): Greener synthesis using Stevia plant and evaluation of its antioxidant properties. *J. Clean. Prod.* **2018**, *208*, 1171–1177. [CrossRef]
38. Reczyńska, K.; Marszałek, M.; Zarzycki, A.; Reczyński, W.; Kornaus, K.; Pamuła, E.; Chrzanowski, W. Superparamagnetic Iron Oxide Nanoparticles Modified with Silica Layers as Potential Agents for Lung Cancer Treatment. *Nanomaterials* **2020**, *10*, 1076. [CrossRef]
39. Raees, K.; Ansari, M.S.; Rafiquee, M.Z.A. Inhibitive effect of super paramagnetic iron oxide nanoparticles on the alkaline hydrolysis of procaine. *J. Nanostructure Chem.* **2019**, *9*, 175–187. [CrossRef]
40. Babu, A.N.; Mohan, G.V.K.; Kalpana, K.; Ravindhranath, K. Removal of Lead from Water Using Calcium Alginate Beads Doped with Hydrazine Sulphate-Activated Red Mud as Adsorbent. *J. Anal. Methods Chem.* **2017**, *2017*, 4650594. [CrossRef]
41. Idris, A.; Ismail, N.S.M.; Hassan, N.; Misran, E.; Ngomsik, A.-F. Synthesis of magnetic alginate beads based on maghemite nanoparticles for Pb(II) removal in aqueous solution. *J. Ind. Eng. Chem.* **2012**, *18*, 1582–1589. [CrossRef]
42. Bilici, Z.; Ozay, Y.; Unal, B.O.; Dizge, N. Investigation of the usage potential of calcium alginate beads functionalized with sodium dodecyl sulfate for wastewater treatment contaminated with waste motor oil. *Water Environ. Res.* **2021**, *93*, 2623–2636. [CrossRef] [PubMed]
43. Khani, R.; Sobhani, S.; Beyki, M.H. Highly selective and efficient removal of lead with magnetic nano-adsorbent: Multivariate optimization, isotherm and thermodynamic studies. *J. Colloid Interface Sci.* **2016**, *466*, 198–205. [CrossRef] [PubMed]
44. Aldavood, S.J.; Abbott, L.C.; Evans, Z.R.; Griffin, D.J.; Lee, M.D.; Quintero-Arevalo, N.M.; Villalobos, A.R.; Quintero, N. Effect of Cadmium and Nickel Exposure on Early Development in Zebrafish (*Danio rerio*) Embryos. *Water* **2020**, *12*, 3005. [CrossRef]
45. Chen, Y.; Li, J.; Zhou, Q.; Liu, Z.; Li, Q. Hexavalent chromium amplifies the developmental toxicity of graphene oxide during zebrafish embryogenesis. *Ecotoxicol. Environ. Saf.* **2021**, *208*, 111487. [CrossRef] [PubMed]
46. Guillén, A.; Ardila, Y.; Noguera, M.J.; Campaña, A.L.; Bejarano, M.; Akle, V.; Osma, J.F. Toxicity of Modified Magnetite-Based Nanocomposites Used for Wastewater Treatment and Evaluated on Zebrafish (*Danio rerio*) Model. *Nanomaterials* **2022**, *12*, 489. [CrossRef]
47. Samrot, A.V.; Shobana, N.; Sathiyasree, M.; Thirugnanasambandam, R.; Visvanathan, S.; Mohanty, B.K.; Sabesan, G.S.; Dhiva, S. Toxicity evaluation of SPIONs on *Danio rerio* embryonic development. *Mater. Today Proc.* **2022**, *59*, 1555–1560. [CrossRef]

toxics

Article

Potential Application of Discarded Natural Coal Gangue for the Removal of Tetracycline Hydrochloride (TC) from an Aqueous Solution

Hongyou Wan [1,2], Chen Wang [1], Lin Gong [1], Xinfeng Zhu [3], Jingwei Yan [1], Jiajia Lu [1] and Wei Zhang [1,2,3,4,5,6,7,*]

1 School of Ecology and Environment, Zhengzhou University, Zhengzhou 450001, China
2 Research Centre of Engineering and Technology for Synergetic Control of Environmental Pollution and Carbon Emissions of Henan Province, Zhengzhou 450001, China
3 Henan Key Laboratory of Water Pollution Control and Rehabilitation Technology, Pingdingshan 467036, China
4 Zhengzhou Key Laboratory of Water Resource and Environment, Zhengzhou 450001, China
5 Yellow River Institute for Ecological Protection and Regional Coordination Development, Zhengzhou University, Zhengzhou 450001, China
6 Henan International Joint Laboratory of Water Cycle Simulation and Environmental Protection, Zhengzhou 450001, China
7 Henan Key Laboratory of Water Resources Conservation and Intensive Utilization in the Yellow River Basin, Zhengzhou 450001, China
* Correspondence: zhangwei88@zzu.edu.cn

Abstract: The generation and accumulation of discarded coal gangue (CG) have severe environmental impacts. CG can adsorb other pollutants in the aquatic environment. However, previous studies have not assessed whether CG can adsorb the emerging contaminant tetracycline hydrochloride (TC). Here, discarded CG taken from a mine was pretreated by crushing, cleaning, and sieving and subsequently applied to the adsorption of TC. The adsorption studies were carried out by batch equilibrium adsorption experiments. Our findings indicated that the adsorption behavior could be accurately described using the quasi-first order kinetic and Langmuir adsorption isotherm models, indicating that monolayer adsorption was the main mechanism mediating the interaction between CG and TC. The adsorption process was classified as a thermodynamic endothermic and spontaneous reaction, which was controlled by chemical and physical adsorption, including electrostatic interaction and cation exchange. The pH of the solution had a great influence on the TC adsorption capacity of GC, with higher adsorption occurring in acidic environments compared to alkaline environments. This was attributed to the changes in CG Zeta potential and TC pKa at different pH conditions. Collectively, our findings demonstrated the potential applicability of discarded CG for the adsorption of TC and provided insights into the adsorption mechanisms.

Keywords: natural coal gangue; spent coal-associated waste; removal; tetracycline hydrochloride; antibiotics pollutant; water environment

Citation: Wan, H.; Wang, C.; Gong, L.; Zhu, X.; Yan, J.; Lu, J.; Zhang, W. Potential Application of Discarded Natural Coal Gangue for the Removal of Tetracycline Hydrochloride (TC) from an Aqueous Solution. *Toxics* **2023**, *11*, 20. https://doi.org/10.3390/toxics11010020

Academic Editors: Yuezhou Wei, Yan Wu and Xinpeng Wang

Received: 11 November 2022
Revised: 18 December 2022
Accepted: 20 December 2022
Published: 26 December 2022

Copyright: © 2022 by the authors. Licensee MDPI, Basel, Switzerland. This article is an open access article distributed under the terms and conditions of the Creative Commons Attribution (CC BY) license (https://creativecommons.org/licenses/by/4.0/).

1. Introduction

The tetracycline hydrochloride (TC) is a kind of broad-spectrum antibiotic that has been widely used to treat infections in both humans and animals in recent years [1,2].

Expired TC drugs are routinely discarded into the environment and the metabolites of this antibiotic are commonly detected in sewage and wastewater due to its extensive use for medical treatment. Therefore, the levels of environmental TC have been steadily increasing [3]. The high levels of TC in aquatic environments have already been linked to adverse outcomes, including the dissemination of genes associated with bacterial antibiotic resistance, in addition to ecological damage [4]. Furthermore, the increasing occurrence of

TC and related emerging pollutants in aquatic environments poses a challenge to municipal wastewater treatment [5,6].

Several methods are commonly used to remove antibiotic pollution from wastewater such as biological processing technology [7], catalytic degradation [8], photocatalytic degradation [9], and adsorption [10,11]. Among these, adsorption is a promising alternative for the removal of emerging pollutants from water environments due to its effectiveness, low cost [12,13], and ease of implementation, in addition to producing fewer by-products than other methods. Several recent studies have assessed the applicability of many kinds of adsorbents for the removal of TC or related emerging pollutants from aqueous solutions, including ferrihydrite and goethite [14], carbon materials [15], biochar [16], and carbon nanotubes [17]. The main adsorption mechanisms of antibiotic pollutants in water environments are physical adsorption, chemical adsorption, and electrostatic interaction [18–20]. As previously reported, modified attapulgite could efficiently remove TC from water environment [21]. Bentonite could also efficiently remove ciprofloxacin hydrochloride (CIP) from aqueous solutions [22]. The potential applications of CG in the removal of traditional pollutants including phosphates [23], metal ions (Cu^{2+}, Co^{2+}) [24], and F^- [25] have been investigated. However, no previous studies have assessed the applicability of discarded CG as an adsorbent material without chemical modification to remove hydrochloride (TC) or related emerging pollutants from an aqueous solution. Natural CG is a widespread by-product generated in the process of coal mining or the washing process in the coal industry [26]. CG is usually a black-gray rock with abundant minerals, organic substances, and stable crystals [27,28]. However, the accumulation of excessive CG in coal related operations can lead to groundwater, air, and land pollution [29,30], and therefore it is crucial to explore new methods for the recycling of CG materials. Natural CG could be used to efficiently remove TC from aqueous solutions due to its high abundance, low cost, high chemical stability, and high ion exchange capacity [13,31].

In this study, the applicability of discarded natural CG for in the removal of TC from an aqueous solution was systematically investigated. The effect of CG dosage, CG particle size, TC concentration, contact time, temperature, and solution pH on TC adsorption by CG were studied. Additionally, adsorption kinetics and adsorption isotherm fitting were performed to interpret the adsorption data. Material characterization routes including FT-IR, XPS, XRD, and SEM were applied to analyze the structural changes of the CG in the adsorption process. This study thus provides innovative insights into the potential applicability of spent CG derived from coal-related waste for the efficient removal of antibiotic contaminants from water.

2. Materials and Methods

2.1. Raw Material and Chemicals

The raw CG (coal gangue) material was directly collected from the fifth mine of Pingdingshan Tian'an Coal Mining in the city of Pingdingshan, Henan Province, China. The treatment process of discarded CG as adsorbate was briefly depicted in Figure 1. The CG samples were firstly dried under 45 °C in a thermostatic oven for 12 h, fractured by jaw breaker, and then finally sieved through the first 100-mesh sieve and the second 120-mesh sieve. The CG samples (with particle-size range of 0.12 to 0.15 mm) were first cleaned with deionized water, filtered by vacuum sucking pump, and then dried at 40 °C in thermostatic oven for 5 h. The CG samples were finally grinded to disperse, and then stored for further use. The elemental compositions of Si, Al, Ca, Fe, and K in the raw CG were determined as 25.0%, 12.7%, 6.7%, 3.4%, and 2.4%, respectively (Table S1). The TC (tetracycline, $C_{22}H_{25}ClN_2O_8$, 96%) was bought from Aladdin Biochemical Co., Ltd., Shanghai, China. The NaOH (sodium hydroxide, 96%) was obtained from Tianjin Sailboat Chemical Reagent Technology Co., Ltd., Tianjin, China, while the H_2SO_4 (sulfuric acid, 98%) was bought from Luoyang Chemical Reagent Factory, Henan, China.

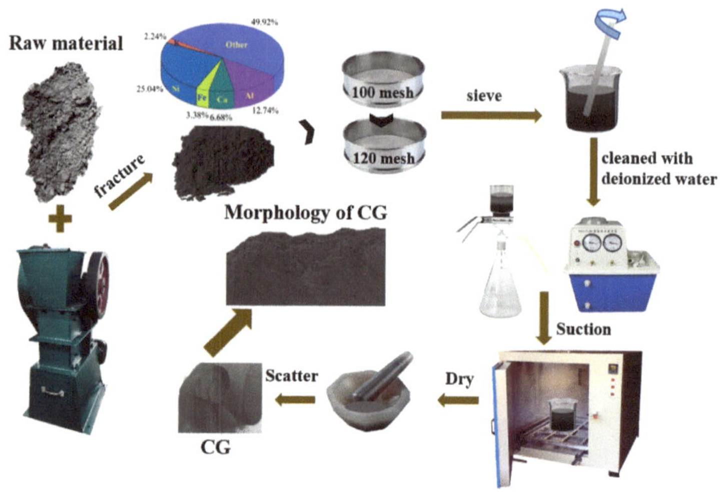

Figure 1. The flow chart for treatment of CG samples derived from the discarded natural coal gangue.

2.2. Adsorption of TC by CG

The stock TC solution (40 mg/L) was prepared by weighing 20 mg TC into a 100 mL beaker, adding a small amount of water to dissolve and pipette into a 500 mL volumetric flask, rinsing the beaker multiple times to pipette, and fixing the volumetric flask to the 500 mL tick line. For each batch experiment, the TC solution (50 mL, 40 mg/L) was mixed with CG (1 g/L) in the conical flask (100 mL) shaking by the thermostatic water bath oscillator (120 rpm) at 298 K until adsorption equilibrium was reached. At the terminal of 3 h, the slurry containing the CG particles and TC was quickly separated through a 0.45-μm syringe filter to get the filtrate. The filtrate after separation was measured by using an Ultraviolet–visible spectrophotometer (UV759CRT, Shanghai, China) at a wavenumber of 357 nm and the proposed process for adsorption of TC by CG was shown in Figure 2a. The ICP (inductively coupled plasma-atomic emission spectroscopy, AAS, AA800, Perkin-Elmer) was used to measure the concentration of iron leaching from CG into the solution during the TC adsorption process. The TC desorption from exhausted CG and CG recycling process were conducted (Texts S1 and S2).

The adsorption capacity (Q, mg/g) and removal efficiency (E, %) of TC by CG were calculated by Equations (1) and (2), respectively.

$$Q = \left[\frac{C_0 - C}{m}\right] \times V \tag{1}$$

$$E = \left[\frac{C_0 - C}{C_0}\right] \times 100\% \tag{2}$$

where C_0 and C (mg/L) represent the initial and equilibrium concentration of TC during the adsorption process, respectively; m (g) is the dosage of CG; V (L) is the volume of TC solution.

2.2.1. Adsorption Kinetics and Isotherms Investigation

In the adsorption kinetics investigation experiments, the slurry of TC solution (50 mL, 40 mg/L) and CG (1 g/L) was mixed in conical flasks (100 mL) by the water bath oscillator at 298 K. The pH of the slurry determined as 6.2 without pH adjustment. After the beginning of the reaction, at given time intervals of 0.17, 0.5, 1, 2, 3, 3.5, 4, 5, 6, 12, and 24 h, the slurry (3 mL) was withdrawn from each conical flask to measure the concentration of TC.

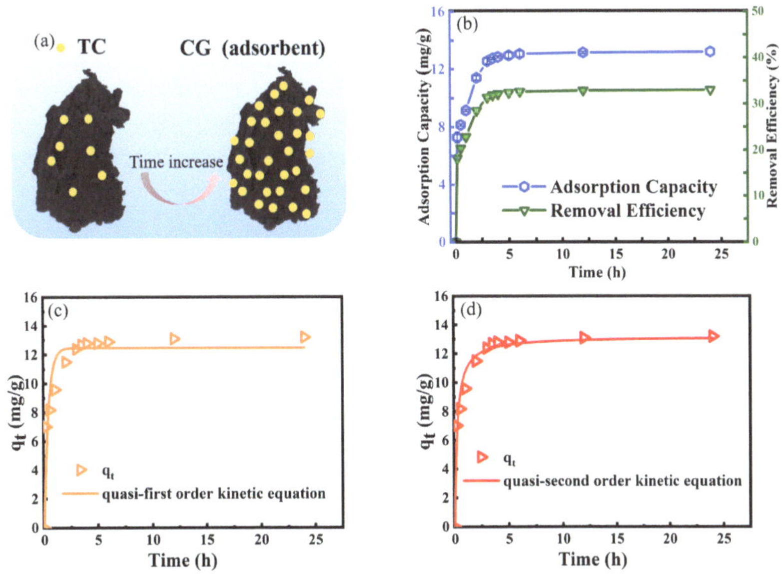

Figure 2. The adsorption kinetic curves of TC by CG: (**a**) the proposed schematic diagram of TC adsorption by CG with increasing contact time; (**b**) effect of contact time on TC adsorption by CG, (**c**) the quasi-first order kinetic model fitting line; (**d**) the quasi-second order kinetic model fitting line. (TC concentration of 40 mg/L, CG dosage of 1 g/L, pH of 6.2, contact temperature of 25 °C, and stirring speed of 120 rpm).

Aiming to analyze the adsorption kinetics of TC by CG particles, the quasi-first-order Equation (3) and quasi-second order kinetic Equation (4) were used to explain adsorption kinetics.

$$q_t = q_e\left(1 - e^{-k_1 t}\right) \tag{3}$$

$$q_t = \frac{k_2 q_e^2 t}{1 + k_2 q_e t} \tag{4}$$

where q_e (mg/g) is the amount of TC adsorbed by CG at equilibrium and q_t is the adsorption amount at certain time of t. The k_1 (h^{-1}) and k_2 (h^{-1}) represent the first order rate constant and the second order rate constant, respectively [32,33].

The adsorption isotherms investigations were conducted by measuring the adsorption capacities of TC by CG at different temperatures of 298 K, 308 K, and 318 K. At the three-mentioned temperatures, three parallel samples (for each temperature) were prepared by mixing TC (50 mL, 20–150 mg/L) solution and CG (1 g/L) into the conical flasks (100 mL) in the water bath oscillators (with oscillating speed of 120 rpm) for 3 h. All samples were conducted in triplicate. At the termination of the setting adsorption equilibrium time, the solution samples were quickly withdrawn, then filtered through a 0.45-μm syringe filter, and finally measured by UV–Vis spectrophotometry.

The Langmuir and Freundlich isotherm models for different temperature (25 °C, 35 °C, 45 °C) adsorption of TC were applied to assess adsorption process.

$$Q = \frac{Q_m C_e K_L}{1 + K_L C_e} \tag{5}$$

$$Q = K_F C_e^{\frac{1}{n}} \tag{6}$$

where Q is the adsorption amount at equilibrium time (mg/g), Q_m is the maximum adsorption capacity at a certain temperature (mg/g), C_e is the concentration of adsorbate in solution at adsorption equilibrium (mg/L), K_L demonstrates Langmuir adsorption equilibrium constant (mg/g), and K_F is Freundlich adsorption equilibrium constant (mg/g), the value of $1/n$ represents an adsorption intensity [34,35].

2.2.2. Effect of CG Particle Size

The effects of CG particle sizes on adsorption performance of TC by CG were performed at ranging particle sizes (0.90–0.45, 0.45–0.30, 0.30–0.15, 0.15–0.12, and ≤0.12 mm). The CG (with different sizes) samples (1 g/L) were mixed with TC solution (50 mL, 40 mg/L) without adjusting the solution pH of 6.2 in the conical flask (100 mL) in the water bath oscillator (120 rpm). The contact time and temperature were 3 h and 298 K, respectively.

2.2.3. Effect of CG Dosage and pH

The effects of CG dosage on TC adsorption were explored at the CG dosage (size of 0.15–0.12 mm) ranged of 0 and 2 g/L, a natural pH of 6.2, and temperature of 298 K under 120 rpm in water bath oscillator. The volume of TC solution (40 mg/L) was 50 mL, with contact time of 3 h. Finally, the effect of pH on the adsorption of TC by CG was investigated in pH ranging from 2 to 12 at the CG dosage of 1 g/L for 3 h. The contact temperature was 298 K in the water bath oscillator (120 rpm).

2.3. Material Characterization

The XRD pattern of CG was obtained by using the D/max 2500 pc (Rigaku, Empyrean, The Netherlands) between the 2θ of 5–90°. The elements composition of the original CG was measured by the XRF (X-ray Fluorescence Spectrometer, ARL, Thermo Scientific, Waltham, MA, USA). The FT-IR spectra (Fourier transform infrared spectroscopy, Vertex 80, Bruker, Germany) was used to investigate the key functional groups in CG before and after the adsorption process experiment. The XPS (X-ray photoelectron spectroscopy, Escalab 250Xi, Thermo Scientific, Waltham, MA, USA) instrument was used to measure the chemical state of main elements of CG. The Zeta potentials of CG in aqueous solution were recorded through a micro-electrophoresis instrument (JS94HM, Shanghai, China). The SEM (scanning electron microscope, Merlin compact, Zeiss, GER) was used to analyze the morphology of CG. The N_2-BET measurement (Brunauer–Emmett–Teller, ASAP 2460, Micromeritics, Norcross, GA, USA) was used to analyze the surface area of CG samples, while the BJH (Barrett–Joyner–Halenda) was used to obtain the pore volume and pore size of CG under N_2–adsorption/desorption process.

3. Results

3.1. Adsorption Kinetic Study

The effect of the reaction time on TC adsorption by CG is shown in Figure 2b. Approximately 33.2% of TC was removed by CG within 24 h, with a TC adsorption capacity by CG of 13.3 mg/g. The TC adsorption capacity by CG was quickly increased within 0–3 h, whereas the TC adsorption capacity was nearly constant within 12–24 h. The quasi-first order equation and quasi-second order kinetic equations were used to fit the TC adsorption kinetic results. The fitting curves (Figure 2c,d) and their detailed parameters (Table S2) indicated that the quasi-second order kinetic equation could better describe the adsorption of TC by CG compared with the quasi-first order equation. Furthermore, the q_e (13.19 mg/g) calculated by the quasi-second order kinetic equation was close to the experimental adsorption capacity (12.74 mg/g), further confirming the suitability of the quasi-second order kinetic equation to fit the TC adsorption by CG in the water environment. The quasi-second order kinetic equation could further indicate that the adsorption performance of TC by CG should be controlled by chemical adsorption [36].

3.2. Adsorption Isotherms Study

The effect of TC concentration on the TC adsorption performance by CG was illustrated in Figure 3a. With the TC concentration ranging from 20 mg/L to 150 mg/L, the TC adsorption capacity was increased from 7.2 mg/g to 24.2 mg/g. With the increasing TC molecules in an aqueous solution, the unit mass of CG would contact and adsorb more TC molecules, thus increasing the adsorption capacity of TC by CG [37]. However, the TC adsorption efficiency was decreased from 36.1% to 16.1% with TC concentration ranging from 20 mg/L to 150 mg/L. The decrease in TC adsorption efficiency was mainly attributed to the adsorption saturation of TC by the limited CG samples in the water environment. With more TC molecules added to the solution, the calculated TC adsorption efficiency by CG would be decreased due to the increased C_0 in Equation (2) [38].

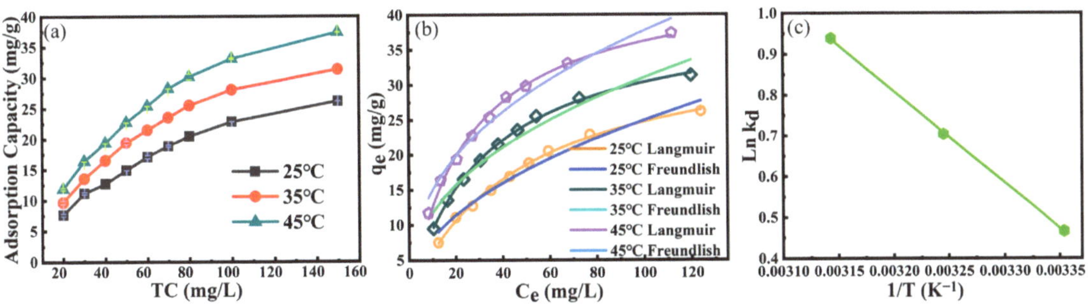

Figure 3. The adsorption isotherms curves of TC by CG: (**a**) impact of initial TC concentration on adsorption capacity under different temperatures (25 °C, 35 °C, 45 °C), (CG dosage of 1 g/L, contact time of 3h, pH of 6.2 and 120 rpm); (**b**) Langmuir and Freundlich adsorption isotherm of TC adsorbed by CG under different temperature (25 °C, 35 °C, 45 °C), (**c**) Plot of LnK_d vs. 1/T for TC adsorption by CG.

The effect of contact temperature (25 °C, 35 °C, and 45 °C) on TC adsorption by CG is shown in Figure 3a. As the contact temperature increased, the adsorption capacity of TC by CG was increased, indicating that the adsorption of TC by CG was an endothermic process [39].

The fitting curves for adsorption isotherms of TC by CG and their related parameters are shown in Figure 3b and Table S2, respectively. Based on the fitting curves at 25 °C, the Langmuir mode (with a correlation coefficient of 0.997) could better describe the TC adsorption by CG, compared with the Freundlich model (with a correlation coefficient of 0.972). These results confirmed that the adsorption of TC by CG was mainly controlled by monolayer adsorption [40].

The thermodynamic parameters, including ΔG, ΔH, and ΔS, were described by Equations (7), (8), and (9), respectively.

$$LnK_d = \frac{q_e}{c_e} \tag{7}$$

$$\Delta G = -RTLnK_d \tag{8}$$

$$LnK_d = \frac{\Delta S}{R} - \frac{\Delta H}{RT} \tag{9}$$

where ΔG (J/mol) is the Gibbs free energy; ΔH (KJ/mol) is the enthalpy change of the adsorption process; ΔS (J/mol·K) is the entropy change between product and reactant; R is the thermodynamic constant, which is approximately equal to 8.314 (J/mol·K); T (K) is the thermodynamic temperature. The values of ΔS and ΔH are computed from the intercept and the slope of the linear plot for LnK_d vs. 1/T [41], which is shown in Figure 3c. The values of ΔG can be calculated from Equation (8). The results are displayed in Table 1.

Table 1. The thermodynamic parameters for TC adsorption by CG.

T (K)	ΔG (J/mol)	ΔH (KJ/mol)	ΔS (J/mol·K)	R^2
298	−1154			
308	−1744	18.635	66.347	0.999
318	−2326			

The positive ΔH (18.63 KJ/mol) (Table 1) further confirmed that the TC adsorption by CG was an endothermic process. The negative values of ΔG indicated that the adsorption process was spontaneous. The positive ΔS value (66.34 J/mol·K) indicated that the freedom and randomness of the adsorption process were higher [39,41,42].

3.3. *Effect of CG Dosage and Particle Size*

Figure 4a illustrates the effect of CG dosage on TC adsorption by CG. With CG dosage ranging from 0.2 g/L to 1.8 g/L, the TC adsorption efficiency increased from 10.8% to 47.7%. The enhancement of TC adsorption efficiency with increasing CG dosage was mainly attributed to the increase in the number of adsorption sites provided by more CG particles (Figure 4b). However, the adsorption efficiency of CG at a dosage of 2 g/L (44.54%) is slightly lower than at a dosage of 1.8 g/L (47.72%), which was mainly due to the insufficient exposure of CG active sites caused by the agglomeration of CG particles [43,44]. However, The TC adsorption capacity was decreased from 22.6 mg/g to 8.79 mg/g with CG dosage ranging from 0.2 g/L to 2 g/L. After increasing the CG in the aqueous solution, more CG particles cannot adsorb the TC (due to the limited TC concentration). Thus, the calculated adsorption capacity was decreased due to the increased m of Equation (1).

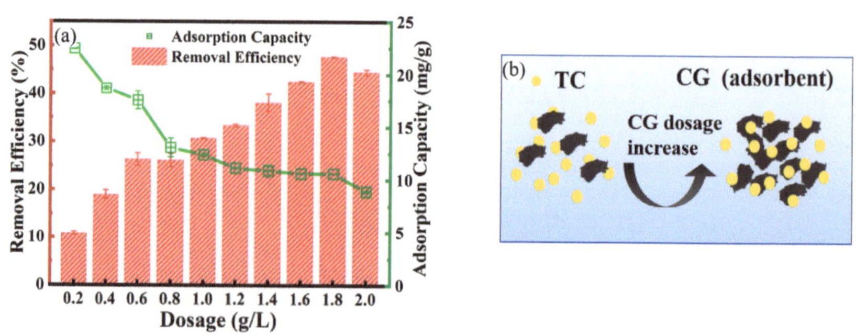

Figure 4. Effect of adsorbent dosage for TC adsorption: (**a**) Effect of CG dosage on removal efficiency and adsorption capacity of TC by CG (contact time of 3 h, TC concentration of 40 mg/L, pH of 6.2 and temperature of 25 °C, 120 rpm); (**b**) the proposed schematic diagram of TC adsorption by CG with increasing CG dosage.

The TC adsorption efficiency was increased from 13.30% to 54.26% with CG (1 g/L) particle size ranging from 0.90–0.45 mm to smaller than 0.12 mm (Figure 5a). The CG with a smaller particle size would provide more specific surface areas to adsorb TC in an aqueous solution, further increasing the TC adsorption efficiency, which is depicted in Figure 5b.

3.4. *Effect of pH*

The effect of pH on TC adsorption by CG in an aqueous solution was shown in Figure 6a. The TC adsorption efficiency by CG in an acidic environment was higher compared with the more alkaline environment [45]. The significant difference in the adsorption efficiency in different pH conditions could be explained by the Zeta potential data (Figure 6b). The dominant phase of TC in the aqueous solution was mainly TCH^{3+} cation (pH < 3), $TC^{2\pm}$ zwitterion species (pH between 3.3 and 7.7), and TCH^-/TC^{2-} (pH > 7.7) [46]. Therefore, the TC molecules were mainly positively charged. With the pH of solution changing from acidic to alkaline, the charge of CG was becoming more negative. Thus, compared with the acidic environment, the electrical repulsion between the CG and TC would be increased in the more alkaline environment, leading to decreased TC adsorption capacity by CG (Figure 6c) [46,47].

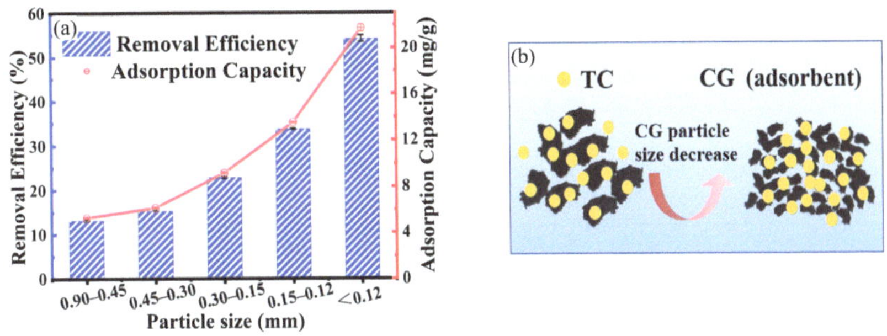

Figure 5. Effect of adsorbent particle size for TC adsorption: (**a**) Effect of CG particle size on removal efficiency and adsorption capacity of TC by CG (contact time of 3 h, TC concentration of 40 mg/L, CG dosage of 1 g/L, pH of 6.2 and temperature of 25 °C, 120 rpm), (**b**) the proposed schematic diagram of TC adsorption by CG with reduced CG particle size.

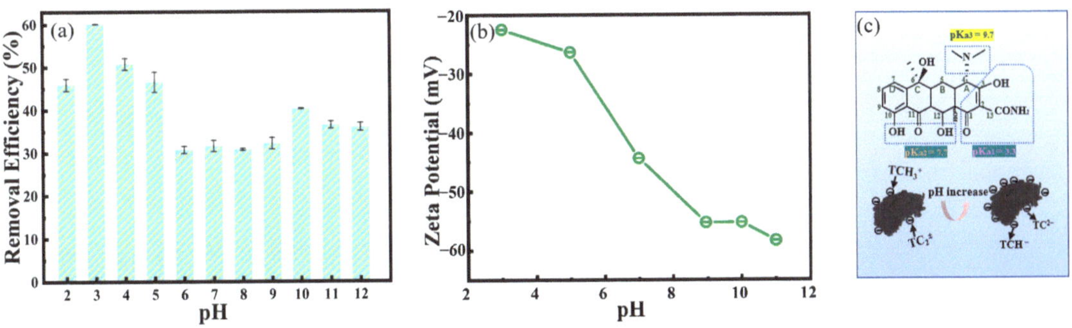

Figure 6. Effect of solution pH for TC adsorption: (**a**) Effect of solution pH on removal efficiency of TC by CG (contact time of 3 h, TC concentration of 40 mg/L, CG dosage of 1 g/L, contact temperature of 25 °C, and stirring speed of 120 rpm); (**b**) the Zeta potential of CG with increasing pH in an aqueous solution, (c) the proposed schematic diagram of TC adsorption by CG with increasing solution pHs.

3.5. TC Desorption and the Reusability of Exhausted CG

The TC desorption rate from exhausted CG was shown in Figure S1, while TC desorption rate from exhausted CG reached only 17.6% at equilibrium time of 1.5 h. This result indicated TC were not easily desorbed from CG, further confirming the relatively stronger force between CG and TC [48,49].

The reusability of exhausted CG in TC removal process was conducted and the TC removal efficiency by used CG could only reach 25.3%. This result illustrated that the exhausted CG was needed further treatment before its reusability.

3.6. Material Characterization of the Natural CG and its Adsorption Mechanism

The SEM image of CG is shown in Figure 7a. The surface of CG is flat and accompanied by flaky kaolin stones, showing a cluttered surface with an irregular arrangement. The XRD patterns of the original CG are shown in Figure 7b. The quartz, kaolinite, quartz, illite, ettringite ($Ca_6Al_2(SO_4)_3(OH)_{12}{\cdot}26H_2O$), and C–S–H ($Ca_4Si_6O_{14}(OH)_4{\cdot}2H_2O$) are the dominant crystal phases in the CG [50]. The chemical state of the main elements in CG was investigated and shown in Figure S2.

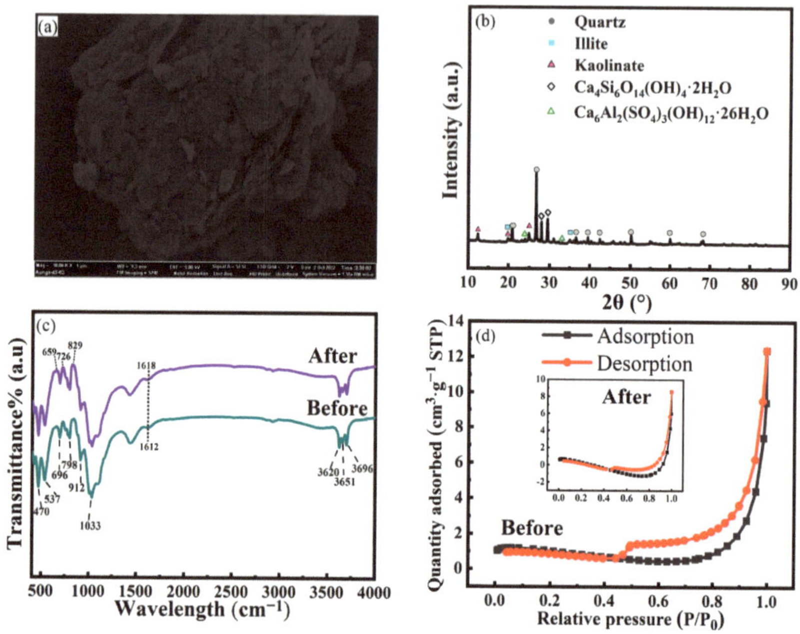

Figure 7. The material characterization results of CG particles: (**a**) SEM images of the CG before adsorption; (**b**) XRD; (**c**) The FT-IR spectra of the CG before and after adsorption; (**d**) the nitrogen adsorption–desorption isotherms before and after reaction.

The detailed BET parameters of CG were shown in Table S3. As depicted in Figure 7d, the isotherm curves are fitted with the IUPAC classification type curve of IV and H_3 type hysteresis loop. Meanwhile, the hysteresis loop closes at a relative pressure of $P/P_0 = 0.4$, indicating the existence of smaller mesopores in the CG sample and a smaller amount of micropores [51,52]. The BET surface area and the total pore volume of CG after the TC adsorptions processes were decreased, and the reasons are illustrated as follows: (1) TC enters the pores of the material and stays in the pores by electrostatic attraction. (2) The active group in the material undergoes ion exchange or complexation reaction with TC, thereby reducing the pore volume and specific surface area.

The FTIR spectra of CG before and after the TC removal process is shown in Figure 7c. The characteristic bands of CG observed at 3696, 3651, and 3620 cm^{-1} are mainly attributed to the hydroxyl stretching vibrations of kaolinite. The peak observed at 470 cm^{-1} is related to the Si-O-Fe, further confirming the existence of Fe. The peaks observed at 694 cm^{-1} and 537 cm^{-1} were representative of the stretching and bending vibrations of Si-O-Al, respectively. The band observed at 912 cm^{-1} is related to the Al-OH bending vibration. The peaks observed at 798 cm^{-1} and 1033 cm^{-1} are mainly attributed to the bending vibration and stretching vibration of Si-O-Si in the natural CG, which are identical with the characteristic absorption peaks of illite, indicating the existence of illite and kaolinite in CG, according to the XRD results [53–57].

The FT-IR spectra show characteristic bands of original CG before and after TC adsorption. Compared with the CG before adsorption, the -OH in absorption peaks at 3651, and 3620 cm^{-1} of CG after adsorption had almost no change. With the peak at 912 cm^{-1} of Al-OH absorption peak weakened after adsorption, suggesting that the decrease in Al-OH may be due to the complexation between Al and TC. The decrease of Si-O-Al absorption peaks at 537 cm^{-1} indicates the damage of kaolinite. This, accompanied by the strengthening of Si-O-Si bending vibration peaks at 798 cm^{-1} and the lower Si-O-Si stretching vibration peaks at 1033 cm^{-1}, indicated the destruction of illite. Nevertheless, the peaks at 659, 726, and 829 cm^{-1} present varying degrees of elevation due to the consequence of TC adsorption onto CG, which represented the characteristic peaks of TC [55,58–60]. Furthermore, the peak observed at 1612 cm^{-1} was mainly attributed to the vibration of amide carbonyl at ring A of the TC (Figure 6c). Therefore, the narrower and higher intensity peak observed at 1618 cm^{-1} of CG after TC adsorption could confirm the deprotonation of TC during the adsorption process [61]. Figure S2 could also support the deprotonation of TC.

Based on the above analysis, the possible adsorption mechanism of TC by CG is shown in Figure 8. The Langmuir model of the TC adsorption process indicated that the physical adsorption process will also work in the TC adsorption by CG. The electrostatic action between TC and CG at varying pH would also affect the adsorption of TC by CG in an aqueous solution [62].

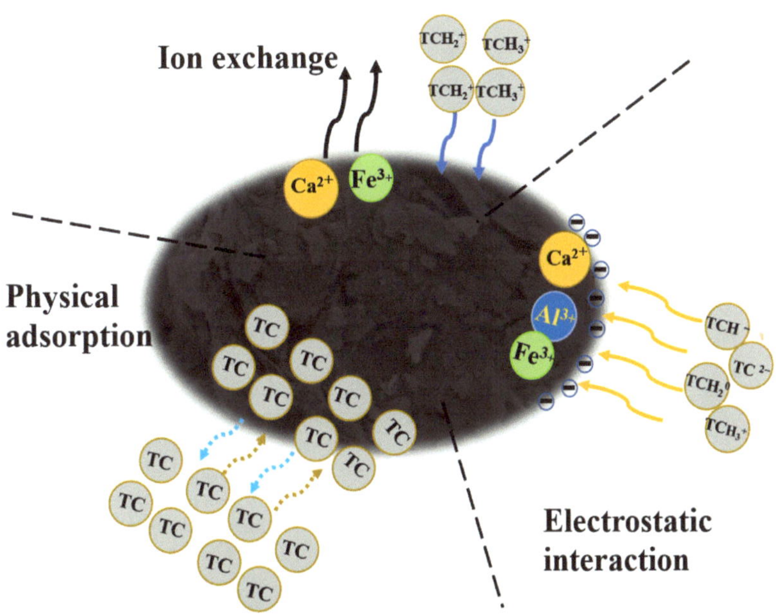

Figure 8. The proposed adsorption mechanism of TC adsorption by CG.

Additionally, the adsorption kinetic analyses confirmed that the adsorption of TC by CG was affected by the chemical adsorption process. Considering that the TC was an amphoteric molecule [46], the TC would occur in the protonation and deprotonation reactions at varying pH conditions. At a pH lower than pK_{a1} (3.3), the TC was dominant with species of TCH^{3+}. With the pH ranging between pK_{a1} and pK_{a2} ($3.3 < pH < 7.7$), the TC was mainly presented with TCH^{2+}. At pH further increasing to a range of ($7.7 < pH < 9.7$) and ($pH > 9.7$), the dominant species of TC were TCH^{-} and TC^{2-}, respectively [47]. The species of TCH^{3+} and TCH^{2+} in acidic and neutral environments would be exchanged with the positively charged ions (Ca^{2+}, Fe^{3+}) in CG. Additionally, the concentrations of Fe and Ca were determined as 140.8 μg/L and 173.2 μg/L, respectively, further confirming the metal element exchange from the CG to the TC solution [63,64]. Therefore, at acidic and neutral pH environments, the cation exchange process would affect the adsorption of TC by the CG samples. However, in the alkaline pH environment, the cation exchange between the TCH^{-}/ TC^{2-} and CG would weaken.

The comparisons of TC adsorption performance by CG and other related minerals were shown in Table 2. Compared with kaolinite [64] and ferrihydrite [14], the natural CG had a higher TC adsorption capacity and a shorter adsorption equilibrium time (3 h). Compared with minerals including, such as illite [65], montmorillonite [62], and bentonite clay [14], natural CG has a lower TC adsorption capacity but reaches adsorption equilibrium within a shorter time of 3 h [66–68].

Table 2. The comparisons of TC adsorption between CG and other minerals.

Used Minerals	Concentration of Pollutants and Adsorption Equilibrium Time	Adsorption Capacity Q_{max} ($mg \cdot g^{-1}$)	References
Natural CG (1 g/L)	TC (40 mg/L), 3 h	24.7 mg/g	This Study
bentonite clays (0.4 g/L)	TC (30 mg/L), 5 h	156.7 mg/g	[11]
Ferrihydrite (10 g/L)	TC (40 mg/L), 96 h	3.2 mg/g	[14]
Montmorillonite (0.2 g/L)	TC (5–100 mg/L), 24 h	250.0 mg/g	[60]
Kaolinite (100 g/L)	TC (480 mg/L), 8 h	4.4 mg/g	[62]
Illite (10 g/L)	TC (200 mg/L), 8 h	32.0 mg/g	[63]

Based on the aforementioned comparison, the CGs used in this study possessed several advantages including a faster equilibrium time, relatively larger adsorption capacity, and easier access to application.

4. Conclusions

In the present study, CG was prepared by simple pretreatment for the removal of TC in solution. The effects of contact time, the particle size of CG, the dosage of CG, different temperatures, and solution pH on the adsorption capacity were analyzed. The adsorption experimental data can be well fitted by the first-order kinetic model and Langmuir adsorption isotherm model. The results showed that the adsorption process can be classified as spontaneous endothermic monolayer adsorption, controlled by physical adsorption and chemical adsorption. The adsorption model, FT-IR, XPS, XRD, and Zeta potential show that the adsorption mechanism of TC by CG mainly includes electrostatic interaction and cation exchange. In addition, the strength of electrostatic attraction is the main reason for the difference in adsorption capacity between acidic and alkaline environments. The comparison between CG and other minerals showed that CG had better adsorption abilities for TC. This study confirmed that CG is a promising and sustainable adsorbent that could be used for the treatment of antibiotic-contaminated wastewater. Therefore, CG can be used to both realize the waste resource reduction of CG and explore the enhanced value of CG in wastewater treatment.

Supplementary Materials: The following supporting information can be downloaded at: https://www.mdpi.com/article/10.3390/toxics11010020/s1, Text S1: TC desorption from the exhausted CG; Text S2: The CG recycling process and the results; Table S1: The elemental compositions of the raw CG; Table S2: The detailed parameters for adsorption kinetics and isotherms of TC by CG; Table S3: BET parameters of CG before and after TC adsorption; Figure S1: The desorption for TC of exhausted CG; Figure S2: The XPS spectra of the raw and used CG.

Author Contributions: Conceptualization, H.W. and W.Z.; methodology, L.G. and X.Z.; validation, C.W., J.Y. and J.L.; data curation, C.W.; writing—original draft preparation, C.W.; writing—review and editing, W.Z.; visualization, W.Z.; supervision, C.W.; project administration, H.W.; funding acquisition, W.Z. All authors have read and agreed to the published version of the manuscript.

Funding: This study was funded by the National Natural Science Foundation of China (grant No. 52000163, and 42277394), the Open fund from Henan Key Laboratory of Water Pollution Control and Rehabilitation Technology (CJSP2022001), the Youth Talent Plan of China Association for Science and

Technology (20220615ZZ07110090), and the Open Research Fund of Henan Key Laboratory of Water Resources Conservation and Intensive Utilization in the Yellow River Basin (grant NO. HAKF202101).

Institutional Review Board Statement: Not applicable.

Informed Consent Statement: Written informed consent has been obtained from the patient(s) to publish this paper.

Data Availability Statement: Not applicable.

Acknowledgments: We thank the Modern Analysis and Gene Sequencing Center of Zhengzhou University and the National Supercomputing Center in Zhengzhou.

Conflicts of Interest: The authors declare no conflict of interest.

References

1. Sarmah, A.K.; Meyer, M.T.; Boxall, A.B.A. A global perspective on the use, sales, exposure pathways, occurrence, fate and effects of veterinary antibiotics (VAs) in the environment. *Chemosphere* **2006**, *65*, 725–759. [CrossRef]
2. Sheng, X.; Wang, J.; Cui, Q.; Zhang, W.; Zhu, X. A feasible biochar derived from biogas residue and its application in the efficient adsorption of tetracycline from an aqueous solution. *Environ. Res.* **2022**, *207*, 112175. [CrossRef]
3. Hao, R.; Xiao, X.; Zuo, X.; Nan, J.; Zhang, W. Efficient adsorption and visible-light photocatalytic degradation of tetracycline hydrochloride using mesoporous BiOI microspheres. *J. Hazard. Mater.* **2012**, *209–210*, 137–145. [CrossRef]
4. Wan, H.; Yan, J.; Guo, C.; Cui, Q.; Zhang, W. Synthesis of core-heteroshell structure for ZIF-67/VTM and its efficient activation of peroxymonosulfate in treatment of levofloxacin from an aqueous solution. *Environ. Res.* **2022**, *204*, 111986. [CrossRef]
5. Ji, L.; Wan, Y.; Zheng, S.; Zhu, D. Adsorption of Tetracycline and Sulfamethoxazole on Crop Residue-Derived Ashes: Implication for the Relative Importance of Black Carbon to Soil Sorption. *Environ. Sci. Technol.* **2011**, *45*, 5580–5586. [CrossRef]
6. Pastor-Navarro, N.; Maquieira, Á.; Puchades, R. Review on immunoanalytical determination of tetracycline and sulfonamide residues in edible products. *Anal. Bioanal. Chem.* **2009**, *395*, 907–920. [CrossRef]
7. Oberoi, A.S.; Jia, Y.; Zhang, H.; Khanal, S.K.; Lu, H. Insights into the Fate and Removal of Antibiotics in Engineered Biological Treatment Systems: A Critical Review. *Environ. Sci. Technol.* **2019**, *53*, 7234–7264. [CrossRef]
8. Cui, Q.; Li, Y.; Chai, S.; Zhang, W.; Zuo, Q.; He, C. Enhanced catalytic activation of H_2O_2 by CNTs/SCH through rapid Fe(III)/Fe(II) redox couple circulation: Insights into the role of functionalized multiwalled CNTs. *Sep. Purif. Technol.* **2022**, *282*, 120000. [CrossRef]
9. Cao, H.-L.; Cai, F.-Y.; Yu, K.; Zhang, Y.-Q.; Lü, J.; Cao, R. Photocatalytic Degradation of Tetracycline Antibiotics over CdS/Nitrogen-Doped—Carbon Composites Derived from in Situ Carbonization of Metal—Organic Frameworks. *ACS Sustain. Chem. Eng.* **2019**, *7*, 10847–10854. [CrossRef]
10. Cui, Q.; Zhang, W.; Chai, S.; Zuo, Q.; Kim, K.-H. The potential of green biochar generated from biogas residue as a heterogeneous persulfate activator and its non-radical degradation pathways: Adsorption and degradation of tetracycline. *Environ. Res.* **2022**, *204*, 112335. [CrossRef]
11. Maged, A.; Iqbal, J.; Kharbish, S.; Ismael, I.S.; Bhatnagar, A. Tuning tetracycline removal from aqueous solution onto activated 2:1 layered clay mineral: Characterization, sorption and mechanistic studies. *J. Hazard. Mater.* **2020**, *384*, 121320. [CrossRef] [PubMed]
12. Imanipoor, J.; Mohammadi, M.; Dinari, M. Evaluating the performance of L-methionine modified montmorillonite K10 and 3-aminopropyltriethoxysilane functionalized magnesium phyllosilicate organoclays for adsorptive removal of azithromycin from water. *Sep. Purif. Technol.* **2021**, *275*, 119256. [CrossRef]
13. Hacıosmanoğlu, G.G.; Mejías, C.; Martín, J.; Santos, J.L.; Aparicio, I.; Alonso, E. Antibiotic adsorption by natural and modified clay minerals as designer adsorbents for wastewater treatment: A comprehensive review. *J. Environ. Manag.* **2022**, *317*, 115397. [CrossRef] [PubMed]
14. Wu, T.; Xue, Q.; Liu, F.; Zhang, J.; Zhou, C.; Cao, J.; Chen, H. Mechanistic insight into interactions between tetracycline and two iron oxide minerals with different crystal structures. *Chem. Eng. J.* **2019**, *366*, 577–586. [CrossRef]
15. Silva, C.P.; Jaria, G.; Otero, M.; Esteves, V.I.; Calisto, V. Adsorption of pharmaceuticals from biologically treated municipal wastewater using paper mill sludge-based activated carbon. *Environ. Sci. Pollut. Res.* **2019**, *26*, 13173–13184. [CrossRef]
16. Liew, R.K.; Nam, W.L.; Chong, M.Y.; Phang, X.Y.; Su, M.H.; Yek, P.N.Y.; Ma, N.L.; Cheng, C.K.; Chong, C.T.; Lam, S.S. Oil palm waste: An abundant and promising feedstock for microwave pyrolysis conversion into good quality biochar with potential multi-applications. *Process Saf. Environ. Prot.* **2018**, *115*, 57–69. [CrossRef]
17. Carrales-Alvarado, D.H.; Leyva-Ramos, R.; Rodríguez-Ramos, I.; Mendoza-Mendoza, E.; Moral-Rodríguez, A.E. Adsorption capacity of different types of carbon nanotubes towards metronidazole and dimetridazole antibiotics from aqueous solutions: Effect of morphology and surface chemistry. *Environmental Sci. Pollut. Res.* **2020**, *27*, 17123–17137. [CrossRef]
18. Nielsen, L.; Biggs, M.J.; Skinner, W.; Bandosz, T.J. The effects of activated carbon surface features on the reactive adsorption of carbamazepine and sulfamethoxazole. *Carbon* **2014**, *80*, 419–432. [CrossRef]

19. Cao, J.; Xiong, Z.; Lai, B. Effect of initial pH on the tetracycline (TC) removal by zero-valent iron: Adsorption, oxidation and reduction. *Chem. Eng. J.* **2018**, *343*, 492–499. [CrossRef]
20. Gu, C.; Karthikeyan, K.G. Interaction of Tetracycline with Aluminum and Iron Hydrous Oxides. *Environ. Sci. Technol.* **2005**, *39*, 2660–2667. [CrossRef]
21. Gao, Y.; Dai, Y.; Zhang, H.; Diao, E.; Hou, H.; Dong, H. Effects of organic modification of montmorillonite on the performance of starch-based nanocomposite films. *Appl. Clay Sci.* **2014**, *99*, 201–206. [CrossRef]
22. Ding, W.; Bai, S.; Mu, H.; Naren, G. Investigation of phosphate removal from aqueous solution by both coal gangues. *Water Sci. Technol.* **2017**, *76*, 785–792. [CrossRef] [PubMed]
23. Lu, X.; Shi, D.; Chen, J. Sorption of Cu^{2+} and Co^{2+} using zeolite synthesized from coal gangue: Isotherm and kinetic studies. *Environ. Earth Sci.* **2017**, *76*, 591. [CrossRef]
24. Jin, Y.; Li, L.; Liu, Z.; Zhu, S.; Wang, D. Synthesis and characterization of low-cost zeolite NaA from coal gangue by hydrothermal method. *Adv. Powder Technol.* **2021**, *32*, 791–801. [CrossRef]
25. Wang, X.-L.; Zhang, Y. Synthesis and Characterization of Zeolite A Obtained from Coal Gangue for the Adsorption of F^- in Wastewater. *Sci. Adv. Mater.* **2019**, *11*, 277–282. [CrossRef]
26. Zhang, W.; Dong, C.; Huang, P.; Sun, Q.; Li, M.; Chai, J. Experimental Study on the Characteristics of Activated Coal Gangue and Coal Gangue-Based Geopolymer. *Energies* **2020**, *13*, 2504. [CrossRef]
27. Li, L.; Long, G.; Bai, C.; Ma, K.; Wang, M.; Zhang, S. Utilization of Coal Gangue Aggregate for Railway Roadbed Construction in Practice. *Sustainability* **2020**, *12*, 4583. [CrossRef]
28. Li, J.; Wang, J. Comprehensive utilization and environmental risks of coal gangue: A review. *J. Clean. Prod.* **2019**, *239*, 117946. [CrossRef]
29. Tang, Y.; Wang, H. Development of a novel bentonite–acrylamide superabsorbent hydrogel for extinguishing gangue fire hazard. *Powder Technol.* **2018**, *323*, 486–494. [CrossRef]
30. Li, S.; Liber, K. Influence of different revegetation choices on plant community and soil development nine years after initial planting on a reclaimed coal gob pile in the Shanxi mining area, China. *Sci. Total Environ.* **2018**, *618*, 1314–1323. [CrossRef]
31. Zhu, Y.; Yang, Q.; Lu, T.; Qi, W.; Zhang, H.; Wang, M.; Qi, Z.; Chen, W. Effect of phosphate on the adsorption of antibiotics onto iron oxide minerals: Comparison between tetracycline and ciprofloxacin. *Ecotoxicol. Environ. Saf.* **2020**, *205*, 111345. [CrossRef] [PubMed]
32. Saghir, S.; Fu, E.; Xiao, Z. Synthesis of CoCu-LDH nanosheets derived from zeolitic imidazole framework-67 (ZIF-67) as an efficient adsorbent for azo dye from waste water. *Microporous Mesoporous Mater.* **2020**, *297*, 110010. [CrossRef]
33. Zhong, X.; Lu, Z.; Liang, W.; Hu, B. The magnetic covalent organic framework as a platform for high-performance extraction of Cr(VI) and bisphenol a from aqueous solution. *J. Hazard. Mater.* **2020**, *393*, 122353. [CrossRef] [PubMed]
34. Santoso, E.; Ediati, R.; Istiqomah, Z.; Sulistiono, D.O.; Nugraha, R.E.; Kusumawati, Y.; Bahruji, H.; Prasetyoko, D. Facile synthesis of ZIF-8 nanoparticles using polar acetic acid solvent for enhanced adsorption of methylene blue. *Microporous Mesoporous Mater.* **2021**, *310*, 110620. [CrossRef]
35. Brigante, M.; Schulz, P.C. Adsorption of the antibiotic minocycline on cerium(IV) oxide: Effect of pH, ionic strength and temperature. *Microporous Mesoporous Mater.* **2012**, *156*, 138–144. [CrossRef]
36. Zhang, L.; Li, Y.; Wang, W.; Zhang, W.; Zuo, Q.; Abdelkader, A.; Xi, K.; Heynderickx, P.M.; Kim, K.-H. The potential of microplastics as adsorbents of sodium dodecyl benzene sulfonate and chromium in an aqueous environment. *Environ. Res.* **2021**, *197*, 111057. [CrossRef]
37. Alimi, O.S.; Farner Budarz, J.; Hernandez, L.M.; Tufenkji, N. Microplastics and Nanoplastics in Aquatic Environments: Aggregation, Deposition, and Enhanced Contaminant Transport. *Environ. Sci. Technol.* **2018**, *52*, 1704–1724. [CrossRef]
38. Lv, B.; Dong, B.; Zhang, C.; Chen, Z.; Zhao, Z.; Deng, X.; Fang, C. Effective adsorption of methylene blue from aqueous solution by coal gangue-based zeolite granules in a fluidized bed: Fluidization characteristics and continuous adsorption. *Powder Technol.* **2022**, *408*, 117764. [CrossRef]
39. Wu, J.; Yan, X.; Li, L.; Gu, J.; Zhang, T.; Tian, L.; Su, X.; Lin, Z. High-efficiency adsorption of Cr(VI) and RhB by hierarchical porous carbon prepared from coal gangue. *Chemosphere* **2021**, *275*, 130008. [CrossRef]
40. Zhang, W.; Lan, Y.; Ma, M.; Chai, S.; Zuo, Q.; Kim, K.-H.; Gao, Y. A novel chitosan–vanadium-titanium-magnetite composite as a superior adsorbent for organic dyes in wastewater. *Environ. Int.* **2020**, *142*, 105798. [CrossRef]
41. Li, H.; Zheng, F.; Wang, J.; Zhou, J.; Huang, X.; Chen, L.; Hu, P.; Gao, J.-m.; Zhen, Q.; Bashir, S.; et al. Facile preparation of zeolite-activated carbon composite from coal gangue with enhanced adsorption performance. *Chem. Eng. J.* **2020**, *390*, 124513. [CrossRef]
42. Bo, L.; Gao, F.; Shuangbao; Bian, Y.; Liu, Z.; Dai, Y. A novel adsorbent Auricularia Auricular for the removal of methylene blue from aqueous solution: Isotherm and kinetics studies. *Environ. Technol. Innov.* **2021**, *23*, 101576. [CrossRef]
43. Li, R.; Zhang, Y.; Deng, H.; Zhang, Z.; Wang, J.J.; Shaheen, S.M.; Xiao, R.; Rinklebe, J.; Xi, B.; He, X.; et al. Removing tetracycline and Hg(II) with ball-milled magnetic nanobiochar and its potential on polluted irrigation water reclamation. *J. Hazard. Mater.* **2020**, *384*, 121095. [CrossRef] [PubMed]
44. Zhao, R.; Wang, B.; Zhang, X.; Lee, X.; Chen, M.; Feng, Q.; Chen, S. Insights into Cr(VI) removal mechanism in water by facile one-step pyrolysis prepared coal gangue-biochar composite. *Chemosphere* **2022**, *299*, 134334. [CrossRef] [PubMed]

45. Saghir, S.; Xiao, Z. Facile preparation of metal-organic frameworks-8 (ZIF-8) and its simultaneous adsorption of tetracycline (TC) and minocycline (MC) from aqueous solutions. *Mater. Res. Bull.* **2021**, *141*, 111372. [CrossRef]
46. Parolo, M.E.; Avena, M.J.; Pettinari, G.R.; Baschini, M.T. Influence of Ca^{2+} on tetracycline adsorption on montmorillonite. *J. Colloid Interface Sci.* **2012**, *368*, 420–426. [CrossRef]
47. Zhao, Y.; Tong, F.; Gu, X.; Gu, C.; Wang, X.; Zhang, Y. Insights into tetracycline adsorption onto goethite: Experiments and modeling. *Sci. Total Environ.* **2014**, *470–471*, 19–25. [CrossRef] [PubMed]
48. Liu, G.; Zhu, Z.; Yang, Y.; Sun, Y.; Yu, F.; Ma, J. Sorption behavior and mechanism of hydrophilic organic chemicals to virgin and aged microplastics in freshwater and seawater. *Environ. Pollut.* **2019**, *246*, 26–33. [CrossRef] [PubMed]
49. Sun, B.; Hu, Y.; Cheng, H.; Tao, S. Releases of brominated flame retardants (bfrs) from microplastics in aqueous medium: Kinetics and molecular-size dependence of diffusion. *Water Res.* **2019**, *151*, 215–225. [CrossRef] [PubMed]
50. Guo, Y.; Yan, K.; Cui, L.; Cheng, F. Improved extraction of alumina from coal gangue by surface mechanically grinding modification. *Powder Technol.* **2016**, *302*, 33–41. [CrossRef]
51. Mei, Y.; Xu, J.; Zhang, Y.; Li, B.; Fan, S.; Xu, H. Effect of Fe–N modification on the properties of biochars and their adsorption behavior on tetracycline removal from aqueous solution. *Bioresour. Technol.* **2021**, *325*, 124732. [CrossRef] [PubMed]
52. Wu, Y.; Yang, M.; Long, D.; Yang, F.; Luo, S. Tetracycline Adsorption on Magnetic Sludge Biochar: Effects of pH, Humic Acid (HA), and Fulvic Acid (FA). *Micromachines* **2022**, *13*, 1057. [CrossRef] [PubMed]
53. Wang, B.; Ma, Y.; Lee, X.; Wu, P.; Liu, F.; Zhang, X.; Li, L.; Chen, M. Environmental-friendly coal gangue-biochar composites reclaiming phosphate from water as a slow-release fertilizer. *Sci. Total Environ.* **2021**, *758*, 143664. [CrossRef] [PubMed]
54. Li, C.; Huang, Y.; Dong, X.; Sun, Z.; Duan, X.; Ren, B.; Zheng, S.; Dionysiou, D.D. Highly efficient activation of peroxymonosulfate by natural negatively-charged kaolinite with abundant hydroxyl groups for the degradation of atrazine. *Appl. Catal. B: Environ.* **2019**, *247*, 10–23. [CrossRef]
55. Zuo, S.; Cao, X.; Liu, W.; Liu, T.; Li, X.; Yao, C.; Xu, R.; Fu, Y. Effective adsorption of zeolite/carbon composite molecular sieve synthesized from spent bleaching earth. *Environ. Sci. Pollut. Res.* **2022**, *29*, 25916–25924. [CrossRef]
56. Ye, T.; Min, X.; Li, X.; Zhang, S.; Gao, Y. Improved holding and releasing capacities of coal gangue toward phosphate through alkali-activation. *Chemosphere* **2022**, *287*, 132382. [CrossRef]
57. Zhao, R.; Zhang, X.; Su, Y.; Liu, Z.; Du, C. Unprecedented catalytic activity of coal gangue toward environmental remediation: Key role of hydroxyl groups. *Chem. Eng. J.* **2020**, *380*, 122432. [CrossRef]
58. Zhang, L.; Zhu, D.; Wang, H.; Hou, L.; Chen, W. Humic acid-mediated transport of tetracycline and pyrene in saturated porous media. *Environ. Toxicol. Chem.* **2012**, *31*, 534–541. [CrossRef]
59. Zhou, L.; Zhou, H.; Hu, Y.; Yan, S.; Yang, J. Adsorption removal of cationic dyes from aqueous solutions using ceramic adsorbents prepared from industrial waste coal gangue. *J. Environ. Manag.* **2019**, *234*, 245–252. [CrossRef]
60. Liu, X.; Zhao, X.; Yin, H.; Chen, J.; Zhang, N. Intermediate-calcium based cementitious materials prepared by MSWI fly ash and other solid wastes: Hydration characteristics and heavy metals solidification behavior. *J. Hazard. Mater.* **2018**, *349*, 262–271. [CrossRef]
61. Leypold, C.F.; Reiher, M.; Brehm, G.; Schmitt, M.O.; Schneider, S.; Matousek, P.; Towrie, M. Tetracycline and derivatives—Assignment of IR and Raman spectra via DFT calculations. *Phys. Chem. Chem. Phys.* **2003**, *5*, 1149–1157. [CrossRef]
62. Zhao, Y.; Gu, X.; Gao, S.; Geng, J.; Wang, X. Adsorption of tetracycline (TC) onto montmorillonite: Cations and humic acid effects. *Geoderma* **2012**, *183–184*, 12–18. [CrossRef]
63. Li, Z.; Chang, P.H.; Jean, J.S. Interaction between tetracycline and smectite in aqueous solution. *J. Colloid Interface Sci.* **2010**, *341*, 311–319. [CrossRef] [PubMed]
64. Li, Z.; Schulz, L.; Ackley, C.; Fenske, N. Adsorption of tetracycline on kaolinite with pH-dependent surface charges. *J. Colloid Interface Sci.* **2010**, *351*, 254–260. [CrossRef] [PubMed]
65. Chang, P.-H.; Li, Z.; Jean, J.-S.; Jiang, W.-T.; Wang, C.-J.; Lin, K.-H. Adsorption of tetracycline on 2:1 layered non-swelling clay mineral illite. *Appl. Clay Sci.* **2012**, *67–68*, 158–163. [CrossRef]
66. Chang, P.-H.; Li, Z.; Jean, J.-S.; Jiang, W.-T.; Wu, Q.; Kuo, C.-Y.; Kraus, J. Desorption of tetracycline from montmorillonite by aluminum, calcium, and sodium: An indication of intercalation stability. *Int. J. Environ. Sci. Technol.* **2014**, *11*, 633–644.
67. Cheng, X.; Zhang, X.; Su, D.; Wang, Z.; Chang, J.; Ma, C. NO reduction by CO over copper catalyst supported on mixed CeO_2 and Fe_2O_3: Catalyst design and activity test. *Appl. Catal. B: Environ.* **2018**, *239*, 485–501. [CrossRef]
68. Chen, Y.; Wang, C.; Wu, Z.; Xiong, Y.; Xu, Q.; Yu, S.; Jiang, H. From Bimetallic Metal-Organic Framework to Porous Carbon: High Surface Area and Multicomponent Active Dopants for Excellent Electrocatalysis. *Adv. Mater.* **2015**, *27*, 5010–5016. [CrossRef]

Disclaimer/Publisher's Note: The statements, opinions and data contained in all publications are solely those of the individual author(s) and contributor(s) and not of MDPI and/or the editor(s). MDPI and/or the editor(s) disclaim responsibility for any injury to people or property resulting from any ideas, methods, instructions or products referred to in the content.

toxics

Article

Advection–Dispersion Behavior for Simulation of H-3 and Pu-238 Transport in Undisturbed Argillaceous Shale of a Near-Surface Repository

Yunfeng Shi [1], Song Yang [1], Enhui Wu [2], Longjiang Wang [1], Wenjie Chen [1], Weijia Xiong [3], Yanna Zhang [1], Aiming Zhang [1,*] and Bing Lian [1,*]

1 Department of Nuclear Environmental Science, China Institute for Radiation Protection (CIRP), Taiyuan 030006, China
2 School of Nuclear Science and Engineering, East China University of Technology, Nanchang 330013, China
3 CNNC Environmental Protection Corporation (ECPC), Beijing 100045, China
* Correspondence: 13834508698@163.com (A.Z.); bennis6723@ecut.edu.cn (B.L.); Tel.: +86-138-0350-8698 (A.Z.); +86-136-0358-4701 (B.L.)

Abstract: In this study, a column experiment was employed to evaluate the nuclide migration behavior in the surrounding rock medium of a near-surface disposal site in China and to investigate the advection–dispersion behavior of tritium (H-3) and plutonium-238 (Pu-238) in highly weathered argillaceous shale. A reasonable numerical model was selected to fit the experimental breakthrough curves (BTCs) and to obtain the relevant migration parameters. The results show the following: (1) the internal structure of the highly weathered argillaceous shale exhibited heterogeneity, and the nuclide migration BTC showed characteristics of a "curve peak moving forward" and a "tail curve trailing"; (2) compared with other models, the stream tube mode could better fit the BTCs and obtain the average dispersion coefficient $\langle D \rangle$, average distribution coefficient $\langle K_d \rangle$, and other parameters; (3) compared to the results of the batch experiment, the distribution coefficient K_d obtained from the column experiment was smaller than that obtained from the batch experiment, which is speculated to be due to the influence of contact time and the contact area between the nuclide and the medium.

Keywords: Pu-238; H-3; undisturbed column experiment; batch experiment; stream tube model

Citation: Shi, Y.; Yang, S.; Wu, E.; Wang, L.; Chen, W.; Xiong, W.; Zhang, Y.; Zhang, A.; Lian, B. Advection–Dispersion Behavior for Simulation of H-3 and Pu-238 Transport in Undisturbed Argillaceous Shale of a Near-Surface Repository. *Toxics* **2023**, *11*, 124. https://doi.org/10.3390/toxics11020124

Academic Editors: Yuezhou Wei, Yan Wu and Xinpeng Wang

Received: 2 January 2023
Revised: 20 January 2023
Accepted: 24 January 2023
Published: 27 January 2023

Copyright: © 2023 by the authors. Licensee MDPI, Basel, Switzerland. This article is an open access article distributed under the terms and conditions of the Creative Commons Attribution (CC BY) license (https://creativecommons.org/licenses/by/4.0/).

1. Introduction

Radioactive waste can be classified as high-level, intermediate-level, and low-level wastes. Currently, the international common disposal method for intermediate-level and low-level wastes is to use near-surface treatment in order to ensure that there are no adverse impacts on biological health within the design lifetime [1]. In the Safety Case concept proposed by the International Atomic Energy Agency (IAEA), near-surface repositories require further research to understand the migration behavior of radionuclides in the surrounding rock in order to determine the key migration parameters [2–4]. Argillaceous shale is a rock type that is widely distributed in China. It is the main wall rock of a near-surface disposal site in China because of its strong adsorption capacity for nuclides and water swelling property [5,6].

Pu (atomic number 94) is mainly available in the forms of Pu-238, Pu-239, and Pu-240 nuclides in low-level and intermediate-level radioactive wastes, in which Pu-238 ($T_{1/2} = 8.77 \times 10^1$ years), mainly from tests of relevant nuclear weapons, exhibits a long life, high activity, and high toxicity, and it is the key nuclide in low- and intermediate-level radioactive wastes [7,8]. There have been many studies on the adsorption and diffusion behaviors of Pu in different media [9]. For example, Xie studied the adsorption dynamic behavior of Pu-239 on sand, highly weathered granite, and silty clay using a batch experiment [10]. Dang studied the diffusion behavior of Pu in granite through a penetration diffusion experiment and discussed the influences of the chemical forms of Pu on the diffusion behavior [11]. In terms of convection dispersion migration behavior, although a lot of studies have reported the convection dispersion behavior of Pu in different media [12–14], most of the media are porous filling media, and there is limited research on the migration behavior of Pu in undisturbed fractured media in the field.

Column experiments can be used to study the advection–dispersion migration behavior of solutes in various solid media [15,16]. According to the type of solid media used, column experiments can be divided into packed column and undisturbed column experiments. Most studies have utilized the undisturbed column experiment, and it has always been favored by scholars because it better reflects the actual situation of the field formation medium. However, it has disadvantages, such as difficulties in obtaining and storing undisturbed media and in eliminating the column wall effect [17–19].

The numerical models that can be used to simulate fractured media include the equivalent porous medium model, the dual porosity/feasibility model, the stream tube model, the discrete frame network model, and the hybrid approach model [20]. The equivalent porous medium and dual porosity/feasibility models are widely used for indoor small-scale column experiments. The internal structure is mostly homogeneous for columns with highly weathered fractured media (or columns filled with fractured media), and a good fitting effect can be achieved when using an equivalent porous media model [21,22]. The weathering degree of fractured media is weak, the internal structure is highly heterogeneous, and the solute breakthrough curves (BTCs) have a clear "preferential flow" phenomenon. Using a dual porosity/feasibility model for such a medium can have a good fitting effect [23,24]. However, for highly weathered fractured media, due to the strong heterogeneity of the internal structure and the obvious tailing phenomenon of the BTCs, research on a numerical model that can be used to better fit the penetration curve is rarely reported.

As mentioned above, at present, the following problems exist when attempting to determine the migration behavior of radionuclides in groundwater environments: (1) in column experiments, the columns are mostly filled with media, and in undisturbed medium experiments, the columns are filled with too little media, and (2) in terms of numerical models, although a lot of numerical models have been established, there are few numerical simulation experiments of strongly weathered fractures. Therefore, in this study, batch and column experiments were used to study the adsorption behavior and convection dispersion migration behavior of H-3 and Pu-238 in the surrounding rock medium of a near-surface repository in China. Different numerical models (the equivalent porous medium model, the two-region nonequilibrium transport model, and the stream tube model) were fitted to the solute penetration curve (BTC), and the key migration parameters were determined. The research results can provide technical support for studying the radionuclide adsorption retardation performance of the surrounding rock medium of low- and intermediate-level solid waste disposal sites.

2. Theory of Advection and Dispersion

2.1. Equivalent Porous Medium Model

The equivalent porous medium model is a numerical model that equates the fractured medium to the porous medium. It can be applied to the numerical simulation calculation of water bodies and solute transport behavior in homogeneous water passages generated by strongly weathered fractures. Its one-dimensional equation is as follows:

$$\frac{\partial}{\partial t}(\theta c_r + \rho_b s) = \frac{\partial}{\partial x}\left(\theta D \frac{\partial c_r}{\partial x} - J_w c\right) - \theta \mu_l c_r - \rho_b \mu_s s + \theta r_l(x) + \rho_b r_s(x) \quad (1)$$

where c_r is the volume-averaged or resident concentration of the liquid phase (ML^{-3}); s is the concentration of the adsorbed phase (MM^{-1}); D is the dispersion coefficient (L^2T^{-1}); θ is the volumetric water content (L^3L^{-3}); J_w is the volumetric water flux density (LT^{-1}); ρ_b is the soil bulk density (ML^{-3}); μ_l and μ_s are the first-order decay coefficients for the degradation of the solute in the liquid and adsorbed phases, respectively (T^{-1}); r_l ($ML^{-3}T^{-1}$) and r_s ($MM^{-1}T^{-1}$) are the zero-order production terms for the liquid and adsorbed phases, respectively; x is the distance (L); and t is the time (T).

The solute adsorption by the solid phase is expressed using a linear isotherm as follows:

$$s = K_d c_r \tag{2}$$

where K_d is an empirical distribution constant ($M^{-1}L^3$). Using (2) and assuming a steady-state flow in a homogeneous soil, (1) may be rewritten as follows:

$$R\frac{\partial c_r}{\partial t} = D\frac{\partial^2 c_r}{\partial x^2} - v\frac{\partial c_r}{\partial x} - \mu c_r + r(x) \tag{3}$$

where v ($v = \frac{J_w}{\theta}$) is the average pore-water velocity; R is the retardation factor given by $R = 1 + \frac{\rho_b K_d}{\theta}$; and μ and r are combined first- and zero-order rate coefficients: $\mu = \mu_l + \frac{\rho_b K_d \mu_s}{\theta}$, $r(x) = r_l(x) + \frac{\rho_b r_s(x)}{\theta}$, respectively.

2.2. Two-Region Nonequilibrium Model

The two-region nonequilibrium model divides the fractured media into mobile (flowing) and immobile (stagnant) regions according to permeability. The model has a good fitting effect for the "preferential flow" generated by large fractures and the tailing phenomenon of the solute BTCs due to the existence of stagnant regions. The two-region solute transport model is given by

$$\begin{gathered}(\theta_m + f\rho_b K_d)\frac{\partial c_m}{\partial t} = \theta_m D_m \frac{\partial^2 c_m}{\partial x^2} - J_w \frac{\partial c_m}{\partial x} - a(c_m - c_{im}) - \\ (\theta_m \mu_{l,m} + f\rho_b K_d \mu_{s,m})c_m + \theta_m \gamma_{l,m}(x) + f\rho_b \gamma_{s,m}(x)\end{gathered} \tag{4}$$

$$\begin{gathered}(\theta_m + (1-f)\rho_b K_d)\frac{\partial c_{im}}{\partial t} = a(c_m - c_{im}) - \\ (\theta_{im}\mu_{l,m} + (1-f)\rho_b K_d \mu_{s,im})c_{im} + \theta_{im}\gamma_{l,m}(x) + (1-f)\rho_b \gamma_{s,im}(x)\end{gathered} \tag{5}$$

where the subscripts m and im refer to the mobile and immobile liquid regions, respectively; $J_w = v\theta = v_m\theta_m$ is the volumetric water flux density (LT^{-1}); f represents the fraction of adsorption sites that equilibrates with the mobile liquid phase; and a is a first-order mass transfer coefficient (T^{-1}) governing the rate of solute exchange between the mobile and immobile liquid regions. Note that $\theta = \theta_m + \theta_{im}$.

2.3. Stream Tube Model

The simulation medium is regarded as a series of mutually independent vertical soil columns. The flow in these columns is regarded as a pipe flow, and the equilibrium model describes the flow process in each pipe. However, the migration parameters in each pipe flow are randomly distributed. The pairs of stochastic parameters in the stream tube model for transport in each stream tube are obtained from a bivariate lognormal joint probability density function. Because of their relatively low coefficient of variation, the same values for θ and ρ_b are used for each stream tube. The joint probability density functions of v, in conjunction with D, K_d, and a, are written as f(v,D), f(v,K_d), and f(v,a), respectively. The general bivariate lognormal joint pdf is defined as follows:

$$f(v,\eta) = \frac{1}{2\pi\sigma_v\sigma_\eta v\eta\sqrt{1-\rho_{v\eta}^2}} exp\left(-\frac{Y_v^2 - 2\rho_{v\eta}Y_v Y_\eta + Y_\eta^2}{2\left(1-\rho_{v\eta}^2\right)}\right) \tag{6}$$

$$Y_v = \frac{ln(v) - \mu_v}{\sigma_v} \tag{7}$$

$$Y_\eta = \frac{ln(\eta) - \mu_\eta}{\sigma_\eta} \tag{8}$$

$$\rho_{v\eta} = \langle Y_v Y_\eta \rangle = \int_0^\infty \int_0^\infty Y_v Y_\eta f(v,\eta) dv d\eta \quad (9)$$

where η denotes D, K_d, or a; μ and σ are the mean and standard deviation of the log-transformed variable, respectively; and $\rho_{v\eta}$ is the correlation coefficient between Y_v and Y_η.

3. Experiment

3.1. Experimental Device

The experimental apparatus shown in Figure 1 was used for the dynamic flow column system. The device mainly includes a high-performance glass cylinder (modified type, Shanghai SuKe Industrial Co., Ltd., Shanghai, China); a high-performance and precise isocratic flow pump (MASTERFLEX L/S, Cole-Parmer Instrument Co., Barrington, IL, USA); an auto-fractional collector (BXZ-100, Shanghai Huxi Industrial Co., Ltd., Shanghai, China); and two reservoirs that contained groundwater (GW), a radiotracer (H-3/Pu-238), and the associated Teflon® (PTFE) tubes and port connectors. The glass geometry is listed in Table 1, and its cuboid consisted of a pressure-resistant glass (<1 MPa) with a length of 360 mm, a width of 120 mm, and a height of 120 mm. It was filled with 12.39 ± 0.50 g of undisturbed rock in a bulk density of 2.39 ± 0.05 g/cm^3. A pressure sensor and switch drain valves were equipped within the flowing column system, providing a constant and stable liquid flow rate.

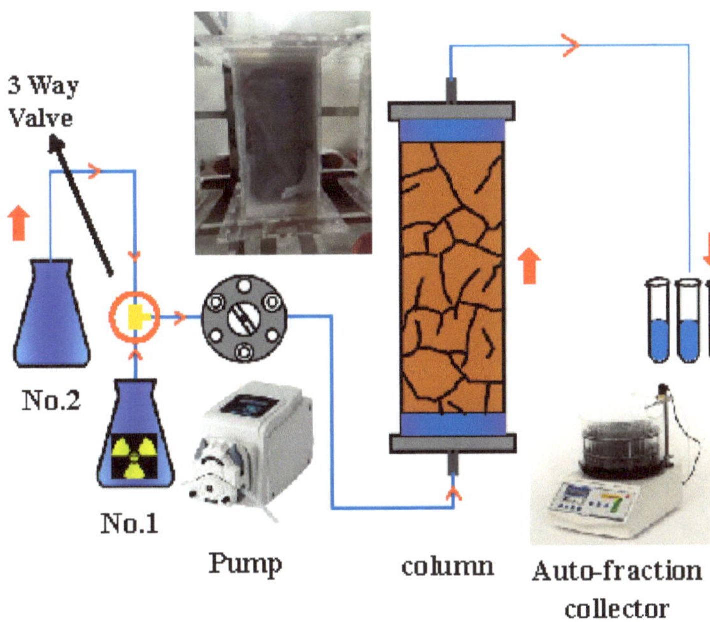

Figure 1. A reliable ADE device with a glass column apparatus in this study.

Table 1. The experimental glass filled and compacted with crushed granite.

Item	Location	Length × Width × Height (cm)	Weight (g)	Density (g/cm^3)	Porosity
Argillaceous shale	Near-surface repository in China	36 × 12 × 12	12.39 ± 0.50	2.39 ± 0.05	0.18 ± 0.03

3.2. Field Sample

The highly weathered argillaceous shale and groundwater samples were collected from a near-surface repository in southern China. The "freezing and grinding" method was used to ensure the collection of undisturbed, highly weathered argillaceous shale samples. After the fragile argillaceous shale was frozen and fixed by liquid nitrogen, it was polished into a cube using an angle grinder and other equipment. The sampling process is shown in Figure 2.

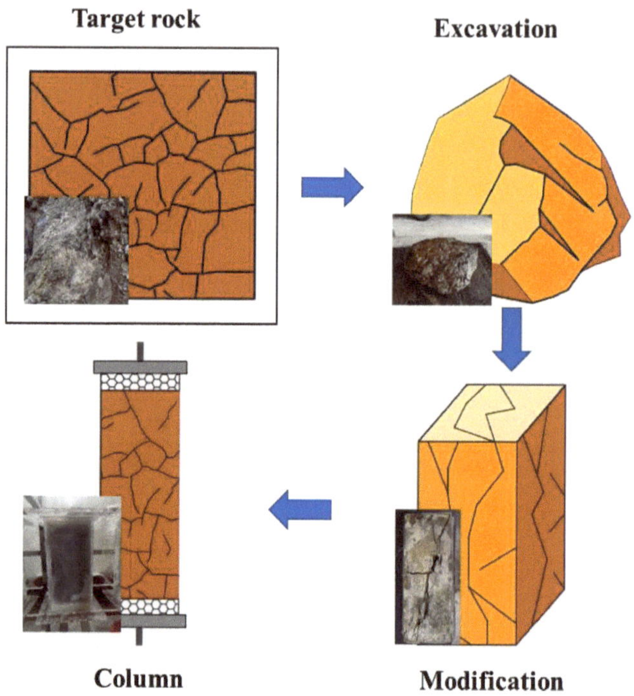

Figure 2. Collection process of shale samples.

3.3. Analysis of Argillaceous Shale and Groundwater

An X-ray diffraction spectrometer (D/max-2500PC), a wavelength-dispersive X-ray fluorescence spectrometer (Axios-mAX), and an ultraviolet–visible spectrophotometer (UV-1100) were used to analyze the mineral composition, chemical composition, organic matter, and cation exchange capacity of the rock samples. The test results are shown in Tables 2 and 3. In addition, a conductivity meter (FE38), an ion chromatograph (ICS-1100), an automatic potentiometric titrator (AT-510), and a plasma mass spectrometer (ELEMENT XR) were used to analyze the chemical composition of the groundwater. The test results are shown in Table 4.

Table 2. Analysis results of argillaceous shale.

Item	SiO_2 (%)	Al_2O_3 (%)	Fe_2O_3 (%)	MgO (%)	CaO (%)	Na_2O (%)
Argillaceous shale	60.02	17.50	6.89	2.06	0.516	0.740
	K_2O (%)	MnO (%)	TiO_2 (%)	P_2O_5 (%)	Loss on ignition (%)	FeO (%)
	3.90	0.09	0.699	0.111	7.42	1.11
	Sr (μg/L)	Ni (μg/L)	Pu (ppm)	Organic matter (g/kg)	Cation exchange capacity (cmol/kg)	
	85.3	41.6	<1	22.1	21.5	

Table 3. Argillaceous shale mineral (%).

Item	Dolomite	Calcite	Quartz	Plagioclase	Potash Feldspar
Argillaceous shale	8	-	28	5	-
	Illite	Amphibole	Chlorite	Pyrite	Montmorillonite
	31	-	2	-	26

Table 4. Analysis results of groundwater.

Item	F^-	Cl^-	NO_3^-	SO_4^{2-}	Na^+	K^+	Mg^{2+}	Ca^{2+}
Groundwater	(mg/L)							
	0.487	1.82	2.23	42.9	24.3	14.5	2.89	19.1
	Sr	Ni	Pu	Conductivity	pH	Turbidity (UNT)		
	μg/L		ppm	μS/cm				
	257	1.1	<1	271	8.19	0.43		

3.4. Batch Sorption Test

The experiment was used to test the adsorption capacity of the highly weathered argillaceous shale on Pu-238 and to determine the distribution coefficient as follows:

$$K_d = \frac{Concentration\ of\ nuclides\ per\ unit\ mass\ of\ soild\ medium\ (\mathrm{Bq/kg})}{Concentration\ of\ nuclides\ per\ unit\ volume\ of\ water\ body\ (\mathrm{Bq/L})} \quad (10)$$

First, the shale was crushed and sieved to achieve a particle size below 75 μm. Then, 1 g of crushed shale was added to 9 mL of groundwater into a 15 mL centrifuge tube, and 1 mL of Pu-238 solution (2000 Bq) was added. After shaking and mixing for 7 days, the test tube was centrifuged for 20 min at 800 r/min in a centrifuge, and the supernatant was extracted to measure the content of Pu-238. The calculated distribution coefficient of the Pu-238 on the strongly weathered argillaceous shale was equal to 2170 L/kg. In addition, the pH = 3.50 ± 0.20 and Eh = 0.24 ± 0.02 V of the liquid sample were measured after the experiment, and Pu(III) was the chemical form according to the pH–Eh diagram [25].

3.5. Water Saturation

A peristaltic pump was opened, and the valve was adjusted to pass the groundwater into the column from the bottom at a flow rate of 3 mL/min in order to ensure that all the pores in the whole column were filled with groundwater. The concentrations of Na, Mg, Ca, and K in the effluent were measured using inductively coupled plasma–optical emission spectrometry (ICP–OES, iCAP 7000, Thermo, Waltham, MA, USA) at various concentrations within 5% of the corresponding liquid-phase concentrations, and it was determined that water saturation had been reached.

3.6. Nonreactive Tests—H-3

The migration behavior of radionuclides can be affected by the structure of the water passage and the chemical reaction simultaneously. To determine the structure of the water passage in the medium, a non-reactive tracer (H-3) was used to perform a migration behavior experiment under advection–dispersion conditions. H-3 (500 Bq/mL) was placed into container 1, and the underground water sample was placed into container 2. The peristaltic pump was opened to inject groundwater into the column at a flow rate of 25 mL/min. After the outflow was stabilized, the valve was adjusted to inject the H-3 solution 1000 Bq (2 mL), and then the valve was quickly adjusted to continue injecting the groundwater. The effluent was collected at the outlet every 1 min, and a liquid scintillator analyzer (LSA3000, Shanghai SIM-MAX Technology Co., Ltd, Shanghai, China) was used to analyze the H-3 and Pu-238 activities in the test solution with a scintillation cocktail. The experiment was concluded when the H-3 content in the effluent was lower than the lower detection limit, and a BTC was obtained.

3.7. Reactive Tests—Pu-238

After the H-3 solution in container 1 was replaced with Pu-238 (2000 Bq/mL), Pu-238 was injected into the column via an instantaneous injection according to the same experimental process. The groundwater flow was maintained at 25 mL/min, and it was continuously injected into the argillaceous shale. When the outflow was stable, the valve was adjusted to inject a 4000 Bq (2 mL) Pu-238 solution, and then it quickly adjusted to continue injecting the groundwater. The outflow was collected at the outflow outlet every 1 min. A liquid scintillator was used to analyze and test the activity of Pu-238 in the solution. The experiment was concluded when the content of Pu-238 in the outflow was lower than the lower limit of detection, and a BTC was obtained.

3.8. Mathematical Model and Parameter Estimations

STudio of ANalytical MODels (STANMOD) is a model package containing several analytical models that solve the solute transport behavior in porous media based on the advection–dispersion equation. It includes a deterministic model and a stochastic model. The deterministic model includes the deterministic equilibrium model and the deterministic nonequilibrium model (the two-site nonequilibrium model and the two-region nonequilibrium model). A stochastic model is a stream tube model.

To determine the fitting effect, the software uses the root mean square error (RMSE), and it is expressed as follows:

$$RMSE = \sqrt{\frac{\sum_{i}^{N}(C^{p} - C^{e})^{2}}{N}} \tag{11}$$

where C^p denotes the results of the numerical fitting, and C^e denotes the experimental data.

4. Results and Discussion

4.1. Breakthrough Curves of H-3

A column experiment was used to obtain the BTC of the advection–dispersion transport of H-3 in highly weathered argillaceous shale. As shown in Figure 3a, the BTC of H-3 showed the phenomena of "peak moving forward" and "tail dragging", which indicates that the water flow resulted in a "preferential flow" inside the column. The equilibrium model, the nonequilibrium model (two-region nonequilibrium model), and the stream tube model were used to fit the BTC of H-3. The fitting effect is shown in Figure 3b–d and Table 5. The equilibrium model (RMSE = 0.930), the two-region nonequilibrium model (RMSE = 0.987), and the stream tube model (RMSE = 0.996) well-reflected the characteristics of the curve. However, the analysis parameters indicated that the dispersion coefficient (D = 19.81 cm^2/min) obtained from the equilibrium model and the dispersion coefficient (D = 17.5 cm^2/min) obtained from the two-region nonequilibrium model were significantly larger, which is inconsistent with the actual situation. Therefore, the stream tube model can better fit the advection–dispersion behavior of H-3 in highly weathered argillaceous shale.

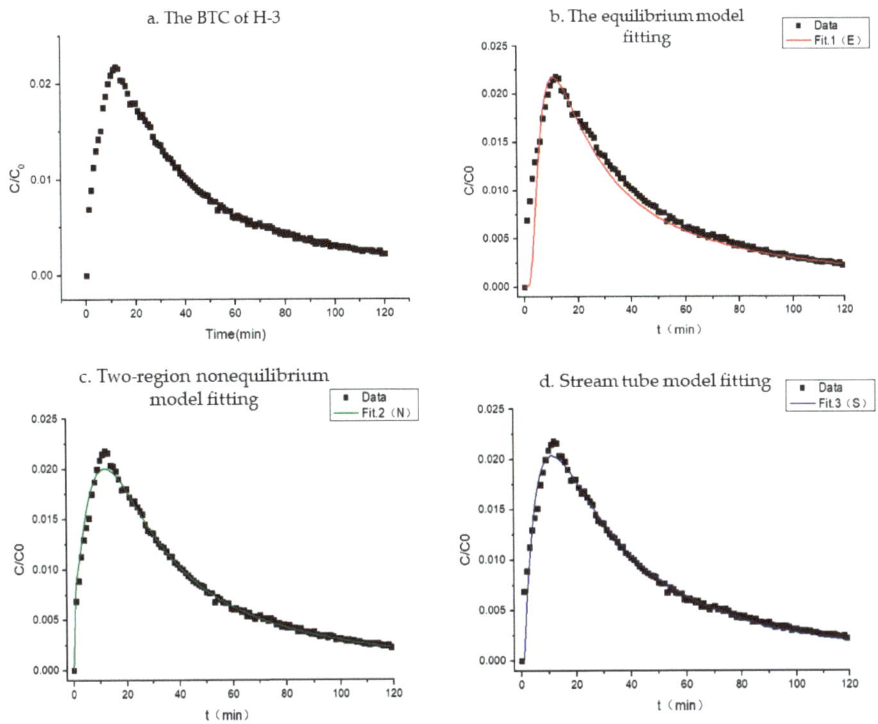

Figure 3. The experimental breakthroughs of H-3 and fitting curves.

Table 5. The parameters for fitting BTCs.

NO	Model	v (cm/min)	D (cm^2/min)	K_d (L/kg)	β ($\frac{\theta_m}{\theta}$)		a		RMSE
FIT1 (H-3)	Equilibrium	0.026	19.812	0	-		-		0.930
FIT2 (H-3)	Two-region non-equilibrium	0.426	17.514	0	0.191		0.832		0.987
NO	model	$\langle v \rangle$ (cm/min)	$\langle D \rangle$ (cm^2/min)	$\langle K_d \rangle$ (L/kg)	σ_V	σ_D	σ_{Kd}	ρ_{Kd*V}	RMSE
FIT3 (H-3)	Stream tube	0.490	0.292	0	1.111	3.120	0		0.996
FIT4 (^{238}Pu)	Stream tube	0.490	0.292	46.924	1.111	3.120	38.512	−1	0.957

4.2. Breakthrough Curves of Pu-238

The BTC of Pu-238 also exhibited the phenomena of "peak moving forward" and "tail dragging". The stream tube model was used to fit the advection–dispersion behavior of Pu-238 in the highly weathered argillaceous shale, and the fitting results are shown in Figure 4 and Table 5. The fitting results of the H-3 'BTC were inputted as fixed values into the model, and other parameters were fitted to obtain the ensemble average distribution coefficient $\langle K_d \rangle$ = 46.9 L/kg and the correlation coefficient between the flow rate and the distribution coefficient $\rho_{Kd\,*\,V} = -1$, indicating that the flow rate was negatively correlated with the distribution coefficient. In addition, the pH = 3.50 ± 0.20 and Eh = 0.24 ± 0.02 V of the liquid sample were measured after the experiment, and Pu(III) was the chemical form according to the pH–Eh diagram.

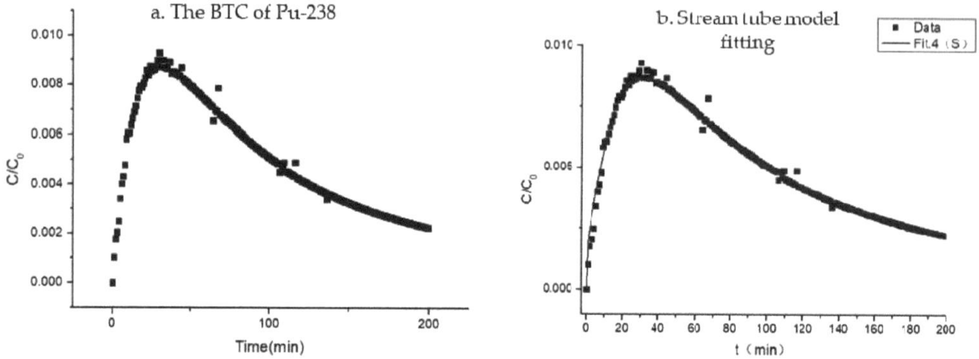

Figure 4. The experimental breakthroughs of Pu-238 and fitting curves.

4.3. Parameter Comparison

The chemical species of Pu is complex, and the chemical species affects the adsorption mechanism of Pu. Therefore, it was necessary to analyze the chemical species of Pu-238 when comparing the distribution coefficients obtained from the batch experiments and the column experiments. According to the measured values of the pH and the Eh of the liquid after the batch and column experiments and the thermodynamic database [25], the chemical species of Pu-238 in the batch and column experiments was determined to be Pu(III). Therefore, the influence of the chemical form of plutonium was not considered when comparing the distribution coefficients obtained from the batch experiments and column experiments.

A comparison of the distribution coefficient of Pu-238 obtained in the batch experiment (K_d = 2170 L/kg) indicated that the distribution coefficient obtained in the column experiment ($<K_d>$ = 46.9 L/kg) was small. The main reason for the difference in the distribution coefficient between the two experiments is the way of sample treatment. Because the argillaceous shale was broken in the batch experiment and the specific surface area of the argillaceous shale was artificially increased, the contact area between the nuclide and the argillaceous shale was larger than that in the column experiment, and Pu-238 could be better adsorbed on the argillaceous shale, so the distribution coefficient obtained in the batch experiment was larger than that obtained in the column experiment. In addition, in order to ensure that the argillaceous shale and the Pu-238 could reach adsorption equilibrium in the batch experiment, the contact time between the argillaceous shale and Pu-238 was extended. The migration of the nuclides in the column experiment was completed under the condition of maintaining a certain flow rate. Therefore, the contact time between the argillaceous shale and Pu-238 was shorter than that in the batch experiment, which also led to the distribution coefficient obtained in the batch experiment being greater than that obtained in the column experiment.

5. Conclusions

This study investigated the advection–dispersion behavior of H-3 and Pu-238 in the fractures of highly weathered argillaceous shale surrounding the rock medium of a near-surface disposal site in China. According to the fitting results of the BTC, an appropriate mathematical model was selected, and the key migration parameters were obtained. The research results are as follows:

1. The advection–dispersion behavior of H-3 and Pu-238 in the highly weathered argillaceous shale showed "preferential flow", reflecting the strong heterogeneity of highly weathered argillaceous shale.
2. The equilibrium model, the nonequilibrium model (in the two-region mode), and the stream tube model were used to fit the BTC. The results show that the stream

tube model can better fit the curve characteristics, indicating that the water-carrying capacity of each channel of media varies greatly and is complex.

3. A comparison of the distribution coefficients obtained from the column and batch experiments indicated that the distribution coefficient obtained from the column experiment was smaller, and an analysis was necessary mainly due to the contact area and contact time.

Author Contributions: Conceptualization, Y.S. and B.L.; methodology, Y.S. and S.Y.; software, Y.S., W.C. and E.W.; validation, B.L. and A.Z.; formal analysis, Y.S., Y.Z. and S.Y.; investigation, W.C. and L.W.; resources, B.L. and A.Z.; data curation, Y.S., W.C. and W.X.; writing—original draft preparation, Y.S. and S.Y.; writing—review and editing, Y.S. and S.Y.; visualization, Y.S. and S.Y.; supervision, Y.S. and B.L.; project administration, Y.S. and B.L.; funding acquisition, A.Z. and B.L. All authors have read and agreed to the published version of the manuscript.

Funding: This project was mainly supported by the "Fundamental Science on Radioactive Geology and Exploration Technology Laboratory" (No.2022RGET02); East China University of Technology, Shanxi Province Basic research Project (No.20210302124176); and The National Natural Science Foundation of China (No. U2267218).

Institutional Review Board Statement: Not applicable for studies not involving humans or animals.

Informed Consent Statement: Not applicable.

Data Availability Statement: Not applicable.

Acknowledgments: The experimentation conducted in this study was assisted by Neng-Chuan Tien, members of the Chemistry and Analysis Division of the Institute of Nuclear Energy Research, Taiwan. Assistance in conducting the experimental and instrumental analyses of this study was provided by the East China University of Technology, Nanchang, Jiangxi, China.

Conflicts of Interest: The authors declare no conflict of interest.

References

1. Zheng, J.-M.; Yin, J.-P.; Chen, H.; Wang, F.-L.; Guo, Y.-F.; Guo, X. Current Status of Foreign Radioactive Waste Management. *Nucl. Electron. Detect. Technol.* **2021**, *41*, 821–827.
2. International Atomic Energy Agency. *Contents and Sample Arguments of a Safety Case for Near Surface Disposal of Radioactive Waste*; IAEA-TECDOC-1814; IAEA: Vienna. Austria, 2017.
3. International Atomic Energy Agency. *The Safety Case and Safety Assessment for the Disposal of Radioactive Waste*; IAEA Safety Standards Series No.SSG-23; IAEA: Vienna, Austria, 2012.
4. International Atomic Energy Agency. *Near Surface Disposal Facilities for Radioactive Waste*; IAEA Safety Standards Serie No. SSG-29; IAEA: Vienna, Austria, 2014.
5. Pan, Z.-Q.; Qian, Q.-H. *Study on Geological Disposal Strategy of High-Level Radioactive Waste*; Atomic Energy Press: Beijing, China, 2009.
6. Zhang, L.-M.; Zhou, Y.; Yue, J.-G. Causes and Regulation Measures of Landslide on High Slope of Feifengshan Low and Medium Level Solid Waste Disposal Site. *Sci. Technol. Eng.* **2020**, *20*, 942–949.
7. Smith, D.K.; Finnegan, D.L.; Bowen, S.M. An inventory of long-lived radionuclides residual from underground nuclear testing at the Nevada test site, 1951–1992. *J. Environ. Radioact.* **2003**, *67*, 35–51. [CrossRef] [PubMed]
8. Woodhead, D. The Radiological Situation at the Atolls of Mururoa and Fangataufa. *J. Radiol. Prot.* **1999**, *19*, 194. [CrossRef]
9. Wang, X.; Liu, C.; Wang, C.; Li, C.; Chen, T. Adsorption and diffusion of some important radionuclides in Beishan granites and Gaomiaozi Bentonites. *Sci. Sin. Chim.* **2020**, *50*, 1585–1599. [CrossRef]
10. Tian, X.I.E.; Chao, C.H.E.N.; Jun, Z.H.U.; Yun-feng, S.H.I.; Ting, L.I.; Ai-ming, Z.H.A.N.G. Adsorption and Kinetics of Uranium, Plutonium and Other Nuclides on Geotechnical Media. *J. Nucl. Radiochem.* **2021**, *43*, 353–361.
11. Haijun, D.; Xiaolin, H.; Wenyuan, L.; Chong LI, U.; Jiang, X.U. Diffusivity of /Sr, I and Pu in Granite with Through-Diffusion Experiment. *J. Nucl. Radiochem.* **2014**, *36*, 53–59.
12. Deng, X.; Tuo, X.; Li, Z.; Leng, Y.; Tian, Q.; Huang, H. Experimental simulation of 238U and 239Pu migration speed in shale and soil medium. *Environ. Chem.* **2012**, *31*, 1948–1952.
13. Chao, C.; Jun, Z.; Aiming, Z.H.A.N.G. The adsorption transfer mechanism of U and 238Pu in sandy aquifers. *Nucl. Tech.* **2019**, *42*, 1–9.
14. Li, Y.; Zhang, A.; Chen, H.; Lian, B.; Zhao, Y. Migration Experiment of 238Pu in Sandy Soil in Vadose Zone of Chinese Arid Area. *Environ. Sci. Technol.* **2017**, *40*, 97–99.

15. Palágyi, Š.; Štamberg, K.; Vopálka, D. A simplified approach to evaluation of column experiments as a tool for determination of radionuclide transport parameters in rock-groundwater or soil-groundwater systems. *J. Radioanal Nucl. Chem.* **2015**, *304*, 945–954. [CrossRef]
16. Palágyi, S.; Štamberg, K. Transport parameters of I- and IO3-determined in crushed rock column and groundwater system under dynamic flow conditions. *J. Radioanal Nucl. Chem.* **2014**, *302*, 647–653. [CrossRef]
17. Lee, C.-P.; Chen, D.; Hu, Y.; Jan, Y.-L.; Shi, Y.; Wang, Z.; Wu, E.; Tien, N.-C.; Sun, Y.; Tsai, S.-C. An Application of Safety Assessment for Radioactive Waste Repository: Non-Equilibrium Transport of Tritium, Selenium, and Cesium in Crushed Granite with Different Path Lengths. *Appl. Sci.* **2021**, *11*, 9750. [CrossRef]
18. Shi, Y.; Lee, C.P.; Yu, H.; Hu, Y.; Liu, H.; Tien, N.C.; Wang, Y.; Liu, W.; Kong, J.; Hua, R.; et al. Study on advection–dispersion behavior for simulation of HTO and Se transport in crushed granite. *J. Radioanal. Nucl. Chem.* **2021**, *328*, 1329–1338. [CrossRef]
19. Shi, Y.; Yang, S.; Chen, W.; Xiong, W.; Zhang, A.; Yu, Z.; Lian, B.; Lee, C.-P. Study on Tritium and Iodine Species Transport through Porous Granite: A Non-Sorption Effect by Anion Exclusion. *Toxics* **2022**, *10*, 540. [CrossRef] [PubMed]
20. Zhang, X.; Ma, F.; Dai, Z.; Wang, J.; Chen, L.; Ling, H.; Soltanian, M.R. Radionuclide Transport in Multi-scale Fractured Rocks: A Review. *J. Hazard. Mater.* **2022**, *424*, 1–23. [CrossRef] [PubMed]
21. Hadgu, T.; Karra, S.; Kalinina, E.; Makedonska, N.; Hyman, J.D.; Klise, K.; Viswanathan, H.S.; Wang, Y. A comparative study of discrete fracture network and equivalent continuum. models for simulating flow and transport in the far field of a hypothetical nuclear waste repository in crystalline host rock. *J. Hydrol.* **2017**, *553*, 59–70. [CrossRef]
22. Jackson, C.P.; Hoch, A.R.; Todman, S. Self-consistency of a heterogeneous continuum porous medium representation of a fractured medium. *Water Resour. Res.* **2000**, *36*, 189–202. [CrossRef]
23. Blessent, D.; Jørgensen, P.R.; Therrien, R. Comparing Discrete Fracture and Continuum Models to Predict Contaminant Transport in Fractured Porous Media. *Groundwater* **2014**, *52*, 84–95. [CrossRef] [PubMed]
24. Reimus, P.; Pohll, G.; Mihevc, T.; Chapman, J.; Haga, M.; Lyles, B.; Kosinski, S.; Niswonger, R.; Sanders, P. Testing and parameterizing a conceptual model for solute transport in a fractured granite using multiple tracers in a forced-gradient test. *Water Resour. Res.* **2003**, *39*, 1–15. [CrossRef]
25. Takeno, N. *Atlas of Eh-pH Diagrams Intercomparison of Thermodynamic Databases, Geological Survey of Japan Open File Report No.419*; National Institute of Advanced Industrial Science and Technology Research Center for Deep Geological Environments: Tokyo, Japan, 2005.

Disclaimer/Publisher's Note: The statements, opinions and data contained in all publications are solely those of the individual author(s) and contributor(s) and not of MDPI and/or the editor(s). MDPI and/or the editor(s) disclaim responsibility for any injury to people or property resulting from any ideas, methods, instructions or products referred to in the content.

toxics

MDPI

Article

Radiation Resistance and Adsorption Behavior of Aluminum Hexacyanoferrate for Pd

Yueying Wen [1], Yan Wu [1,*] and Lejin Xu [2]

[1] School of Nuclear Science and Engineering, Shanghai Jiao Tong University, Shanghai 200240, China
[2] School of Energy and Power Engineering, Huazhong University of Science and Technology, Wuhan 430074, China
* Correspondence: wu_yan@sjtu.edu.cn

Abstract: Irradiation resistance is important for adsorbents used in radioactive environments such as high-level liquid waste. In this work, a silica-based composite adsorbent ($KAlFe(CN)_6/SiO_2$) was synthesized and γ-irradiated from 10 to 1000 kGy. The angles of the main X-ray diffraction peaks slightly decreased with the increase in irradiation dose, and a minor decomposition of CN^- occurred after irradiation to 1000 kGy, indicating that the $KAlFe(CN)_6/SiO_2$ adsorbent could preserve structural integrity with a dose below 100 kGy. In 1 to 7 M HNO_3, the adsorption ability of the irradiated $KAlFe(CN)_6/SiO_2$ remained performant, with a higher K_d than 1625 cm^3 g^{-1}. The adsorption equilibrium of Pd(II) in 3 M HNO_3 was attained within 45 min before and after irradiation. The maximal adsorption capacity Q_e of the irradiated $KAlFe(CN)_6/SiO_2$ on Pd(II) ranged from 45.1 to 48.1 mg g^{-1}. A 1.2% relative drop in Q_e was observed after 100 kGy irradiation, showing that γ-irradiation lower than 100 kGy insignificantly affected the adsorption capacity of $KAlFe(CN)_6/SiO_2$. Calculating and comparing the structures and free energies of different adsorption products via the density functional theory (DFT) method showed that $KAlFe(CN)_6/SiO_2$ was more inclined to completely adsorb Pd(II) and spontaneously generate $Pd[AlFe(CN)_6]_2$.

Keywords: γ irradiation; aluminum hexacyanoferrate; palladium; adsorption site

Citation: Wen, Y.; Wu, Y.; Xu, L. Radiation Resistance and Adsorption Behavior of Aluminum Hexacyanoferrate for Pd. *Toxics* **2023**, *11*, 321. https://doi.org/10.3390/toxics11040321

Academic Editor: Tiziana Missana

Received: 17 March 2023
Revised: 26 March 2023
Accepted: 27 March 2023
Published: 29 March 2023

Copyright: © 2023 by the authors. Licensee MDPI, Basel, Switzerland. This article is an open access article distributed under the terms and conditions of the Creative Commons Attribution (CC BY) license (https://creativecommons.org/licenses/by/4.0/).

1. Introduction

The development of nuclear energy has brought evident benefits in reducing carbon emissions and promoting human health [1,2], although the disposal of spent highly radioactive fuel discharged from nuclear power plants is complex and challenging [3–5] partly because of the limited miscibility of platinum group metals (PGMs) [6]. The plutonium uranium extraction (PUREX) process is commonly used to reprocess spent fuel and produces approximately 500 L of radioactive high-level liquid waste (HLLW) per ton of spent fuel. HLLW is dangerous, and has a complex chemical composition and radiation levels of up to 1000 Ci/L [7]. If HLLW is released into the biosphere, radioactive and heavy-metal elements may migrate to the soil and plants, destroying the genetic material of organisms and human organs [8,9]. Therefore, the proper disposal of HLLW, such as by separating high-heat radionuclides, is necessary to protect humans and the biosphere [10–12].

Glass solidification is widely used to form a solidified product for HLLW that can envelop radioactive elements and reduce waste volume. This is a type of technology that dissolves the ions in HLLW into amorphous borosilicate glass or phosphate glass to form a new, stable structure. Some studies indicated that the separation of Pd is beneficial for the safety of glass solidification [13–15]. Palladium has a low solubility of 0.3 mg g^{-1} in glass, facilitating precipitation during glass solidification [16]. The precipitated substance has high density, viscosity, and electric conductivity, leading to problems such as blocking the outlet and short-circuiting the electrodes in the melter, threatening production safety. In addition, palladium is a precious strategic resource with distinctive properties that is widely used in areas such as catalytic converters in vehicles and high-pressure turbine blades in jet engines [17,18]. Many scholars have investigated recovering valuable resources such as Pd in HLLW [19]. The long-lived radioisotope of Pd in HLLW is ^{107}Pd, which has a low beta decay energy of 0.033 MeV and is essentially safe for most industrial applications [20]. Adsorbents such as hexacyanoferrate [14], nitrogen-donor-type adsorbents (R-Amine) [21], amino acid-functionalized cellulose microspheres [22], magnetic cross-linking chitosan nanoparticles [23], quaternary ammonium group ion exchange resin [24], bayberry tannin immobilized collagen fiber membranes [25], and ligand-immobilized meso-adsorbents [26] were studied to remove Pd. Among these adsorbents, hexacyanoferrate is expected because of its excellent adsorption capacity and easy accessibility [14,27]. Compared with expensive organic extractants, e.g., TODGA, costing CNY 150–500 (about USD 22–73) per gram, hexacyanoferrate is economical as a commercially available conventional reagent, costing CNY 0.1–0.3 (about USD 0.015–0.045) per gram.

Hexacyanoferrate powder is not suitable for the column, and a solid support is required for better surface area. SiO_2 and PAN are commonly used as solid supports for these kinds of systems [28–30], as they have better mechanical and hydraulic properties, and a greater surface area. HLLW is a radioactive and heat-releasing material with high requirements for material irradiation resistance. Generally, inorganic adsorbents have better radiation resistance than that of equivalent organic solvent-extraction and ligand systems [31,32]. However, the effects of irradiation on hexacyanoferrate and palladium removal, despite being important for the industry, have rarely been discussed in existing research.

In this work, silica-based aluminum hexacyanoferrate was selected as the adsorbent. In addition, we study the γ-irradiation resistance and theoretical Pd(II) adsorption products of $KAlFe(CN)_6/SiO_2$. The $KAlFe(CN)_6/SiO_2$ adsorbent was prepared and irradiated with ^{60}Co. The effects of different γ-irradiation doses on the crystal structure of adsorbent were analyzed. The adsorption performance of irradiated $KAlFe(CN)_6/SiO_2$ on Pd(II) at different HNO_3 concentrations was investigated. Adsorption kinetics and adsorption isotherms before and after γ-irradiation were explored. Adsorption products with different adsorption rates and Pd(II) binding sites were calculated by using the DFT method, and their structural characteristics and free energy were compared.

2. Materials and Methods

2.1. Materials

Aluminum chloride hexahydrate ($AlCl_3 \cdot 6H_2O$) with 99% purity and potassium hexacyanoferrate trihydrate ($K_4Fe(CN)_6 \cdot 3H_2O$) with 99.5% purity were purchased from Shanghai Macklin Inc, Shanghai, China. Palladium nitrate dihydrate ($Pd(NO_3)_2 \cdot 2H_2O$) with 99% purity was obtained from Beijing Innochem, Beijing, China. Nitric acid (HNO_3) was obtained from Shanghai Sinopharm, Shanghai, China and was of analytical grade. Silica (SiO_2) was purchased from Fuji Silysia Chemical Ltd., Kasugai, Japan. The conductivity of deionized water used in the experiments was less than 5 $\mu S\ cm^{-1}$.

2.2. Preparation of $KAlFe(CN)_6/SiO_2$

The $KAlFe(CN)_6/SiO_2$ adsorbent was synthesized by using a two-step loading method. First, the silica was washed and dried at 90 °C for 1 day. The 1 M $AlCl_3$ solution and 1 M $K_4Fe(CN)_6$ solution were successively sucked into the silica pores at a phase ratio of 10 $cm^3\ g^{-1}$. Then, the filter flask was left at 25 °C for 1 day to fully react. When the solution had been layered and there was no newly generated precipitate, the product was removed, filtered, washed, and dried at 90 °C for 1 day. According to EDS mapping analysis, the $KAlFe(CN)_6$ accounted for about 30% of the $KAlFe(CN)_6/SiO_2$ composite adsorbent in the mass.

2.3. Irradiation Treatment and Characterization

The $KAlFe(CN)_6/SiO_2$ adsorbent was γ-irradiated in an air atmosphere at 25 °C. The irradiation source was ^{60}Co (10^5 Ci), and the irradiation doses were 10, 50, 100, and 1000 kGy calibrated with a Frick dosing agent at a rate of 5 kGy/h. After irradiation, $KAlFe(CN)_6/SiO_2$ was characterized. Scanning electron microscopy (SEM) was performed with a Mira 3 (TESCAN ORSAY HOLDING a.s., Brno, Czech Republic). An X-ray diffractometer (XRD, D8 Advance, Brucker Corporation, Billerica, MA, USA) and a Fourier transform infrared spectrometer (FT–IR, Nicolet 6700, Thermo Fisher Scientific, Waltham, MA, USA) were used to analyze the crystal structure and chemical bonds of the adsorbents.

2.4. Batch Experiments

Batch experiments were conducted at 25 °C. In a clean vial, a Pd(II) solution (V = 5 mL) was mixed with $KAlFe(CN)_6/SiO_2$ (m = 0.05 g). The supernatant inside the vial was filtered and diluted. The concentration of Pd(II) in the filtrate was obtained by using an inductively coupled plasma (ICP) spectrometer (Shimadzu Corporation, Kyoto, Japan, ICP-7500).

The distribution coefficient of Pd(II) (K_d) was calculated as follows:

$$K_d = [(C_0 - C_e)/C_e] \times (V/m), \quad (1)$$

where C_0 is the concentration of Pd(II) in the preadsorption solution, C_e represents the equilibrium concentration, V is the volume of the Pd(II) solution, and m is the mass of the $KAlFe(CN)_6/SiO_2$ adsorbent.

The experimental adsorption capacity of Pd(II) (Q_e) is defined as follows:

$$Q_e = (C_0 - C_e) \times (V/m). \quad (2)$$

2.5. DFT Calculations

In exploring the adsorption mechanism of Pd(II) by $KAlFe(CN)_6$, the optimized structures and free energies of the adsorbent and adsorption products were calculated on the basis of DFT. Calculations were performed via VASP 6.2. The modeling and visualization of the hexacyanoferrates were conducted in VESTA [33]. Considering that the adsorption capacity may have been different, the Pd(II) products with adsorption rates of 100% ($Pd[AlFe(CN)_6]_2$) and 50% ($Pd_{0.5}K[AlFe(CN)_6]_2$) were calculated. The crystal parameters and free-energy changes caused by the difference in Pd(II) adsorption sites in $Pd_{0.5}K[AlFe(CN)_6]_2$ were compared. The generalized gradient approximation (GGA) method was selected as the wavefunction optimization potential. Calculations were performed in the inverted space by using the Perdew–Burke–Ernzerhof (PBE) functional [34].

3. Results and Discussion

3.1. XRD Analysis before and after Irradiation

In studying whether irradiation changes the crystal structure of $KAlFe(CN)_6$, XRD characterization was performed on the $KAlFe(CN)_6$ powder sample after irradiation (Figure 1). Table 1 shows the angle of incidence and lattice parameters of the $KAlFe(CN)_6$ powder before and after irradiation.

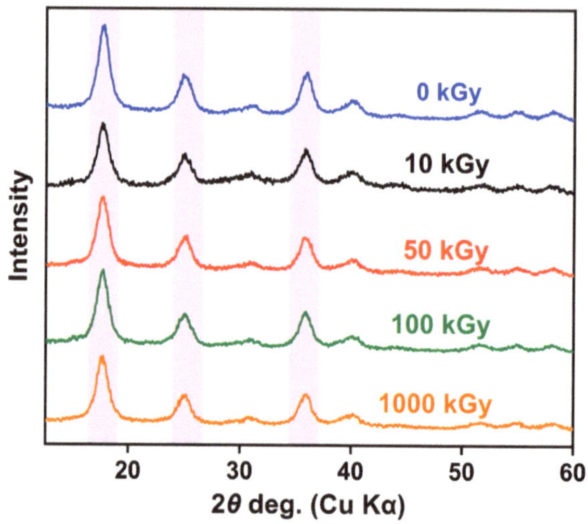

Figure 1. XRD patterns of $KAlFe(CN)_6$ before and after irradiation.

Table 1. Angles of incidence of main diffraction peaks and lattice parameters of $KAlFe(CN)_6$ before and after irradiation.

Irradiation Dose (kGy)	Angle of Incidence 2θ (°)			Lattice Parameters $a = b = c$ (Å)
	Peak 1	Peak 2	Peak 3	
0	17.672	25.028	36.051	10.0155
10	17.661	24.997	35.923	10.0246
50	17.642	24.958	35.884	10.0268
100	17.582	24.938	35.765	10.0323
1000	17.482	24.840	35.746	10.0464

As shown in Figure 1 and Table 1, the lattice of $KAlFe(CN)_6$ expanded slightly after irradiation for 10, 50, 100, or 1000 kGy. The angles of the three main diffraction peaks slightly decreased with the increase in irradiation dose from 17.672°, 25.028°, and 36.051° to 17.482°, 24.840°, and 35.746°, respectively. Nevertheless, the strength and shape of the peaks in the spectrum remained similar throughout irradiation, indicating that irradiation did not cause a notably change in the substance, and the structure of the crystal remained stable. The results can be explained by the action mechanism of irradiation. The effect of γ rays is identical with that of β rays except for the thermal effect [35]. Under γ-ray irradiation, free electrons were formed in the $KAlFe(CN)_6$ structure because of the photoelectric and Compton effects. The electrons interacted with the adsorbent and caused atomic displacements, forming defects such as dislocation loops [36,37].

Further analysis showed that, after irradiation at 1000 kGy, the lattice parameter of the $KAlFe(CN)_6$ adsorbent increased slightly from 10.0155 to 10.0464 Å, and the unit cell remained face-centered cubic. Thus, the irradiation dose from 10 to 1000 kGy hardly affects the lattice structure and irregularity of $KAlFe(CN)_6$, and the radiation resistance of the adsorbent was excellent.

3.2. SEM and FT–IR Analysis before and after Irradiation

The SEM images of $KAlFe(CN)_6/SiO_2$ before and after irradiation are shown in Figure 2. The adsorbent particle was spherical with a diameter of about 100 μm before irradiation. It maintained that shape and surface morphology with some minor cracks forming after 50 and 1000 kGy of γ irradiation. The adsorbents did not collapse or disintegrate after irradiation, indicating that $KAlFe(CN)_6/SiO_2$ had good radiation resistance.

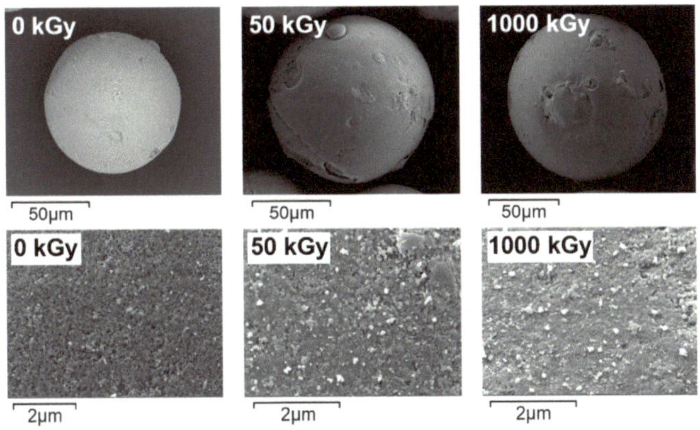

Figure 2. SEM images of KAlFe(CN)$_6$/SiO$_2$ before and after (left to right) 0, 50, and 1000 kGy irradiation.

The FT–IR spectra of KAlFe(CN)$_6$/SiO$_2$ before and after irradiation are shown in Figure 3. The peak at 3430 cm^{-1} was attributed to the O–H stretching vibration peak, and 1621 cm^{-1} was the H–O–H bending vibration peak [38]; both came from crystal water. The peaks at 1110 and 801 cm^{-1} were assigned to the Si–O stretching vibration in SiO$_2$, and 473 cm^{-1} corresponds to the inplane bending vibration of O–Si–O [38]. The peak at 554 cm^{-1} was assigned to the C–Fe stretching vibration [39]. The peaks at 2126 and 2078 cm^{-1} correspond to the C≡N stretching vibration [38,40], indicating the presence of hexacyanoferrate. The peak of CN$^-$ weakened with the increase in irradiation dose, which indicates that the γ ray destroyed a minor portion of CN$^-$ [41,42]. The above analysis shows that KAlFe(CN)$_6$/SiO$_2$ could preserve structural integrity under up to 1000 kGy of irradiation.

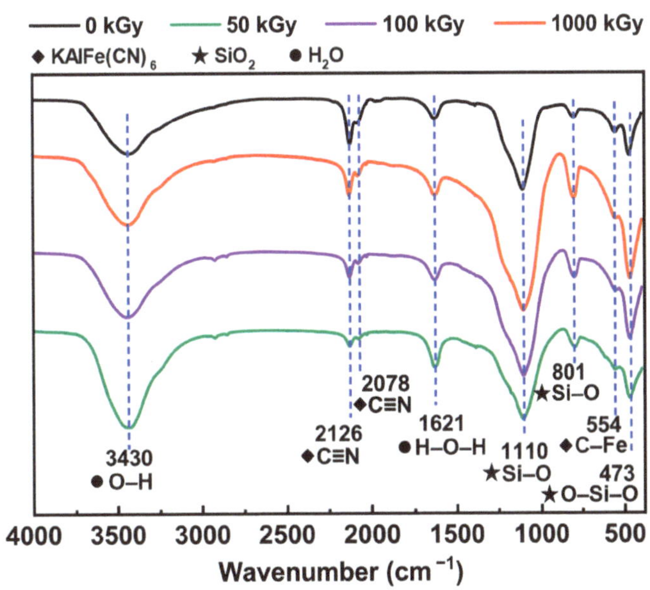

Figure 3. FT–IR patterns of KAlFe(CN)$_6$/SiO$_2$ before and after (top to bottom) 0, 50, 100, and 1000 kGy irradiation.

3.3. Adsorption Properties in Nitric Acid before and after Irradiation

The adsorption performance of the irradiated $KAlFe(CN)_6/SiO_2$ was investigated. Figure 4 shows the adsorption effects of different concentrations of HNO_3 on Pd(II) after receiving 10 to 1000 kGy of irradiation. Distribution coefficient K_d was slightly reduced after irradiation, and the magnitude of the drop decreased with the increase in HNO_3 concentration. In 5 and 7 M HNO_3 concentration, K_d remained almost the same from 0 to 100 kGy. In 1 to 3 M HNO_3 concentration, K_d was highly sensitive to the adsorption rate because the rate was high. In addition, K_d remained high after 1000 kGy of irradiation, ranging from 1625 to 38,445 $cm^3\ g^{-1}$, indicating that the adsorption ability of the irradiated $KAlFe(CN)_6/SiO_2$ remained performant in 1 to 7 M HNO_3.

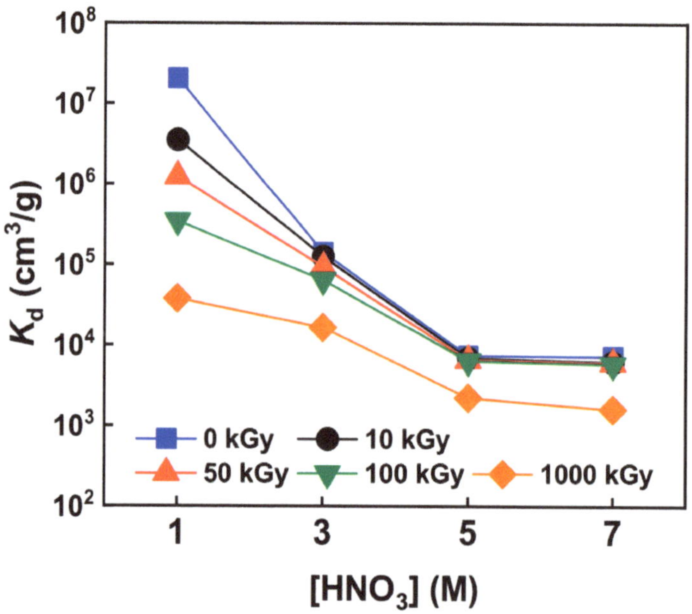

Figure 4. K_d of $KAlFe(CN)_6/SiO_2$ for Pd(II) before and after irradiation in 1–7 M HNO_3; phase ratio = 100 $cm^3\ g^{-1}$, [Pd] = 200 mg L^{-1}, temperature = 25 °C, and contact time = 24 h.

The adsorption rate is an important factor in judging adsorption performance. Figure 5 shows the effects of contact time on Pd(II) adsorption. The adsorption ratios of the $KAlFe(CN)_6/SiO_2$ adsorbent for Pd(II) rapidly increased with time. When the contact time reached 15 min, the adsorption ratios of Pd(II) were over 87%, 86%, and 81% for 0, 50, and 1000 kGy irradiated adsorbents, respectively, indicating excellent adsorption kinetics. Subsequently, the adsorption rate of Pd(II) slowed down because the sites on $KAlFe(CN)_6/SiO_2$ tended to be saturated. After 45 min, the adsorption of Pd(II) by $KAlFe(CN)_6/SiO_2$ reached equilibrium, and the equilibrium adsorption rates were 99.9%, 99.8%, and 99.8%, respectively.

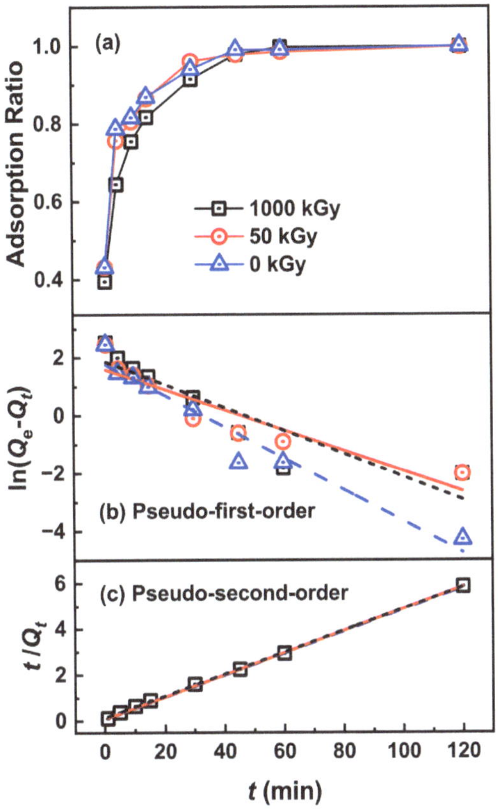

Figure 5. (**a**) Adsorption ratio of Pd(II) by $KAlFe(CN)_6/SiO_2$ versus adsorption time with a phase ratio = 100 cm^3 g^{-1}, [Pd] = 200 mg L^{-1}, temperature = 25 °C, and contact time = 24 h; (**b**) pseudo-first-order adsorption kinetic model fitting; (**c**) pseudo-second-order adsorption kinetic model fitting.

Further, the pseudo-first-order kinetic equation (Equation (3)) and pseudo-second-order kinetic equation (Equation (4)) were used for fitting to explore the adsorption rate control steps of $KAlFe(CN)_6/SiO_2$ on Pd(II) [43].

$$ln(Q_e - Q_t) = ln(Q_e) - (k_1 t), \quad (3)$$

$$t/Q_t = 1/\left(k_2 Q_e^2\right) + t/Q_e, \quad (4)$$

where k_1 and k_2 are fitted parameters related to the adsorption rate.

The pseudo-second-order equation fitted better than the pseudo-first-order equation (Figure 5). For the pseudo-first-order equation, the correlation coefficients were lower than 0.95; the values of the pseudo-second-order equation, on the other hand, were close to 1, indicating good linearity (Table 2). Moreover, the Q_e values of $KAlFe(CN)_6/SiO_2$ for Pd(II) obtained with the pseudo-second-order equation were 20.95, 20.82, and 20.61 mg g^{-1} for the irradiated adsorbents with 0, 50, and 1000 kGy, respectively, which values were close to the experimental data. k_2 shows that the adsorption rate remained approximately equal before and after 50 kGy irradiation. These results indicate that the pseudo-second-order equation could describe the adsorption kinetics of $KAlFe(CN)_6/SiO_2$ well at different nitric acid concentrations, and the adsorption type of Pd(II) through the $KAlFe(CN)_6/SiO_2$ adsorbent was controlled via chemical adsorption.

Table 2. Adsorption kinetic parameters of Pd(II) by $KAlFe(CN)_6/SiO_2$ before and after irradiation.

Adsorption Kinetic Models	Parameters			
		0 kGy	50 kGy	1000 kGy
Pseudo-first-order kinetic model	k_1 (10^{-3} mg g^{-1} min^{-1})	54.51 ± 5.06	35.26 ± 5.73	32.99 ± 7.34
	Q_e (mg g^{-1})	6.158 ± 1.29	5.015 ± 1.34	5.001 ± 1.46
	R^2	0.9427	0.8405	0.8038
Pseudo-second-order kinetic model	k_2 (10^{-3} mg g^{-1} min^{-1})	25.36 ± 1.71	24.83 ± 1.83	17.48 ± 2.87
	Q_e (mg g^{-1})	20.95 ± 0.12	20.82 ± 0.10	20.61 ± 0.19
	R^2	0.9998	0.9999	0.9994
Experimental data	Q (mg/g)	20.60 ± 0.8	20.48 ± 0.8	20.45 ± 0.8

3.4. Adsorption Isotherms before and after Irradiation

In investigating the capacity and mechanism of the adsorption of the $KAlFe(CN)_6/SiO_2$ adsorbent on Pd(II), adsorption isotherms were obtained from the experiments, and fitted using the Langmuir (Equation (5)) [44], Freundlich (Equation (6)) [45], and Redlich–Peterson (Equation (7)) [46] models.

$$Q_e = (Q_{max}K_LC_e)/(1 + K_LC_e), \tag{5}$$

$$Q_e = Q_{max}C_e^{1/n}, \tag{6}$$

$$Q_e = (K_RC_e)/\left(1 + K_PC_e^g\right), \tag{7}$$

where Q_{max} is the maximal adsorption capacity predicted by the model; K_L, $1/n$, K_R, K_P, and g are the fitting parameters.

As shown in Figure 6 and Table 3, the coefficients of determination R^2 of the Langmuir and Redlich–Peterson models were higher than 0.96, and the fitted parameter g in the Redlich–Peterson model was between 0.96 and 0.995, indicating that the Redlich–Peterson model could be converted into the Langmuir model in these cases. With a smaller irradiation dose than 1000 kGy, the adsorption of $KAlFe(CN)_6/SiO_2$ on Pd(II) corresponded to the Langmuir model, suggesting that the adsorption sites on $KAlFe(CN)_6/SiO_2$ were on a single layer, and the adsorbed Pd(II) had no internal interaction [44].

Table 3. Fitted parameters of adsorption isotherms before and after irradiation.

Isotherm Models	Parameters	Irradiation Dose (kGy)				
		0	10	50	100	1000
Langmuir model	Q_{max} (mg g^{-1})	48.3 ± 0.3	48.1 ± 1.5	47.8 ± 0.9	47.7 ± 1.4	45.1 ± 1.2
	K_L (L mg^{-1})	60.8 ± 2.5	15.5 ± 3.2	41.9 ± 5.9	28.9 ± 6.5	82.2 ± 13.6
	R^2	0.9989	0.9724	0.9685	0.9707	0.9769
Freundlich model	Q_{max} (mg g^{-1})	41.8 ± 2.8	39.8 ± 2.3	40.3 ± 2.4	41.7 ± 2.4	38.8 ± 2.5
	$1/n$	0.14 ± 0.04	0.17 ± 0.04	0.14 ± 0.03	0.16 ± 0.03	0.15 ± 0.03
	R^2	0.8305	0.8683	0.8698	0.8686	0.8627
Redlich–Peterson model	K_R (L mg^{-1})	3001.8 ± 14.2	976.8 ± 30.8	2474.7 ± 48.0	1922.4 ± 57.8	4119.1 ± 81.1
	K_P (L mg^{-1})	62.43 ± 3.18	21.1 ± 7.4	53.3 ± 11.3	41.6 ± 13.3	92.7 ± 19.8
	g	0.995 ± 0.006	0.96 ± 0.04	0.972 ± 0.018	0.96 ± 0.03	0.98 ± 0.02
	R^2	0.9991	0.9775	0.9909	0.9801	0.9814
Experimental data	Q_e (mg g^{-1})	48.4 ± 0.8	48.2 ± 0.8	48.1 ± 0.8	47.8 ± 0.8	45.5 ± 0.8

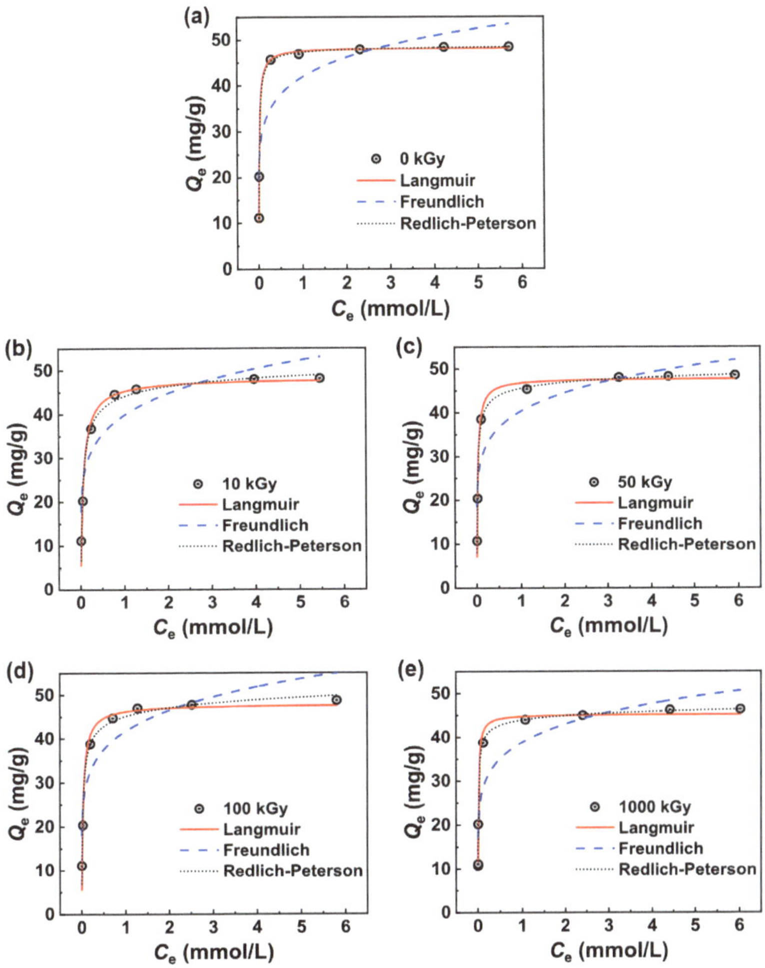

Figure 6. Adsorption isotherms and model fitting of $KAlFe(CN)_6/SiO_2$ for Pd(II) before and after irradiation in 3 M HNO_3 with a phase ratio = 100 cm^3 g^{-1}, temperature = 25 °C, and contact time = 24 h. (**a**) before irradiation; after (**b**) 10 kGy, (**c**) 50 kGy, (**d**) 100 kGy, and (**e**) 1000 kGy irradiation.

The fitted results of Equation (5) show that, after irradiation with 10, 50, 100, and 1000 kGy, the maximal adsorption capacities Q_e of $KAlFe(CN)_6/SiO_2$ on Pd(II) were 48.1, 47.8, 47.7, and 45.1 mg g^{-1}, respectively. The relative decrease in the unirradiated sample was only 0.4%, 1.0%, 1.2%, and 6.2%, respectively. The results indicate that the adsorption capacity of $KAlFe(CN)_6/SiO_2$ was insignificantly influenced by the irradiation dose and remained virtually unchanged after no more than 100 kGy of irradiation. Thus, $KAlFe(CN)_6/SiO_2$ could be used for the separation and recovery of Pd(II) under radioactive conditions.

As shown in Table 4, compared with other adsorbents, the adsorption capacity of $KAlFe(CN)_6/SiO_2$ was higher with shorter equilibrium time.

Table 4. Comparison of adsorption capacity and equilibrium time on Pd(II) with different adsorbents

Adsorbent	Solution	Q_e (mg g^{-1})	T_e * (min)	Reference
Thiourea–formaldehyde resin	pH = 4	31.9	–	[47]
Bayberry tannin immobilized collagen fiber	pH = 4	33.4	240	[25]
PMA–SNP	pH = 3	53.6	480	[48]
Commercial activated carbon pellets	2 M HCl	27.2	–	[49]
CA–BOPhen@SiO_2–P	3 M HNO_3	35.0	10	[50]
$K_2NiFe(CN)_6/SiO_2$	3 M HNO_3	41.0	60	[14]
$KAlFe(CN)_6/SiO_2$	3 M HNO_3	48.3	45	This work

* Equilibrium time.

3.5. DFT Calculations

The DFT method was applied to analyze the adsorption products of Pd(II) by $KAlFe(CN)_6$. Stable structures with different adsorption rates and positions of adsorption sites were calculated. Figure 7a illustrates the optimized structure of the $Pd[AlFe(CN)_6]_2$ (S1) product of saturated adsorption. Figure 7b–f show the optimized structures of the $Pd_{0.5}K[AlFe(CN)_6]_2$ product with an adsorption rate of 50%, namely, S2-1 to S2-5, for different adsorption sites. Table 5 lists the lattice parameters of these optimized structures, proving that all six types of products had FCC cells with similar volumes, where the cell of the saturated adsorption of product S1 was a cube, whereas unsaturated product S2 was a cuboid because of its asymmetry in the position of K and Pd atoms.

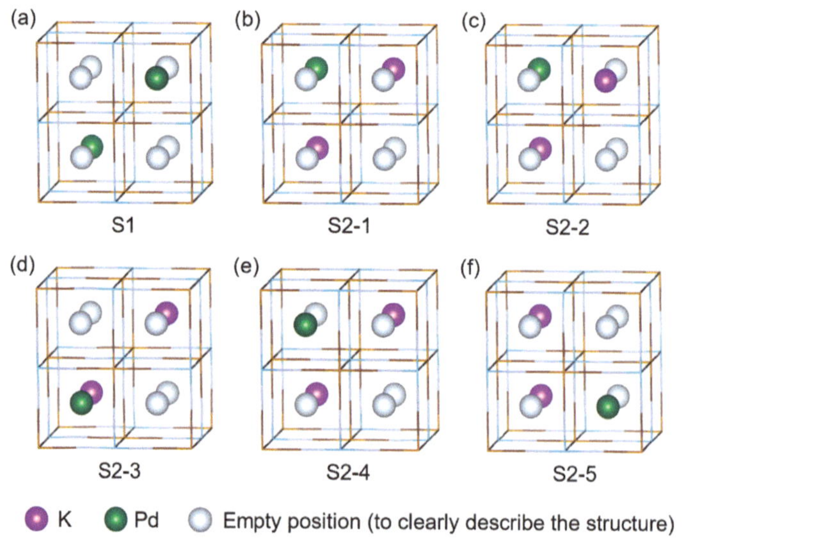

Figure 7. Optimized stable structures of (**a**) $Pd[AlFe(CN)_6]_2$ and (**b–f**) $Pd_{0.5}K[AlFe(CN)_6]_2$.

Table 5. Structural parameters of $Pd[AlFe(CN)_6]_2$ and $Pd_{0.5}K[AlFe(CN)_6]_2$ with different adsorption positions.

Structure	a (Å)	b (Å)	c (Å)	$\alpha = \beta = \gamma$ (°)	V (Å^3)
S1	10.0347	10.0347	10.0347	90	1010.44
S2-1	10.0375	10.0375	10.0421	90	1011.75
S2-2	10.0415	10.0466	10.0466	90	1013.53
S2-3	10.0440	10.0432	10.0432	90	1013.09
S2-4	10.0297	10.0327	10.0327	90	1009.54
S2-5	10.0617	10.0617	10.0531	90	1017.76

On the basis of the optimized stable structures before and after adsorption, the Helmholtz free energy (F) values were calculated. Thermodynamic calibration was performed via Phonopy [51] and VASPKIT [52], and the change in free energy (ΔF) was calculated as

$$\Delta F = F_{\mathrm{Pd}_m\mathrm{K}_n[\mathrm{AlFe(CN)}_6]_2} + nF_{\mathrm{K}^+} - 2F_{\mathrm{KAlFe(CN)}_6} - mF_{\mathrm{Pd}^{2+}}, \quad (8)$$

where F represents the free energy, m is the number of Pd atoms, and n is the number of K atoms.

The results are illustrated in Figure 8, where the free-energy changes of S1 are in the line chart and those of S2s are in the column chart. The ΔF of S1 was lower than that of S2, from 273.15 to 373.15 K, showing that S1 was spontaneously generated. The $KAlFe(CN)_6$ adsorbent tended to completely absorb Pd(II) and become $Pd[AlFe(CN)_6]_2$.

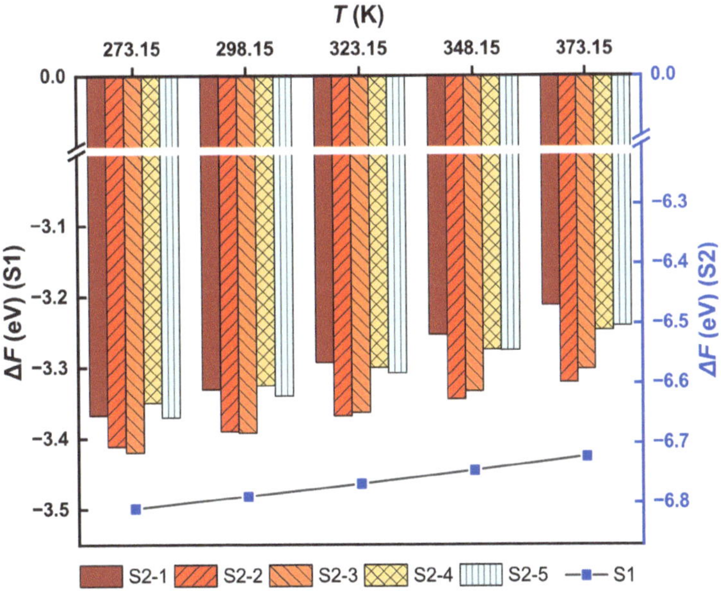

Figure 8. Calculated free-energy changes of different reactions.

S2 structures, on the other hand, were intermediate products of adsorption. The system free-energy changes indicate that S2-2 and S2-3 were likely generated more spontaneously than the others because of the low ΔF. In addition, the free-energy changes varied with temperature, that is, S2-3 was slightly superior, below 323.15 K, whereas S2-2 changed to the dominant structure of above 323.15 K.

4. Conclusions

The crystal structure and adsorption ability of gamma-irradiated $KAlFe(CN)_6/SiO_2$ on Pd(II) were investigated. SEM, XRD, and FT–IR characterizations showed that the peak positions and peak intensities of the irradiated adsorbent changed slightly, but its crystal structure was stable and not significantly damaged. After irradiation, the K_d of Pd(II) dropped slightly, and the magnitude of the drop decreased with the increase in HNO_3 concentration. For the irradiation dose of 100 kGy, K_d on the adsorbent was higher than 5846 $cm^3\ g^{-1}$, and Q_e was 48.1 mg g^{-1}, which was only 1.2% lower than that of the unirradiated samples. The adsorption equilibrium was attained within 45 min, and the adsorption of Pd(II) followed the pseudo-second-order kinetic equation, which corresponds to chemical adsorption. The adsorption results indicated that $KAlFe(CN)_6/SiO_2$ achieved excellent adsorption performance on Pd(II) with 100 kGy irradiation. The DFT calculations showed that the stablest theoretical Pd(II) adsorption product of $KAlFe(CN)_6$ was $Pd[AlFe(CN)_6]_2$, and this adsorption process was spontaneous. When the temperature was lower than 323.15 K, the most dominant intermediate product was $Pd_{0.5}K[AlFe(CN)_6]_2$ with a cell volume of 1013.53 Å^3, followed by $Pd_{0.5}K[AlFe(CN)_6]_2$ with a cell volume of 1013.09 Å^3. Therefore, $KAlFe(CN)_6/SiO_2$ could be used to separate and recover Pd(II) under radioactive conditions for use in other industries. Because of the low beta decay energy, the recovered Pd is essentially safe for industrial applications. Further study into recovering Pd from adsorbents is required, and there are several ongoing works on this topic.

Author Contributions: Investigation, visualization, and writing—original draft preparation, Y.W. (Yueying Wen); writing—review and editing and funding acquisition, Y.W. (Yan Wu); resources, L.X. All authors have read and agreed to the published version of the manuscript.

Funding: This research was funded by the National Natural Science Foundation of China (grant numbers 12175143 and 11975082).

Institutional Review Board Statement: Not applicable.

Informed Consent Statement: Not applicable.

Data Availability Statement: Not applicable.

Conflicts of Interest: The authors declare no conflict of interest.

References

1. IAEA. *Climate Change and Nuclear Power 2018*; International Atomic Energy Agency: Vienna, Austria, 2018.
2. Kharecha, P.A.; Hansen, J.E. Prevented Mortality and Greenhouse Gas Emissions from Historical and Projected Nuclear Power. *Environ. Sci. Technol.* **2013**, *47*, 4889–4895. [CrossRef] [PubMed]
3. Zou, S.; Kuang, Y.; Tang, D.; Guo, Z.; Xu, S. Risk analysis of high level radioactive waste storage tank based on HAZOP. *Ann. Nucl. Energy* **2018**, *119*, 106–116. [CrossRef]
4. Wu, Y.; Lee, C.-P.; Mimura, H.; Zhang, X.; Wei, Y. Stable solidification of silica-based ammonium molybdophosphate by allophane: Application to treatment of radioactive cesium in secondary solid wastes generated from fukushima. *J. Hazard. Mater.* **2018**, *341*, 46–54. [CrossRef] [PubMed]
5. Wu, Y.; Zhang, X.-X.; Wei, Y.-Z.; Mimura, H. Development of adsorption and solidification process for decontamination of Cs-contaminated radioactive water in Fukushima through silica-based AMP hybrid adsorbent. *Sep. Purif. Technol.* **2017**, *181*, 76–84. [CrossRef]
6. Laurin, C.; Régnier, E.; Gossé, S.; Laplace, A.; Agullo, J.; Mure, S.; Brackx, E.; Toplis, M.; Pinet, O. Redox behavior of ruthenium in nuclear glass melt: Ruthenium dioxide reduction reaction. *J. Nucl. Mater.* **2021**, *545*, 152650. [CrossRef]
7. Wei, Y.; Wu, Y.; Li, H. *The Newest Nuclear Fuel Cycle*; Shanghai Jiao Tong University Press: Shanghai, China, 2016; pp. 99–125, 225–268, 313–360.
8. Hassan, I.A.; Saleh, H.M. Biological Effects of Hazardous Waste: Threshold Limits of Anomalies and Protective Approaches. *Hazard. Waste Manag. Health Risks* **2020**, *1*, 62. [CrossRef]
9. Choppin, G.; Liljenzin, J.-O.; Rydberg, J.; Ekberg, C. Chapter 15—Radiation Biology and Radiation Protection. In *Radiochemistry and Nuclear Chemistry*, 4th ed.; Choppin, G., Liljenzin, J.-O., Rydberg, J., Ekberg, C., Eds.; Academic Press: Cambridge, MA, USA, 2013; pp. 445–491.
10. Abdelhamid, A.A.; Badr, M.H.; Mohamed, R.A.; Saleh, H.M. Using Agricultural Mixed Waste as a Sustainable Technique for Removing Stable Isotopes and Radioisotopes from the Aquatic Environment. *Sustainability* **2023**, *15*, 1600. [CrossRef]
11. Dawoud, M.M.A.; Hegazi, M.M.; Saleh, H.M.; El Helew, W.K. Removal of stable and radio isotopes from wastewater by using modified microcrystalline cellulose based on Taguchi L16. *Int. J. Environ. Sci. Technol.* **2023**, *20*, 1289–1300. [CrossRef]
12. Forsberg, C.W. Rethinking High-Level Waste Disposal: Separate Disposal of High-Heat Radionuclides (90Sr and 137Cs). *Nucl. Technol.* **2000**, *131*, 252–268. [CrossRef]
13. Sawai, M. Rokkasho Reprocessing Plant: Vitrification Problems Continue Nuke Info Tokyo No. 125. Available online: https://cnic.jp/english/?p=1233 (accessed on 19 October 2022).
14. Wang, Q.; Sang, H.; Chen, L.; Wu, Y.; Wei, Y. Selective separation of Pd(II) through ion exchange and oxidation-reduction with hexacyanoferrates from high-level liquid waste. *Sep. Purif. Technol.* **2020**, *231*, 115932. [CrossRef]
15. Xu, K. International research progress on nuclear waste vitrification: A review. *Mater. China* **2016**, *35*, 481–488+517. [CrossRef]
16. Lu, J.; Lin, P.; Li, L.; Zhao, X.; Wang, Y.; Weng, H.; Lin, M. Solubility of typical PGMs in HLLW glass and influence of PGMs on the properties of HLLW glass. *Radiat. Prot.* **2023**, *43*, 17–30.

17. Yadav, M.; Misra, A.; Malhotra, A.; Kumar, N. Design and analysis of a high-pressure turbine blade in a jet engine using advanced materials. *Mater. Today Proc.* **2020**, *25*, 639–645. [CrossRef]
18. Yakoumis, I. PROMETHEUS: A Copper-Based Polymetallic Catalyst for Automotive Applications. Part I: Synthesis and Characterization. *Materials* **2021**, *14*, 622. [CrossRef]
19. Holdsworth, A.F.; Eccles, H.; Sharrad, C.A.; George, K. Spent Nuclear Fuel—Waste or Resource? The Potential of Strategic Materials Recovery during Recycle for Sustainability and Advanced Waste Management. *Waste* **2023**, *1*, 249–263. [CrossRef]
20. Venkatesan, K.A.; Selvan, B.R.; Antony, M.P.; Srinivasan, T.G.; Rao, P.R.V. Extraction of palladium from nitric acid medium by commercial resins with phosphinic acid, methylene thiol and isothiouronium moieties attached to polystyrene-divinylbenzene. *J. Radioanal. Nucl. Chem.* **2005**, *266*, 431–440. [CrossRef]
21. Suzuki, T.; Otsubo, U.; Ogata, T.; Shiwaku, H.; Kobayashi, T.; Yaita, T.; Matsuoka, M.; Murayama, N.; Narita, H. Selective adsorption of Pd(II) over Ag(I) in nitric acid solutions using nitrogen-donor-type adsorbents. *Sep. Purif. Technol.* **2023**, *308*, 122943. [CrossRef]
22. Peng, L.; Zhang, M.; Dong, Z.; Qi, W.; Zhai, M.; Zhao, L. Efficient and selective adsorption of Pd(II) by amino acid-functionalized cellulose microspheres and their applications in palladium recovery from PCBs leaching solution. *Sep. Purif. Technol.* **2022**, *301*, 122037. [CrossRef]
23. Zhou, L.; Xu, J.; Liang, X.; Liu, Z. Adsorption of platinum(IV) and palladium(II) from aqueous solution by magnetic cross-linking chitosan nanoparticles modified with ethylenediamine. *J. Hazard. Mater.* **2010**, *182*, 518–524. [CrossRef]
24. Els, E.R.; Lorenzen, L.; Aldrich, C. The adsorption of precious metals and base metals on a quaternary ammonium group ion exchange resin. *Miner. Eng.* **2000**, *13*, 401–414. [CrossRef]
25. Ma, H.-W.; Liao, X.-P.; Liu, X.; Shi, B. Recovery of platinum(IV) and palladium(II) by bayberry tannin immobilized collagen fiber membrane from water solution. *J. Membr. Sci.* **2006**, *278*, 373–380. [CrossRef]
26. Awual, M.R.; Khaleque, M.A.; Ratna, Y.; Znad, H. Simultaneous ultra-trace palladium(II) detection and recovery from wastewater using new class meso-adsorbent. *J. Ind. Eng. Chem.* **2015**, *21*, 405–413. [CrossRef]
27. Tang, Y.; Li, W.; Feng, P.; Zhou, M.; Wang, K.; Jiang, K. Investigation of alkali-ion (Li, Na and K) intercalation in manganese hexacyanoferrate KxMnFe(CN)6 as cathode material. *Chem. Eng. J.* **2020**, *396*, 125269. [CrossRef]
28. Hamoud, M.A.; Allan, K.F.; Sanad, W.A.; Saad, E.A.; Mahmoud, M.R. Synthesis of PAN/ferrocyanide composite incorporated with cetrimonium bromide and its employment as a bifunctional adsorbent for coremoval of Cs^+ and $HCrO_4^-$ from aqueous solutions. *J. Radioanal. Nucl. Chem.* **2020**, *324*, 647–661. [CrossRef]
29. Nilchi, A.; Saberi, R.; Moradi, M.; Azizpour, H.; Zarghami, R. Adsorption of cesium on copper hexacyanoferrate–PAN composite ion exchanger from aqueous solution. *Chem. Eng. J.* **2011**, *172*, 572–580. [CrossRef]
30. Holdsworth, A.F.; Eccles, H.; Rowbotham, D.; Brookfield, A.; Collison, D.; Bond, G.; Kavi, P.C.; Edge, R. The Effect of Gamma Irradiation on the Physiochemical Properties of Caesium-Selective Ammonium Phosphomolybdate–Polyacrylonitrile (AMP–PAN) Composites. *Clean Technol.* **2019**, *1*, 294–310. [CrossRef]
31. Holdsworth, A.F.; Eccles, H.; Rowbotham, D.; Bond, G.; Kavi, P.C.; Edge, R. The Effect of Gamma Irradiation on the Ion Exchange Properties of Caesium-Selective Ammonium Phosphomolybdate-Polyacrylonitrile (AMP-PAN) Composites under Spent Fuel Recycling Conditions. *Separations* **2019**, *6*, 23. [CrossRef]
32. Tranter, T.J.; Herbst, R.S.; Todd, T.A.; Olson, A.L.; Eldredge, H.B. Evaluation of ammonium molybdophosphate-polyacrylonitrile (AMP-PAN) as a cesium selective sorbent for the removal of 137Cs from acidic nuclear waste solutions. *Adv. Environ. Res.* **2002**, *6*, 107–121. [CrossRef]
33. Momma, K.; Izumi, F. VESTA 3 for three-dimensional visualization of crystal, volumetric and morphology data. *J. Appl. Crystallogr.* **2011**, *44*, 1272–1276. [CrossRef]
34. Perdew, J.P.; Burke, K.; Ernzerhof, M. Generalized Gradient Approximation Made Simple. *Phys. Rev. Lett.* **1996**, *77*, 3865–3868. [CrossRef]
35. Mao, J.; Wang, T.; Zhang, X.; Sun, M.; Wang, T.; Peng, H. Equivalence analysis of electron and γ-ray irradiation effects on borosilicate glass. *Mod. Appl. Phys.* **2021**, *12*, 116–122.
36. Was, G.S. *Fundamentals of Radiation Materials Science*; Springer: New York, NY, USA, 2017; pp. 167–181.
37. Jing, T. Study on radiation damage of electron and γ-rays and mechanism of nuclear hardening. *China Nucl. Sci. Technol. Rep.* **2000**, *00*, 1284–1294.
38. Nakamoto, K. *Infrared and Raman Spectra of Inorganic and Coordination Compounds*, 4th ed.; Chemical Industry Press: Beijing, China, 1991.
39. Roque-Malherbe, R.; Carballo, E.; Polanco, R.; Lugo, F.; Lozano, C. Structure and adsorption properties of a porous cooper hexacyanoferrate polymorph. *J. Phys. Chem. Solids* **2015**, *86*, 65–73. [CrossRef]
40. Nishioka, H.; Nishikawa, M.; Katagi, M.; Tatsuno, M.; Tsuchihashi, H. Analysis of Cyanide Compounds Using Fourier Transform Raman Spectroscopy. *Jpn. J. Sci. Technol. Identif.* **2000**, *4*, 61–66. [CrossRef]
41. Hu, Y.; Hu, H.; Yang, M.; Huang, H.; Feng, H. Radiolysis of copper cyanide solution by electron-beam irradiation. *Environ. Pollut. Prev.* **2009**, *31*, 31–34.
42. Liu, Y.; Hu, H.; Yang, M.; Liu, J.; Dang, J.; Wu, Y. Study on the degradation of zinc-cynide complex in gold-leaching wastewater by the E-beam radiolysis. *Ind. Saf. Environ. Prot.* **2011**, *37*, 4–6.
43. Ho, Y.S.; McKay, G. Pseudo-second order model for sorption processes. *Process Biochem.* **1999**, *34*, 451–465. [CrossRef]

44. Langmuir, I. The constitution and fundamental properties of solids and liquids. Part I. Solids. *J. Am. Chem. Soc.* **1916**, *38*, 2221–2295. [CrossRef]
45. Laskar, I.I.; Hashisho, Z. Insights into modeling adsorption equilibria of single and multicomponent systems of organic and water vapors. *Sep. Purif. Technol.* **2020**, *241*, 116681. [CrossRef]
46. Redlich, O.; Peterson, D.L. A Useful Adsorption Isotherm. *J. Phys. Chem.* **1959**, *63*, 1024. [CrossRef]
47. Muslu, N.; Gülfen, M. Selective separation and concentration of Pd(II) from Fe(III), Co(II), Ni(II), and Cu(II) ions using thiourea-formaldehyde resin. *J. Appl. Polym. Sci.* **2011**, *120*, 3316–3324. [CrossRef]
48. Zhao, J.; Wang, C.; Wang, S.; Zhang, L.; Zhang, B. Augmenting the adsorption parameters of palladium onto pyromellitic acid-functionalized nanosilicas from aqueous solution. *Colloids Surf. A Physicochem. Eng. Asp.* **2019**, *578*, 123581. [CrossRef]
49. Kasaini, H.; Goto, M.; Furusaki, S. Selective Separation of Pd(II), Rh(III), and Ru(III) Ions from a Mixed Chloride Solution Using Activated Carbon Pellets. *Sep. Sci. Technol.* **2000**, *35*, 1307–1327. [CrossRef]
50. Xu, L.; Zhang, A.; Pu, N.; Xu, C.; Chen, J. Development of Two novel silica based symmetric triazine-ring opening N-donor ligands functional adsorbents for highly efficient separation of palladium from HNO3 solution. *J. Hazard. Mater.* **2019**, *376*, 188–199. [CrossRef]
51. Togo, A.; Tanaka, I. First principles phonon calculations in materials science. *Scr. Mater.* **2015**, *108*, 1–5. [CrossRef]
52. Wang, V.; Xu, N.; Liu, J.-C.; Tang, G.; Geng, W.-T. VASPKIT: A user-friendly interface facilitating high-throughput computing and analysis using VASP code. *Comput. Phys. Commun.* **2021**, *267*, 108033. [CrossRef]

Disclaimer/Publisher's Note: The statements, opinions and data contained in all publications are solely those of the individual author(s) and contributor(s) and not of MDPI and/or the editor(s). MDPI and/or the editor(s) disclaim responsibility for any injury to people or property resulting from any ideas, methods, instructions or products referred to in the content.

MDPI AG
Grosspeteranlage 5
4052 Basel
Switzerland
Tel.: +41 61 683 77 34

Toxics Editorial Office
E-mail: toxics@mdpi.com
www.mdpi.com/journal/toxics

Disclaimer/Publisher's Note: The title and front matter of this reprint are at the discretion of the Guest Editors. The publisher is not responsible for their content or any associated concerns. The statements, opinions and data contained in all individual articles are solely those of the individual Editors and contributors and not of MDPI. MDPI disclaims responsibility for any injury to people or property resulting from any ideas, methods, instructions or products referred to in the content.

www.ingramcontent.com/pod-product-compliance
Lightning Source LLC
Chambersburg PA
CBHW041415210926
51850CB00013B/3275